MICROACTUATORS

Electrical, Magnetic, Thermal, Optical, Mechanical, Chemical & Smart Structures

THE KLUWER INTERNATIONAL SERIES IN:

ELECTRONIC MATERIALS: Science and Technology

Series Editor
Harry L. Tuller
Massachusetts Institute of Technology

Other books in the series:

DIAMOND: Electronic Properties and Applications
Lawrence S. Pan and Don R. Kania, Editors
ION IMPLANTATION: Basics to Device Fabrication
Emanuele Rimini, Author
SOLID STATE BATTERIES: Materials Design and Optimization
C. Julien and G. Nazri, Authors
SOL-GEL OPTICS: Processing and Applications
L.C. Klein, Editor
PHOTOREFRACTIVE EFFECTS AND MATERIALS
David Nolte, Editor
MATERIALS FOR OPTOELECTRONICS
Maurice Quillec, Editor
PIEZOELECTRIC ACTUATORS AND ULTRASONIC MOTORS
Kenji Uchino, Author
WIDE-GAP LUMINESCENT MATERIALS: Theory and Applications
Stanley R. Rotman, Editor
THIN FILM FERROELECTRIC MATERIALS AND DEVICES
R. Ramesh, University of Maryland, Editor

The Series **ELECTRONIC MATERIALS: Science and Technology** will provide publication with an interdisciplinary approach in the following topic areas:

- Sensors and Actuators
- Electrically Active Ceramics & Polymers
- Structure-Property-Processing-Performance Correlations in Electronic Materials
- Electronically Active Interfaces
- High Tc Superconducting Materials
- Optoelectronic Materials
- Composite Materials
- Defect Engineering
- Solid State Ionics
- Electronic Materials in Energy Conversion-Solar Cells, High Energy Density Microbatteries, Solid State Fuel Cells, etc.

MICROACTUATORS

Electrical, Magnetic, Thermal, Optical, Mechanical, Chemical & Smart Structures

by

Massood Tabib-Azar
Case Western Reserve University

KLUWER ACADEMIC PUBLISHERS
Boston / Dordrecht / London

Distributors for North, Central and South America:
Kluwer Academic Publishers
101 Philip Drive
Assinippi Park
Norwell, Massachusetts 02061 USA
Telephone (781) 871-6600
Fax (781) 871-6528
E-Mail <kluwer@wkap.com>

Distributors for all other countries:
Kluwer Academic Publishers Group
Distribution Centre
Post Office Box 322
3300 AH Dordrecht, THE NETHERLANDS
Telephone 31 78 6392 392
Fax 31 78 6546 474
E-Mail <orderdept@wkap.nl>
Electronic Services <http://www.wkap.nl>

Library of Congress Cataloging-in-Publication Data

Tabib-Azar, Massood.
Microactuators: electrical, magnetic, thermal, optical,
Mechanical, chemical, and smart structures / Massood Tabib-Azar.
p. cm. -- (Kluwer international series in --electronic
materials, science and technology)
Includes bibliographical references and index.
ISBN 0-7923-8089-4 (alk. paper)
1. Microactuators I. Title II. Series
TJ223.A25T33 1997
629.8'815--dc21 97-42709
CIP

Printed on acid-free paper.

Printed in the United States of America

This printing is a digital duplication of the original edition.

Table of Content

Chapter 3 Magnetic Microactuators

Chapter 4 Thermal and Phase-Transformation Microactuators

Chapter 5 Optical Microactuators

Chapter 8 Smart Structures

Micro-Actuators

(Electrical, Magnetic, Thermal, Optical, Mechanical, and Chemical)

It has become quite apparent that sensors and actuators are the main bottleneck of the modern information processing and control systems. Microprocessors and computers used to be the main limiting element in most information processing systems. But thanks to the enormous progress in the microelectronics industry, most information analysis tasks can be processed in real time. The data has to be acquired by the processor in some form and processed and used to produce some useful function in the real world. One may come up with a block diagram shown in figure 1, where a nearly autonomous system acquires the data, turns it into useful information, and uses it to alter the source of the data in a manner consistent with a given goal. I call this a nearly autonomous system since "some biological" entity with a goal (usually survival in different forms) has to be present to impart the system with its goal. At the present time the weakest link in this system is the actuator. Sensors used to be quite weak but no more.

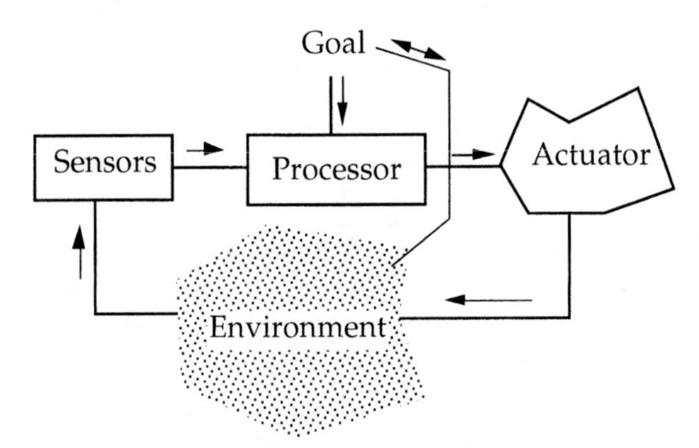

Figure 1 The different parts of a nearly autonomous system.

This is a research oriented book that covers the emerging topic of micro-actuators that have attracted much attention in recent years. Largely owing to the microfabrication methods used in the microelectronics industry, microactuators are being developed at a very fast rate. Although some excellent review articles cover parts of this important field, there is not a single book devoted to its comprehensive coverage.

This book covers the fundamentals of actuation in a textbook manner and it exposes the reader to some research examples. In combining fundamentals with the latest reported actuators, this book distinguishes itself from other monograms or textbooks.

The main intended audiences of this book are academic and industrial researchers and graduate students interested in initiating projects in microactuators. It can also be used as a textbook for a senior/graduate level course in the general area of sensors and actuators.

As in my previous book (Integrated Optics, Microstructures, and Sensors, published by Kluwer), I start with the fundamentals of each actuation mechanism. This is followed by an implementation example which is followed by a literature survey.

Like any other researcher with a healthy respect for writing a book, I asked myself many times whether this is an undertaking that I should be commiting myself to? Two main reasons drove me to write this book. First, I was interested to find out what is out there in this exciting and enabling field. Second, I could not find a book that covered the basics as well as the up-to-date research ideas in the same place. In each of the sub-areas of microactuators there are many well-written review articles that clearly explain the design issues as well as the state-of-the-art devices. But there is not any one article transcending the individual sub-areas to give an overall picture. Probably I should have included more design issues in each chapter, but I think this work does a pretty good job of covering different microactuators in a logical manner.

I am indebted to my parents for their encouragement and continued support of my activities in pursuit of knowledge. I owe it to my father who seeded in me the desire to do more. I dedicate this work to them and to my wife who suffered throughout this work and kept up with my late work and lack of presence. She also contributed to Chapter 7 and my understanding of physiology. I am also thankful of my past and present students who helped me in different ways. Specially to Boonsong Suttupan who proofread the manuscript and generated the index for this book. I am also indebted to Mrs. Cinthia Azim for correcting the text and her thorough and dedicated criticism of my phrases. My only hope is that my readers will have as much fun reading this book as I had in writing it. Please do not hesitate to drop me a line or two regarding your opinion/criticism of this work. My e-mail address is mxt7@po.cwru.edu.

Fall of 1997
Massood Tabib-Azar
Cleveland, Ohio

MICROACTUATORS

Electrical, Magnetic, Thermal, Optical, Mechanical, Chemical & Smart Structures

Actuator Performance and Preliminaries

1.1 Introduction

Actuators perform useful work on the environment in response to a command or a control signal. The amount of work that they perform and the energy expenditure that they require to do the desired work depend drastically on the method of actuation. These methods can be divided into six categories: electrical, magnetic, thermal/phase, optical, mechanical/acoustic, and chemical/biological.

The electrical actuation method relies on the coulombic attraction/repulsion between charged bodies. Electrical actuators include a variety of actuation schemes, or actuation means, as discussed in Chapter 2. Electrostatic actuation uses attraction between dissimilar charges to exert a force on a deformable/mobile body. Piezoelectric actuation, on the other hand, relies on the change in the dimension(s) of piezoelectric materials when a voltage is applied across them.

Magnetic actuators include electro-magnetic schemes, such as the scheme encountered in electromotors, and magneto-elastic and magneto-restrictive schemes, where the material dimensions change when a magnetic field is applied, as discussed in Chapter 3.

Thermal actuators rely on the difference in the thermal expansion coefficients of different materials. The thermal expansion coefficient may undergo a drastic change when the phase of the material changes as a function of temperature. Certain materials known as "martensites", such as perm-alloys, have the interesting property of "remembering" their shape after plastic deformation and upon heating they return to their pre-deformation shape. These methods are being exploited in large displacement/force microactuators. Thermal actuators are discussed in Chapter 4.

Optical schemes are divided into direct and indirect optical methods, as discussed in Chapter 5. Direct optical methods use light to interact with the active parts of the actuator and cause actuation. Indirect optical methods take advantage of the heating power of the light or its ability to generate electrical current or to change the resistivity in photo-responsive materials.

Mechanical schemes use one type of mechanical motion, such as linear displacement, to generate another type of physical motion, such as rotation. As discussed in Chapter 6, mechanical actuators are extensively used in everyday life, and they cover a wide range of actuators, such as those in water faucets, where the rotational motion is converted to a linear displacement that causes the opening of the valve. Mechanical methods may also rely on levers and gear boxes to amplify small displacements. In fluidic systems, a rich variety of structures exist that take advantage of channels with different cross -sections to achieve amplification and switching.

Acoustic methods, also discussed in Chapter 6, take advantage of the momentum and displacement that can be generated by acoustic waves and mechanical vibrations in gases and solids. In these actuators, either a mechanical rectifier is used to rectify the displacement and cause it to generate a linear displacement or the acoustic energy is concentrated and used to vaporize liquids that, upon condensation, constitute the output of the liquid pump.

Actuators that employ chemical reactions to generate force and displacement are numerous. Almost all the mechanical energy that biological (life) organisms generate uses chemical reactions, as discussed in Chapter 7. The car engine is a classic example of harnessing a chemical reaction, i.e., oxidation of the fossil fuel, to generate force and displacement (discussed in Chapter 4). Explosives are another example of chemical actuators since they rely on the generation of pressure waves and large forces by spontaneous chemical reactions.

Chemical reactions are not extensively exploited in microactuators at present, but because of the large output energy density of these reactions, they constitute a very useful source of actuation power and they will be employed extensively in future microactuators.

Table 1.1 Examples of different actuators.

<table>
<tr><th>Output / Input</th><th>Viscosity</th><th>Displacement</th><th>Force/ Pressure</th></tr>
<tr><td>Heat</td><td>Liquids</td><td colspan="2">Bi-metals
Shape memory alloys</td></tr>
<tr><td>Chemical reactions</td><td>Gels</td><td>Muscle
Car Engine</td><td>Muscle
Car Engine</td></tr>
<tr><td>Force</td><td></td><td>Lever</td><td>Solid link</td></tr>
<tr><td>Displacement</td><td></td><td>Gear</td><td>Lever</td></tr>
<tr><td>E-field</td><td>Electro-rheological</td><td colspan="2">Electrostatics, Piezoelectrics</td></tr>
<tr><td>H-field</td><td>Magneto-rheological</td><td colspan="2">Magneto-elastic &
Magneto-restrictive
Ferromagnetics</td></tr>
<tr><td>E&M</td><td></td><td colspan="2">Molecular motor</td></tr>
<tr><td>Light</td><td>Gels</td><td>Photo-electrons
Photo-thermal
Crookes radiometer</td><td>Photo-electrons
Photo-thermal
Crookes radiometer</td></tr>
</table>

The field of microactuators is in its infancy and many lessons are to be learned from biological actuators, which are much more efficient than man-made actuators. Biological actuators, such as muscle, usually involve a chemical component that is used to generate the necessary forces and displacements. In these actuators, there is a fatigue factor that is related to the rate at which the chemicals (also called nutrients) can be replenished and by-products can be removed and discarded. Such a consideration does not exist in man-made actuators. The second issue in biological actuators is the "mode" of operation or performance that is usually related to the rate and duration of work extraction. At slower rates, aerobic behavior is usually observed, while as the rate increases, an anaerobic regimen sets in and energy stored in the actuator itself is taken up. As we move towards more complex systems with a multitude of actuators, such a distributed and bi-modal operation scheme may prove quite useful.

The scope of this book is to describe and study the underlying principles and implementations of different actuation methods. In this chapter, we present criteria that can be used to evaluate and study different types of microactuators (section 1.2). Next, we discuss how an actuator can be decomposed to a structural shell and an actuation mean, and define a figure of merit to enable comparison between different actuation means (section 1.3). To give an example of how the figure of merit can be used to compare different actuators. In addition, we briefly discuss methods that are used in fabricating these devices (section 1.4). Since the field of microactuators is experiencing an explosive growth at the time of the writing of this book, we will be seeing many new and different fabrication methods in the near future. Thus, the fabrication methods discussed here should be treated as introductory and are included to give the essentials of various methods.

1.2 Performance Measures

In the remainder of this section we discuss evaluation criteria that apply equally to both man-made and biological actuators. In the next section, additional criteria are discussed. There are different attributes, very similar to those of sensors [1,2], that can be used to evaluate actuators:

Repeatability	Linearity	Accuracy
Precision	Resolution	Sensitivity
Speed	Threshold	Span
Conformance	Hysteresis	Scaling
Power efficiency	Noise	Instability and drift
Minimum Inducible Output		
Overall performance		
Load-bearing Capability and Stiffness		

Referring to figures 1.1 and 1.2, we define the above attributes. In these figures, the actuator input and output are respectively denoted by X and Y(X), and to make the figure more general, X and Y are expressed in terms of percent of their full scale (FS). Reaching Y from smaller (larger) values is denoted by Y^- (Y^+). For a single valued function Y(X), $Y^- = Y(X^-)$, and $Y^+ = Y(X^+)$.

Repeatability: How repeatable the behavior of an actuator is, is referred to as its repeatability (**R**). Actuators' output behavior may change due to internal relaxations, frictions, and other instabilities that occur in their structures. These "relaxations" may result in drifts and variations in the actuator output as a function of time and over cycles of operation. **R** is defined as:

$$\mathbf{R} = Y_i(X) - Y_k(X) \tag{1.1}$$

where subscripts i and k refer to the ith and k^{th} cycles of operations. Over only two cycles of operation, i=1 and k=2, and R from figure 1.1 at X=80% FS is 10% FS. When the output of an actuator is monitored through many cycles of operation, a worst-case repeatability can be defined as $R_m = Y_{max}(X) - Y_{min}(X)$, where X is either X^- or X^+, whichever results in the largest possible R.

Linearity: Linearity of an actuator (**L**) refers to the linearity of its output as a function of its input and is expressed as a percent of its full-scale output. To measure the linearity of an actuator, a reference linear line is needed and this reference line is taken either as a best-fit straight line or as a line drawn between the minimum and maximum output (terminal) values as shown in figure 1.1. Denoting the reference line as $Y_r(X)$, **L** is given by:

$$\mathbf{L} = |Y(X)-Y_r(X)|_{max} \tag{1.2}$$

The linearity of the device shown in figure 1.1 is 4.5% full-scale if the best-fit is chosen and it is 9% full-scale if the second reference line based on the terminal values is taken.

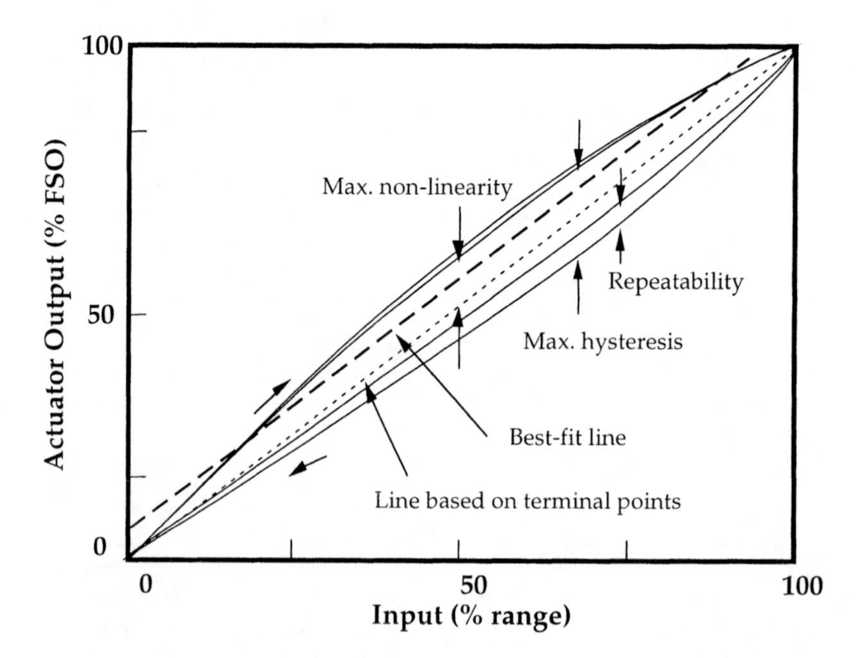

Figure 1.1 The output of an actuator as a function of its input, showing hysteresis and non-linearity.

Precision: How exactly and reproducibly an unknown value is measured is called precision in sensors. In actuators, precision is referred to as how exactly and reproducibly a desired actuation is executed. Precision has nothing to do with how accurately the actuator's output represents a calibrated scale. Precision does not imply accuracy and accuracy without precision is not meaningful.

Accuracy: A measure of how closely the output of the actuator approximates a desired calibrated scale is called accuracy. For example, a linear displacement actuator may produce a displacement of 0.09 μm instead of 0.1 μm (desired). Measured against a calibrated scale, such an actuator has an accuracy of 100*(0.09-0.1)/0.1=10 %. We can write:

$$\varepsilon_a\,(\%) = 100\,(Y_a - Y_t)/Y_t \qquad (1.3)$$

where Y_t is the true value of the desired actuation, and Y_a is the actuated value. In practice, the inaccuracy is expressed as a percentage of full-scale output (FSO):

$$\varepsilon_{FSO}\,(\%) = (Y_a - Y_t)/Y_{FSO} \qquad (1.4)$$

It can be shown that $|\varepsilon_{FSO}| \leq |\varepsilon_a|$. Y_t is either measured independently using other measurements, or it can be obtained in the form of calibrated scales from the National Bureau of Standards (NBS). Since for a repeated number of actuations different actuator outputs may be obtained, it is useful to define an error bar denoting the maximum range of Y_a. Error bars can be generalized to introduce an error band in the case of an experimental curve.

Resolution: The smallest increment in the value of the input that results in a detectable actuation is called the resolution of the actuator. For example, if a displacement sensor yields a displacement increment of δ in response to a practical voltage input of ΔV, then the maximum resolution (R_{max}) is the smallest δ that can be induced (δ_{min}):

$$R_{max}\,(\%) = 100 * (\delta_{min})/(\delta_{max} - \delta_{min}) \qquad (1.5)$$

The average resolution is given by the average of R over the range of the actuator output.

Sensitivity: The ratio of the actuator output (ΔY) to an incremental change in its input (ΔX) is called the sensitivity of the actuator:

$$S = \Delta Y / \Delta X \qquad (1.6)$$

Sensitivity of actuators may vary as a function of temperature and other environmental parameters. Usually, actuator sensitivity is not linear over the range of the actuator output.

Smallest Inducible Output: The smallest change in the actuator output that can be induced and detected is called the smallest inducible output (sIO). sIO is determined by the actuation mechanism and the noise at the actuator output. The sIO is shown by letter "d" in figure 1.2. For example, sIO for piezoelectric actuators is very small, limited only by the thermal vibration noise. In the TiNi shape memory alloy and magnetic actuators, sIO is quite large because these actuators tend to be bistable. In electrostatic microactuators, sIO is small only over a limited range of actuation.

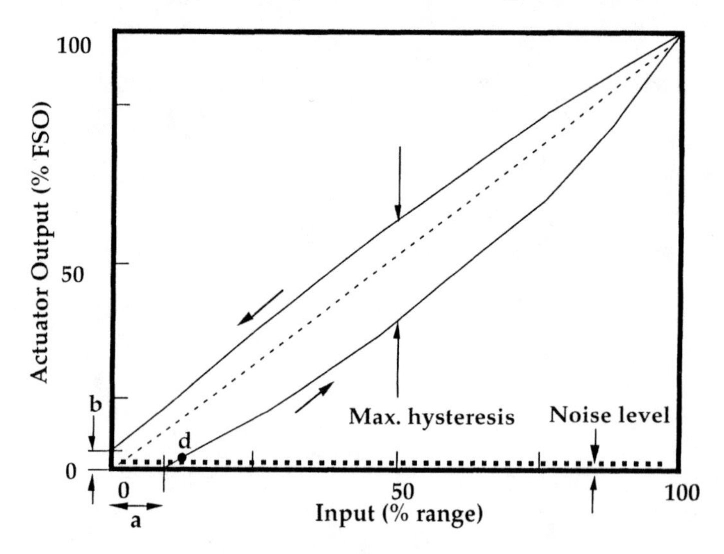

Figure 1.2 The output of an actuator as a function of its input. Threshold is shown by letter "a", smallest inducible output (sIO) is shown by letter "d", and backlash is shown by letter "b".

Threshold: Starting from zero input, the smallest initial increment in the input that results in a detectable actuator output is the threshold. Threshold is usually due to the actuator non-linearity and it is different from sIO. In figure 1.2 the threshold is shown by letter "a".

Conformance: The closeness of the experimental actuator output to a theoretical curve or to curves obtained using least-squares, or other fits, is called conformance. It is expressed in % FSO at any given value of the actuator output.

Hysteresis: The difference in the actuator output Y when Y is reached from two opposite directions, i.e., from Y^- and Y^+ as shown in figure 1.2, is called hysteresis. Hysterisis in the actuators is usually caused by a lag in the action of their deformable parts. In magnetic actuators, hysterisis is caused by the lag in the alignment of magnetic moments in response to the actuating magnetic field. Backlash error, which is usually observed in actuators that employ gears, should not be confused with hysterisis. Backlash is shown by letter "b" in figure 1.2.

Instability and Drift: Changes in actuator output (with zero input) with time, temperature and any other parameter is called drift.

Load-bearing Capability and Stiffness: Actuators behave differently when they are connected to a load. Thus, to determine the actuator output, the mechanical characteristics of the load should be known [21-24]. Let us consider a simple example of a linear piezoelectric actuator connected to a load that can be modeled by a spring with a Hooke's constant of k. The linear actuator generates force and displacement when a voltage is applied across it. When not loaded, the force exerted by the actuator is zero. Therefore, the actuator induces the maximum possible displacement for a given voltage. On the other hand, when the load is rigid and un-yielding, the displacement introduced by the actuator is zero and the force that it generates is maximum. Thus, we can construct a force-displacement (f-x) curve for the actuator, as shown in figure 1.3.

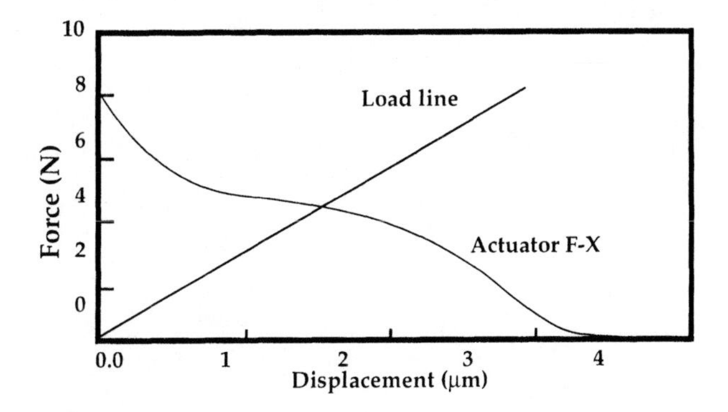

Figure 1.3 The actuator output and the influence of the load on its behavior.

The equilibrium position of the actuator is determined when the sum of the forces exerted by the actuator and the reaction force of the load is zero. Thus, the intersection between the f-x curve of the actuator and the load-line gives the equilibrium position of the actuator displacement under static conditions. When the actuator is excited by a step input, its output response can be quite complex. But under small signal conditions, i.e., when the magnitude of the step input is small compared with the actuator output range, the actuator output response can be determined from the static curves (figure 1.3). In general, dynamic load lines and f-x curves are required to find the actuator output behavior when the input excitation varies rapidly (at a time scale comparable to the natural response of the system) as a function of time.

Span: The full-scale operating range of the actuator output is called its span.

Speed: Actuator speed (v) is defined as the speed at which its output (Y) can be changed by an incremental amount.

$$v = dY/dt \tag{1.7}$$

Experimentally, it is usually easier to determine the actuator speed as the ratio of change in its output over one cycle, which includes a turn-on (τ_{on}) time and a turn-off (τ_{off}) time. Thus, the actuator speed is usually expressed as: $v = \Delta Y/(\tau_{on} + \tau_{off})$.

Step-Response: Because of the inertia and restoring elastic forces associated with the active region of actuators, their output does not change abruptly in response to a step input. The behavior of the actuator response can be quite complicated, exhibiting oscillations when under-damped or critically damped. When over-damped, the output usually executes a "saturation" behavior, as shown in figure 1.4. Usually this saturation behavior is not exponential. However, if it is exponential, a simple time constant would suffice to characterize the actuator output.

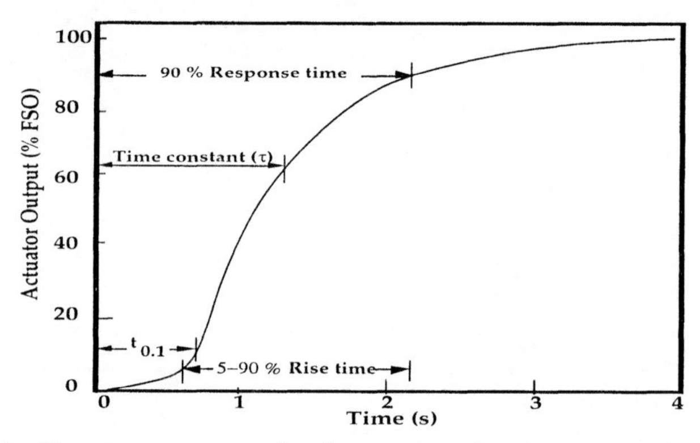

Figure 1.4 The step response of a linear piezoelectric actuator.

Power Efficiency: Actuators can be treated as active transformers that transform one type of energy into another form with a transfer function. Four different sources of energy or work are involved in a given actuator: input (P_{in}) and output (P_{out}) powers, power drawn from the power supply (P_s), and internal power consumption or "wasted" power (P_w). The conservation of energy requires:

$$P_{in} + P_s = P_{out} + P_w \tag{1.8}$$

Clearly, an optimum actuator will have nearly zero wasted power and the largest output power for a given input and source powers. Noting that the power efficiency (η_p) is directly proportional to the output power, while being inversely proportional to the input power, the wasted power and the power withdrawn from the power supply, we can write:

$$\eta_p = P_{out}/(P_{in} + P_W + P_s) \tag{1.9}$$

While one can view electronic devices as actuators, the output powers that are encountered in actuators in most cases are non-electrical. They are mechanical in nature. Mechanical power can be viewed as the force exerted by the actuator multiplied by the displacement that it induces. In general, the relationship between the force and the displacement is quite non-linear, depending on many factors, including method of deposition and fabrication of the actuator, powering mechanism, etc. A general relationship between force and displacement, therefore, is of little use, and actuators should be treated individually.

The power efficiency of biological actuators (i.e., muscle) ranges between 0.25-0.50, which can vary considerably, depending on the load conditions, the condition of the muscle, diet, temperature, and other biological parameters.

Noise: Assuming that the actuator input is noiseless, fluctuations in the actuator output (for example, displacement or force) are usually manifestations of mechanical or electrical fluctuations occurring in the actuator. The actuator noise is directly determined by its actuation mechanism(s) and the fabrication methods that are used to deposit the active materials, etc. In magnetic actuators, for example, the domain wall motions (Barkhausen effect) inside the ferromagnetic material introduce fluctuations in the force/displacement generated by the actuator. The size of the domains in these materials is determined by how they are deposited and by what annealing sequences are used to impart their desired magnetization behavior. The spectrum and magnitude of the noise generated by the domain wall movement also depends on the excitation magnitude and its time dependence.

Scaling: Reducing the size of an actuator may be of great value if it results in a better yield and performance. Not all actuators can be scaled down, however. Thus, it is important to have a "scaleability" measure to evaluate different actuation methods. For example, electrostatic actuators are completely scaleable, and reducing their size improves their performance, while magneto-static actuators are difficult to scale down. We introduce a scaleability measure as:

$$Sc = -\ (d\eta_{\pi}/dV) \tag{1.10}$$

where η_{π} is the power efficiency and V is the volume of the active parts of the actuator. Thus, if a method is highly scaleable, *Sc* may even increase as the volume is decreased. Even if *Sc* does not increase and stays constant at smaller volumes, it would indicate that the method is scaleable. It is also possible to define scaleability as the ratio of an ultimate actuator, such as an amoeba, and the actuator volume.

Overall Performance: Two methods can be used to evaluate the overall performance of an actuator: (i) the worst case approach, which assumes that all the errors add in the same direction and determines the overall error as the linear sum of the performance error; (ii) the root-mean-square (RMS) approach, which uses the root-mean-square of all the performance errors. RMS does not assume all errors are in one direction, which makes it more reasonable than the worst-case approach.

1.3 Actuation Means, Actuator Shells and Figure of Merit

As in sensors [3], one can divide a given actuator into the two parts of actuator shell and actuation method or mean [3]. The actuator shell is usually the edifice of the actuator and it may contain deformable parts and mechanical links, etc. Its main function is to provide a mechanism for the actuation method to produce useful work. Hence, a given actuator shell may be used with different actuation methods. The actuation method (or mean) is the method by which a control signal is converted to a force that is applied to the actuator shell. In the actuator example shown in figure 1.5a, the actuation method is electrostatic and the actuator shell is a microbridge. The microbridge can be used with different actuation methods as shown in figure 1.5b, where the difference in the thermal expansion coefficient of the metal and Si causes the microbridge to bend when the silicon is heated by passing a current through it.

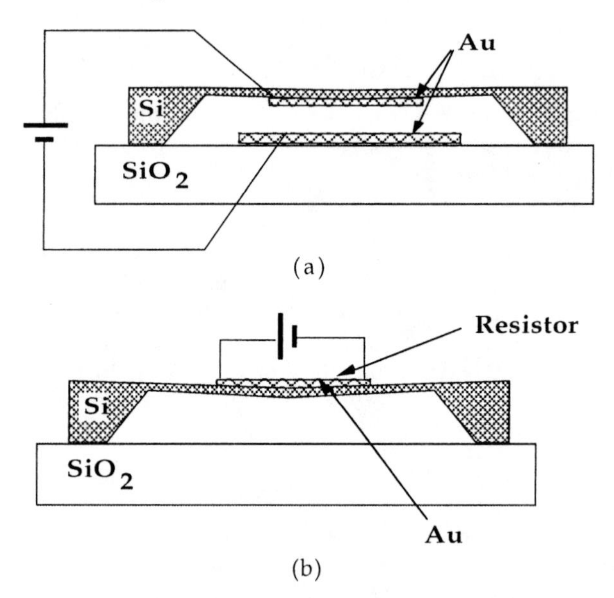

Figure 1.5 An example to illustrate actuator shell and actuation method (mean). In (a) the electrostatic attraction between two metallic plates deforms the microbridge, while in (b) the microbridge is deformed because of the non-zero difference between the thermal expansion coefficients of the silicon and the metal.

Decomposing actuators into a shell and a mean enables comparison between different shells and actuation methods. In the following section we discuss common actuator shells that can be used with a variety of actuation methods.

1.3.1 Deformable Shells

In this section we will briefly discuss the mechanics of small deformable micromachined elements. Most of the treatment that is presented here can be found in many textbooks. However, it is important to discuss some cases that are often used in the literature and to point out some differences that are encountered in the elastic behavior of micromachined mechanical elements when compared to the others. Starting with a simple cantilever beam, we discuss its static and dynamic behavior. Then we proceed to cover microbridges, diaphragms, and torsional mirrors. These simple mechanical elements are used in the majority of actuators, and other more elaborate structures can be analyzed using them.

1.3.2 Elastic Constants of Micromachined Structures

In table 1.2 we present the important elastic constants of materials that are used in microstructures. Most often the Young's modulus of the deposited or micromachined film depends on the deposition method [4,5]. Thus, the values given in table 1.3 are typical. Organic materials, like polyimide, tend to be more sensitive to deposition conditions than others. Young's modulus (Y) is defined as [6]:

$$T=YS \tag{1.11}$$

where T is the stress, and S is the strain. Different methods are used to measure the Young's modulus [5-7].

Table 1.2 Young's modulus of materials used in micromachining [4,5].

Material	Deposition Method	Young's Modulus (10^{10} dynes/cm^2)	Material Density (g/cm^3)
SiO_2	fused silica	73	2.2
	thermal-wet	57	2.2
	thermal-dry	67	2.25
	sputtered	92	2.2
Si_3N_4	crystalline	385	
	bulk	150	3.1
	CVD	146	3.1
	sputtered	130	3.1
polysilicon	CVD	165	2.33
p^+ silicon	diffused	140	2.33
bulk silicon	slightly doped	190	2.33
Aluminum	evaporation	70	2.7
Gold	evaporation	80	19.3
Ir	spun cast	520	2.25

Static methods include slowly varying the stress and measuring strain [6,7]. The slope of the stress-strain curve is the Young's modulus. Clearly, since both stress and strain are second-rank tensors, the Young's modulus is not the same in all directions and itself is a second-rank tensor.

Alternatively, instead of measuring the Young's modulus using the static method, dynamic measurements can be performed [8-18]. In these measurements a cantilever beam, for example, is fabricated using the material of interest, and its resonance behavior is examined, as discussed in the next section. This method yields a dynamic Young's modulus that may be different from the static Young's modulus; in polymers, the static and dynamic Young's moduli are usually different.

1.3.3 Microcantilever Beams

The microstructure is shown in figure 1.6. The Si cantilever beam has a thickness $t<<L$ and is rigidly mounted at one end. Assuming uniformity in the y-direction, the bending of the cantilever is a function only of x, the distance along its length.

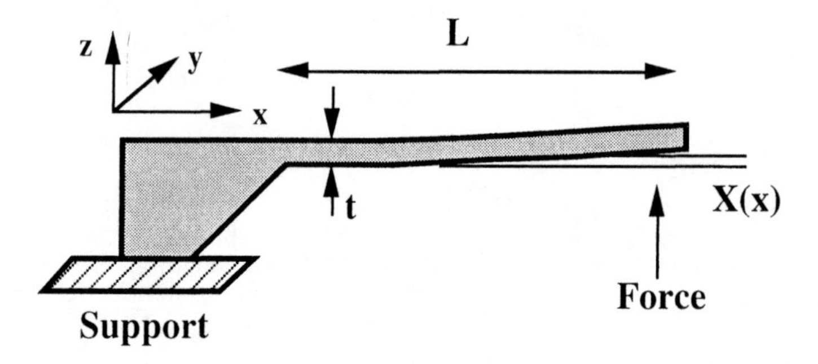

Figure 1.6 The geometry of a micromachined Si cantilever beam.

Small deflections $w(x)$ in the z-direction of the point (x,y) on the cantilever beam in the presence of viscous damping ($-\beta\, dw/dt$) satisfy the differential equation [8,9]:

$$\frac{D}{t\rho}\frac{d^4 w(x,t)}{dx^4} + \frac{d^2 w(x,t)}{dt^2} - \beta\frac{d\,w(x,t)}{dt} = 0 \tag{1.12}$$

In the above equation, the flexural rigidity (D) is $Yt^3/12(1-v^2)$, where Y is Young's modulus (scalar in this case) and ν is Poisson's ratio. Classical boundary conditions apply at the clamped and free ends.

Using the separation of variables $w(x,t)=X(x)T(t)$, we find the following equations for the independent variables x and t:

$$\frac{d^4 X}{dx^4} - \frac{t\rho}{D}\omega_i^2 X = 0 \tag{1.13}$$

$$\frac{d^2 T}{dt^2} - \beta\frac{d\,T}{dt} + \omega_i^2 T = 0 \tag{1.14}$$

For free vibrations, the spatial solution is of the form:

$$X_i(x) = c_1\cos\lambda_i x + c_2\sin\lambda_i x + c_3\cosh\lambda_i x + c_4\sinh\lambda_i x \quad (1.15)$$

where $\lambda_i^4 = (t\rho/D)\omega_i^2$, and the following boundary conditions should be satisfied:

$$X_i(0) = X'_i(0) = 0;\ X''_i(L) = X'''_i(L) = 0 \quad (1.16)$$

where prime denotes a derivative with respect to position "x". These boundary conditions constrain the eigenvalues λ_i such that $\cos(\lambda_i L)\cosh(\lambda_i L) = -1$, hence forming an infinite ordered set of eigenvalues [9]:

$$\lambda_o L = 1.875,\ \lambda_1 L = 4.694,\ \text{etc.} \quad (1.17)$$

The temporal equation has a general solution:

$$T_i(t) = d_1 e^{m_1 t} + d_2 e^{m_2 t} \quad (1.18)$$

where $m_{1,2} = \beta/2 \pm i[\omega_i^2 - \beta^2/4]^{1/2}$. For forced vibrations, detailed spatial dependence of the applied force along the cantilever beam is required. Assuming that the force is applied at the tip of the cantilever beam and that it only excites the lowest mode of vibration, the temporal equation becomes:

$$\frac{d^2T}{dt^2} - \beta\frac{dT}{dt} + \omega_0^2 T = f(t) = B\cos\omega t \quad (1.19)$$

The solution of equation 1.19 is of the form $T(t) = A\cos(\omega_o t + \phi)$, where the force amplitude F and the oscillation amplitude A are related by the well known response or transfer function:

$$\frac{T(\omega)}{F(\omega)} = \frac{a^2K^{-1}}{1-(\omega/\omega_0)^2 + j(\omega/Q\omega_0)} \quad (1.20)$$

where the quality factor Q is just $1/\beta$, "a" is related to the amplitude of the oscillation and can be approximated as the average of $X(x)$ over the tip section of the cantilever, and the effective spring constant k is related to the mass of the beam (M) by definition: $k=\omega_o^2 M$.

Clearly, the amplitude A reaches its peak when $\omega = (\omega_o^2 - \beta^2/2)^{1/2}$. For small displacement, the displacement of the tip of the cantilever is related to the applied force (F) through Hooke's law $F= -kX(L)$, where k is related to Y, I and L through:

$$k=3YI/L^3. \quad (1.21)$$

Typical values for the silicon cantilever beam dimensions are beam width in the y-direction (b) 50 μm, beam length (L) 500 μm, and thickness (t) 5 μm which result in an effective Hooke's constant of 1-2 N/m [3].

1.3.4 Microbridges

A variety of materials, including silicon, polysilicon, and Si_3N_4 have been used to fabricate microbridges for micromachined sensors. Figure 1.7 shows an example of a microbridge having the dimensions of L in the x direction, b in the y-direction, and thickness t.

The behavior of the free oscillation of the above bridge, with the total axial load P, is given by [10,11]:

$$YI\frac{\partial^4 w(x,t)}{\partial x^4} + \rho A\frac{\partial^2 w(x,t)}{\partial t^2} - P\frac{\partial^2 w(x,t)}{\partial x^2} = 0 \quad (1.22)$$

where $w(x,t)$ is the displacement of the neutral axis of the microbridge from equilibrium, Y is Young's modulus, A is the cross-sectional area, and I is the moment of inertia.

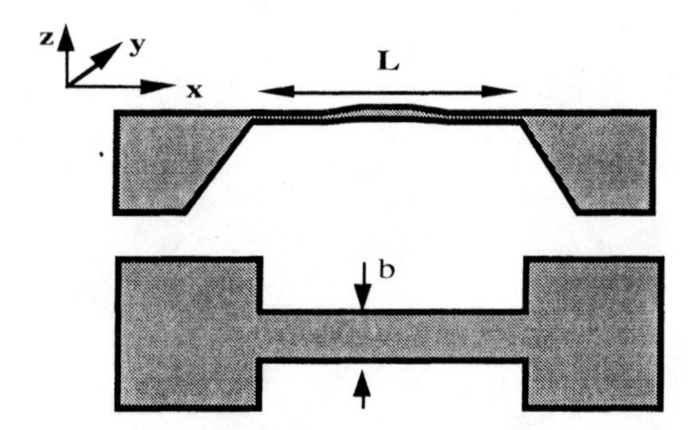

Figure 1.7 A micromachined microbridge.

The spatial deflection of the bridge ($X(x)$) can be approximated, using the Raleigh-Ritz approximation, as [10,11]:

$$X(x) = Cx^2(x-L)^2 \qquad \text{for } 0 \le x \le L \quad (1.23)$$

For a sinusoidal excitation, the temporal solution is of the form $A\cos(\omega_o t+\phi)$, where ω_0 is given by [10]:

$$\omega_0^2 = (E\,t^2/12L^4)\frac{\int_0^L b(x)\left(\frac{d^2X(x)}{dx^2}\right)dx}{\int_0^L \rho(x)b(x)X^2(x)dx} \quad (1.24)$$

where X is given by equation 1.23, $b(x)$ is the width of the microbridge, and ρ its density. In general, both ρ and b may depend on x. For uniform b and ρ, we find the lowest oscillation frequency as:

$$\omega_o = \frac{42Yt^2}{\rho L^4}\,(1+\frac{2L^2}{7t^2}\,S_T) \quad (1.25)$$

where the total strain S_T is the sum of the built-in strain and the applied strain. The transfer function relating the average deflection of the center of the microbridge to the total applied force is of the form given in equation (1.20).

The quality factor, when damping by the air underneath the bridge is taken into account, is given by:

$$Q^{-1} = \mu(Y\rho)^{-1/2}t^{-2}d^{-3}b^2L^2/4 \quad (1.26)$$

where μ is the viscosity of the air. Quality factors as high as 35,000 are obtained in polysilicon microbridges (L=200 μm, b=45 μm, t=2.2 μm) in vacuum [10,11]. Typical values of 10-1000 are common in air.

1.3.5 Diaphragms

Diaphragms and plates are extensively treated in the literature [33,34]. Here we present a very simplified treatment to give an essence of what has been done. Two simple cases of square and circular diaphragms are considered.

The shape of the square diaphragm is approximately given by [13]:

$$X(x) = a\cos(\frac{\pi x}{L})\cos(\frac{\pi y}{L}) \quad (1.27)$$

where a is the displacement of the diaphragm in the center from its equilibrium position. The shape of the circular diaphragm is approximately a semi-spherical cap:

$$X(x) = a - R + (R^2 - r^2)^{1/2} \quad (1.28)$$

where r is the distance from the center of the diaphragm and R is the radius of the curvature of the deflected diaphragm:

$$R = \frac{a^2 + L^2/4}{2a} \quad (1.29)$$

In both cases, the relationship between the applied pressure (P) and the displacement is given by:

$$P = \frac{4C_1t}{L^2}\,\sigma_o\,a + \frac{16C_2f(\nu)t}{L^4}\,\frac{Y}{1-\nu}\,a^3 \quad (1.30)$$

where S_o is the residual stress, $Y/(1-\nu)$ is the bi-axial modulus, and t is the thickness of the diaphragm. The dimensionless C_1, C_2 and $f(\nu)$ depend on the geometry of the diaphragm. For circular diaphragms, f(ν) is 1, C_1=4, and C_2

$=2.67$, while for square diaphragms $f(\nu)$ is $1.075\text{-}0.292\nu$, $C_1=3.04$, and $C_2=1.37$. A more accurate finite element calculation yields $f(\nu)$ of $(1.026 + 0.233\nu)^{-1}$, $C_1=4$, and $C_2=2.67$, and $f(\nu)$ of $1.446\text{-}0.427\nu$, $C_1=3.41$, and $C_2=1.37$, respectively, for the circular and the square diaphragms [13].

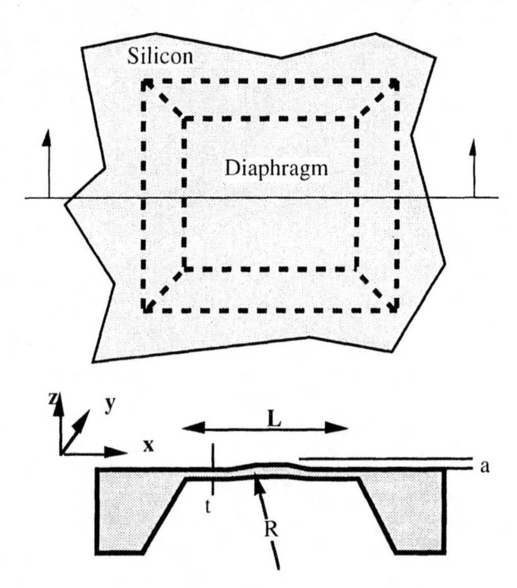

Figure 1.8 Micromachined silicon diaphragm.

For small oscillations, the temporal behavior is also described by equation (1.20). The first mode oscillation of a square diaphragm is [14]:

$$f_o = \frac{35.99}{2\pi L^2}\left[\frac{Yt^2}{12\rho(1\text{-}\nu)^2}\right]^{1/2} \tag{1.31}$$

For a silicon diaphragm of 15 μm-thick ($=t$), 2 mm-width ($=L$), $Y=1.4x10^{12}$ dynes/cm^2, and $\nu=0.278$, f_o is 12.1 kHz.

1.3.6 Torsional Mirrors

Torsional mirrors may be of some interest in integrated optic sensors. They can be used in scanning applications, and, because of their relatively higher oscillation frequency, they may be used as a microbalance. Figure 1.9 shows a schematic of a simple torsional structure that can be fabricated using silicon micromachining [15-17,19].

Of interest are the relationship between the twisting angle (θ) and applied force (static) and the oscillation behavior of the structure, which includes the resonance frequency (f_c) and a quality factor (Q). It can be shown that θ is given by [19]:

$$\theta = \frac{h\tau}{k_1 G g t^3} \tag{1.32}$$

where t is the applied torque, $k_1 = 1/3$, g and t are the structure dimensions shown in figure 1.9, and G is Lamb's elastic coefficient given by:

$$G = \frac{Y}{[2(1+\nu)]} \tag{1.33}$$

Usually, force (F) is applied at the edge of the structure using electrostatic fields. In that case $t = Fg/2$, and we can derive a Hooke-type relationship $F = k\theta$, where k is $2Gt^3/3h$ or $Yt^3/[3(1+\nu)]$.

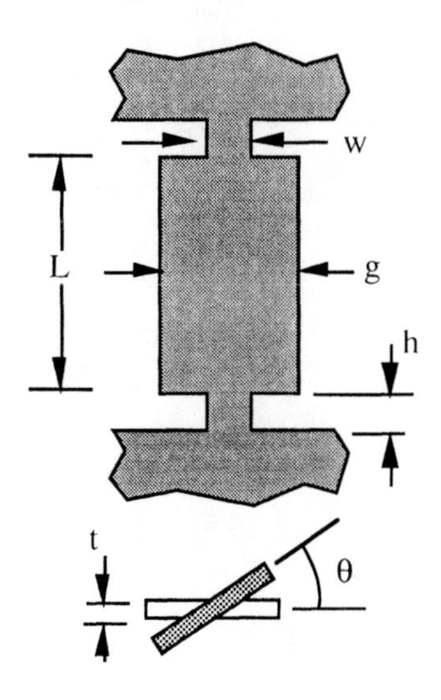

Figure 1.9 A micromachined silicon torsional structure.

The equation of motion for small oscillation amplitudes is given by:

$$\ddot{\theta} = \frac{-2k\theta}{Ih} \tag{1.34}$$

where I is the moment of inertia ($h >> w, t$) given by:

$$I = \frac{M}{12}(g^2 + t^2) = \frac{\rho L t g}{12}(g^2 + t^2) \tag{1.35}$$

where L is shown in figure 1.9. The resonance frequency (ω_0), therefore, is:

$$\omega_o = \sqrt{\frac{Gt^2}{h} \frac{72}{\rho L(g^2 + t^2)}} \tag{1.36}$$

For example, a silicon torsional structure with the following dimensions: $t = 3$ µm, $h = 20$ µm, $g = 50$ µm, and $L = 250$ µm, will exhibit an effective Hooke's constant of 0.22 N/° and a resonance frequency of 18400 rad/s [19].

1.3.7 Pivoted/Rotational Shells

The most notable rotational shell is the rotor of the variable capacitance micromotor discussed in the next chapter. Given an electromotive or any other force causing rotation F, we would like to find the torque in the presence of different frictional and viscous (air or liquid damping) forces.

It is well-known that the torque, when no friction is present, is given by rXF, and since in these motors F is always perpendicular to r, torque is simply rF. τ, in the case of the electrostatic motors discussed in the next chapter, is given by the derivative of the electrostatic co-energy with respect to the rotational angle: $\tau(\theta)=(1/2)V^2dC(\theta)/d\theta$, where $C(\theta)$ is the capacitance between the stator and the rotor poles and V is the excitation voltage. The power is simply $P=\tau\omega$, where ω is the angular frequency.

The equation of motion of the rotor is [41,42]:

$$I\frac{d^2\theta}{dt^2} + C_v\frac{d\theta}{dt} + C_d g m r_c + \tau(\theta) = 0 \qquad (1.37)$$

where I is the moment of inertia of the rotor (approximately in the range of 10^{-20}-10^{-19} kg m^2 for the polysilicon rotors of 125-200 μm-diameter and 3-4 μm-thick discussed in Chapter 2), θ is the rotation angle ($\omega(t)= d\theta/dt$), C_v is the coefficient of viscous torque, C_d is the dynamic coefficient of friction, m is the rotor mass, g is the gravitational constant and r_c is the radius of bushing rim on the bottom of the rotor (figure 1.10).

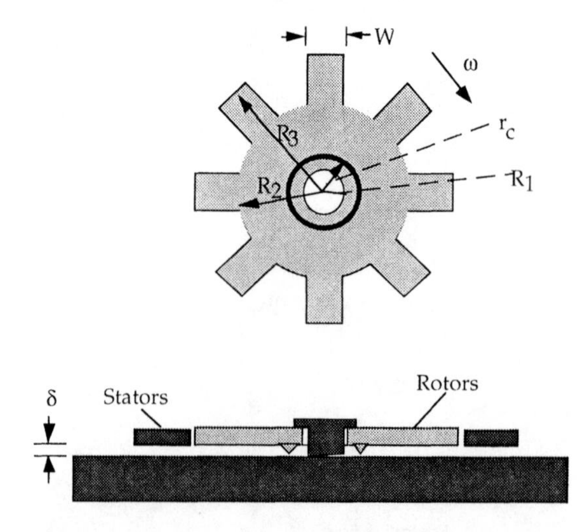

Figure 1.10 A schematic of a variable capacitance micromotor used as an example of pivoted and rotational shell. The bottom drawing is the cutaway side view of the micromotor, while the top view shows the rotor structure.

The coefficient of viscous torque is given by [41,42]:

$$C_v = \frac{2\pi\mu}{4\delta}(R_2^4 - R_1^4) + N\frac{\mu W}{3\delta}(R_3^3 - R_2^3) \tag{1.38}$$

where δ is the separation between the rotor and the substrate plate, μ is the dynamic viscosity of the fluid (typically around 1.7×10^{-5} kg/m s), W is the width of the individual teeth or blades and various radii (R_1-R_3) are shown in figure 1.10. Typical values of C_v and C_d are around 9×10^{-7} N m s and 0.28, respectively.

1.3.8 Figure of Merit

Figure of merits are usually defined to facilitate comparison among different types of devices and actuators. Denoting it by μ_A, we define a figure of merit for actuators that is related to the actuator power efficiency, its scaleability, the amount of time it takes to be turned off (τ_{off}) and turned on (τ_{on}), its smallest inducible output (sIO), its maximum output span (Y_{max}), and its sensitivity (S). The figure of merit is also related to the ratio of the mechanical work output ($\Delta F\ \Delta x$) during one actuation cycle to the energy required to control one cycle ($(\tau_{ov}+\tau_{off})P_{in}$):

$$\mu_A = \eta_p . Sc . S . \frac{\Delta F . \Delta x}{(\tau_{on} + \tau_{off})P_{in}} . \frac{Y_{max}}{sIO} \tag{1.39}$$

Since we are interested in microactuators, we take the scaleability to be 100%, or 1, when the size of the actuator can be reduced to 1000 μm^3. At first glance, this may look quite arbitrary, but it is based on the size of an amoeba, which is around 10-100 μm in diameter. Thus, we take the stand-alone biological system as an example of the ultimate in microactuators.

1.4 Microfabrication Methods

We divide the present chapter into three main parts. The first part deals with steps that are common to both bulk and surface micromachining. The second and third parts deal with the bulk and surface micromachining fabrication steps separately. We almost exclusively discuss silicon-based microstructures, and we assume that some familiarity already exists with silicon crystallography, silicon growth , and its epitaxy.

1.4.1 Common Fabrication Steps

Common to the fabrication of both surface and bulk micromachined structures are the following:

1) Deposition of thin films (metallic, semiconducting, insulating, and protective)
2) Patterning techniques
3) Etching

4) Dicing, milling, and cutting methods
5) Packaging

Deposition Methods. Deposition of thin layers is usually carried out by the following different methods:

i) Thermal growth
ii) Thermal evaporation, electron beam-assisted evaporation, and laser ablation
iii) Sputtering
iv) Ion-beam method
v) Sol-gel method
vi) Metal-organic chemical vapor deposition
vii) Molecular beam evaporation methods

i) Thermal Growth. Thermal oxides and nitrides usually have superior electrical, optical, and mechanical qualities compared to other films [22]. However, in a thermal growth process, the deposition rate is usually much smaller than it is in other techniques, such as the chemical vapor deposition method; and, more often, thermal growth is carried out at higher temperatures. Nonetheless, to impart acceptable electronic quality to the interface between the semiconductor and an insulator, an intervening layer of thermally-grown oxide is very often necessary. This situation arises in optoelectronic circuits where optical components are integrated with electronic devices. On the other hand, optical components require relatively thick buffer layers that, if grown thermally, would adversely affect the minority carrier lifetime in the semiconductor, resulting in unwanted diffusion of impurities away from the device regions. Thus, the initial thermally-grown oxide is usually covered by a thick insulator that is deposited using the methods discussed in the next chapter.

ii) Evaporation. In evaporation methods, which are commonly used in IC fabrication [23], the substrate is placed inside a bell jar and, using roughing and diffusion pumps, a vacuum of 10^{-5}-10^{-7} Torr is created. A suitable material, such as a metal, etc., is placed on heating elements usually situated underneath the substrate. Upon heating, the material melts and then evaporates onto the substrate surface. While in optical devices trace impurities and ions are not important, in electronic devices great care should be taken to insure the cleanliness of the heating element and of the starting materials. In some cases, an electron beam is used to melt the material. Electron beam heating is usually required when depositing refractory metals and ceramics like SiO_2 and Si_3N_4 [45].

iii) Sputtering. Sputtering is performed using an apparatus identical to the one used in evaporation, except that instead of using thermal energy to evaporate the material, an electric field is used to bombard the material surface with noble gases and to knock atoms out of the target material. These atoms are subsequently collected on the substrate, forming the desired film [5,25]. Sputtering is avoided in some cases because of the potential damage of the substrate surface by the energetic atoms. It is, however, the only possible deposition method in depositing some ceramics.

iv) Ion-beam Method. The main idea is to form ions of the desired material and to direct them onto a substrate to form a uniform film. This method usually has a small growth rate and deposits high-quality electronic films. Its application to integrated optics and microstructures may be limited to AlGaAs/GaAs microstructure systems, discussed at the end of the next chapter. For a description of ion-beam deposition of materials, please refer to [26].

v) Sol-gel Method. The sol-gel method is extensively used in fabricating piezoelectric actuators. Sol-gel is an inexpensive deposition method process based on combining liquid reactants and the subsequent solidification of the resultant solution into an amorphous oxide gel [27]. The porous oxide gel is then heated to give densified glasses or polycrystalline solids. Since the heating can be done locally by a laser or an electron beam, the sol-gel method is a good candidate for direct patterning using lasers. It also lends itself naturally to doping and incorporation of impurities and formation of composites. This last attribute is quite important in regards to formation of waveguides and multilayer structures. Moreover, due to the softness of the oxide gel, embossing can be used to form gratings and other devices [28-31].

To form SiO_2 films, tetraethoxy silane (TEOS) is dissolved in an alcoholic solution and then hydrolyzed (by adding H_2O). After the hydrolization, the reaction starts and an alkoxy group of one TEOS molecule reacts with the OH group of an adjacent molecule to form an Si-O-Si bond. This process then repeats itself, resulting in an effective polymerization of the solution and in an increase in its viscosity. The polymerization process can even occur at room temperature, making the sol-gel technique quite desirable.

To form other oxides, such as TiO_2, the silicon in the tetraethoxy silane is replaced by another element, such as Ti. Various oxides and composites, including ZrO_2, GeO_2, $BaTiO_3$, ZnO and PZT, are formed and deposited using the versatile sol-gel method [28-31].

vi) Metal-Organic Chemical Vapor Deposition. The metal-organic chemical vapor deposition method is reviewed in [32]. It consists of using a precursor that is heated at some temperature T_p, corresponding to its evaporation temperature. A carrier neutral gas, such as argon, nitrogen or xenon, carries the vapors downstream over a desired substrate that is kept at temperature T_s, which is higher than T_p and which corresponds to the decomposition of the metal-organic precursor. Thus, the metal-organic vapor decomposes on the substrate, depositing the desired film. This method potentially has a very high deposition rate at relatively low temperatures. Thus, it is quite desirable in depositing high-quality thick insulator and semiconductor films. It can also be used to deposit a variety of different materials (this requires synthesis of the desired metal-organics). Thus, this method is quite versatile and is used extensively to deposit both electronic- and optical-quality films.

vii) Molecular Beam Epitaxy Method. The molecular beam epitaxy method is an ultra-high vacuum evaporation technique where the base pressure is less than 10^{-10} Torr. It results in the highest possible quality electronic films. Its growth rate is very small and it is used to deposit a few hundred atomic layers. This method results in very sharp interfaces between different layers and is used to fabricate quantum wells, superlattices, and heterostructures [33].

Patterning. Patterning is usually performed by photolithography or electron beam lithography methods. This process is very similar to photography, where the exposed areas either protect (when negative photoresist is used) or expose (when positive photoresist is used) the underlying material to the subsequent processing, such as etching. An excellent review of these methods along with the description of some common methods and photoresists is given in [22].

Etching. Etching of silicon is usually carried out in an electrolyte with two components: one of the components oxidizes the silicon and the other component dissolves the resulting oxide [22,34,35]. Strong oxidants, such as nitric acid, are commonly used as the first component. Buffered hydrofluoric acid is usually used as the oxide remover.

Anisotropic Etching. There are well-known etchants that etch the crystalline silicon anisotropically. They usually etch silicon much slower in the <111> direction than in other directions, resulting in delineation of these planes. Among these anisotropic etchants are KOH and ethylenediamine- base solution (EDP) [15,16,35-38]. Table 1.3 shows the etch rates of these etchants in SiO_2 and silicon having different crystal directions [35].

Table 1.3 Etch rates of anisotropic etchants of Si, SiO_2, and Si_3N_4 [35].

Material	EDP $R(T)=R(\infty)e^{-Ea/KT}$ $R(\infty)$, E_a	EDP @ 85°C (μm/h)	20% KOH @ 85°C (μm/h)	42% KOH @ 85°C (μm/h)
(110) Si*	7.3×10^5, 0.33	2.4		
(100) Si*	5.7×10^6, 0.40	2	55	45
(111) Si*	7.3×10^6, 0.52	0.4		
(100) 10^{16} B^+-Si	6.95×10^6, 0.4	≈2	55	45
3.3×10^{19} B-Si	5.8×10^7, 0.5	8		
4.4×10^{19} B-Si	1.18×10^7, 0.48	2.5	20	18
7×10^{19} B-Si	3.46×10^6, 0.5	0.5		
9×10^{19} B-Si	9.9×10^5, 0.49	0.15	2	4.5
Si_3N_4	very low	very low rate	very low rate	very low rate
SiO_2	9.94×10^7, 0.80	8×10^{-4}	0.5	0.7

* Low doping concentrations.

† Boron doped.

Typical etching profiles of silicon that are obtained using isotropic and anisotropic etchants are shown in figure 1.10.

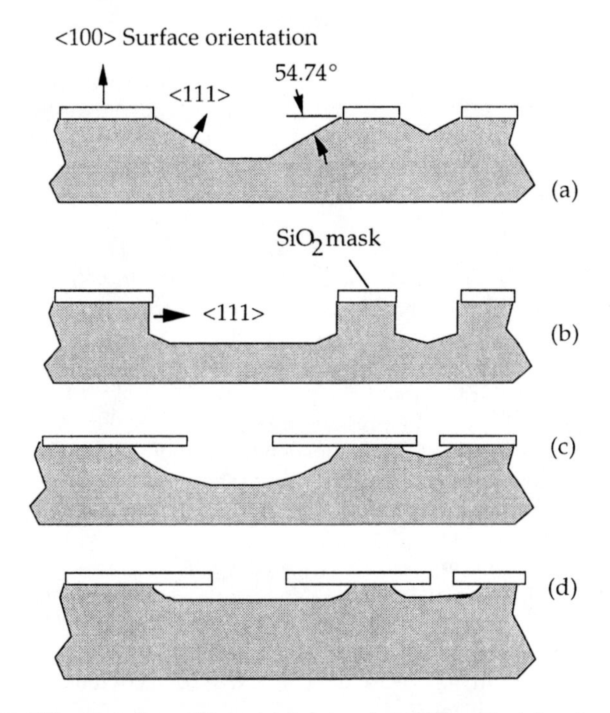

Figure 1.10 Silicon etch profiles obtained using different etchants. (a) and (b) Anisotropic etching on (100) and (110) surfaces respectively. (c) Isotropic etching with agitation. (d) Isotropic etching without agitation [16,17].

Etch-stops. Heavy boron doping is commonly used to stop the etching of silicon by KOH and EDP. The etch rate drops by a factor of 10 when the boron concentration increases from 10^{19} cm^{-3} to $4x10^{9}$ cm^{-3}. Thus, heavily doped regions remain nearly intact [16,17,35-38].

Anisotropic etchants also etch silicon at much faster rates than they etch SiO_2 [35]. At 30 °C, the ratio of EDP's etch rate of (100) silicon over SiO_2 was $2.5x10^{5}$. Using 10% KOH, the same ratio was $2.5x10^{3}$.

The space charge region in a p-n junction can also be used to stop etching of silicon [36]. This etch-stop technique takes advantage of the anodic passivation of silicon. In this scheme, an electrochemical cell is constructed with three electrodes. Connected to the anode is the n-side of the silicon p-n junction. A positive voltage is applied directly to the n-type silicon, and the electrical contact to the p-type silicon is accomplished through the electrolyte, which is either KOH or EDP. Since the majority of the voltage drop is across the reverse biased p-n junction, the p-type silicon essentially remains at small potentials and etches. With the complete removal of the p-type silicon, the voltage drop appears across the n-type silicon and the electrolyte. This anodic voltage results in oxidation of the n-type silicon which prevents it from etching. Using this method, n-type silicon mechanical structures were fabricated [36].

Dry Etchants. These etchants are extremely attractive because highly anisotropic etching can be achieved by manipulating the momentum of the reactive gases. Thus, photoresist undercutting can be virtually eliminated and very high aspect ratios can be achieved. Another important feature of the dry etching process is the lack of liquid contaminants in the etching process. On the other hand, some explosive and dangerous gases are used in this method despite the fact that their concentration is usually quite low. Therefore, dry etching, which includes plasma etching and reactive ion etching, is quite suitable in the fabrication of microstructures [22, 37].

Ion Milling. Ion milling is very similar to sputtering since both methods use momentum transfer for the removal of material. The only difference is that a separate source is used to prepare the argon ions which are used for milling. Thus, in ion milling, both the density and energy of the ions can be controlled independently. A typical milling rate for silicon is 215-500 Å/min when 500 eV ions are used (see page 527 in [22]).

Dicing and Cutting. Dicing is usually accomplished by inscription, using a diamond tip. In the case of crystalline material, great care should be given to crystallographic planes. The cleavage plane of silicon is {111} and that of GaAs is {110}. To cut silicon, either an SiC or a diamond saw is used. Usually wire diamond saws produce the fewest defects, but they are quite slow. A discussion of dicing and cutting of silicon is given in [34].

Packaging. The packaging of integrated optics and microstructures is particularly challenging. In these applications, the package not only provides a mechanical support and electrical connections to the chip, but it also provides optical windows. The packaging of integrated optic sensors is not well-developed, and much work is needed to develop reliable packaging for these devices. The packaging of integrated circuits is extensively discussed in the literature [22] and [34].

1.4.2 Bulk Micromachining

Bulk micromachining refers to fabrication of microstructures from the bulk of the silicon wafer. This is usually accomplished using anisotropic etchants and etch stops as discussed in the previous sections. Cantilever beams, diaphragms, and microbridges are all fabricated using bulk micromachining techniques. These structures are discussed in the next chapter.

Figures 1.11 and 1.12 show the essential steps of bulk micromachining. The process starts with a choice of an appropriate silicon substrate (figure 1.11a), followed by patterning and photholithography to define regions of silicon dioxide that act as a diffusion mask (figure 1.11b). Next, boron is diffused into the silicon (figure 1.11c), and oxidation and photolithography are used to open windows in the oxide on both sides of the silicon substrate (figure 1.11d) to enable the anisotropic etchant to react and remove the silicon (figure 1.11e).

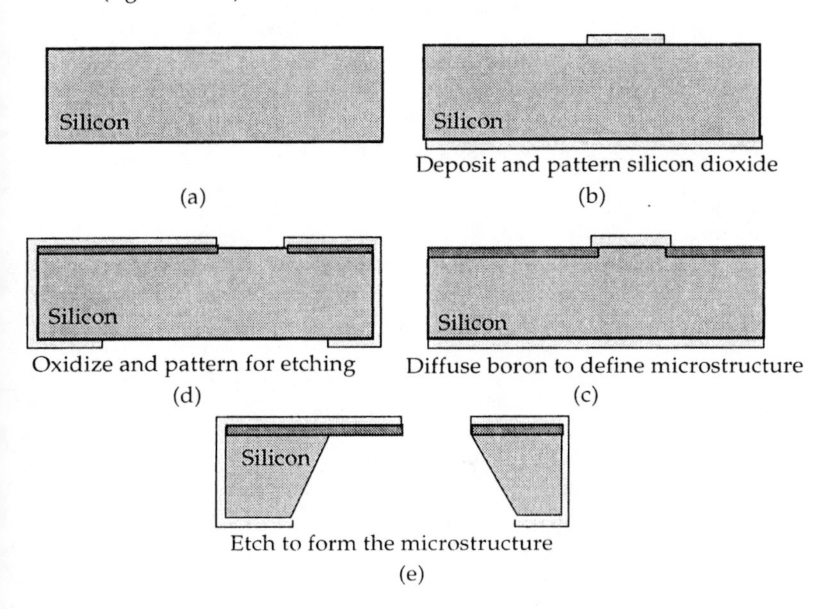

Figure 1.11 Important steps used in bulk micromachining of silicon.

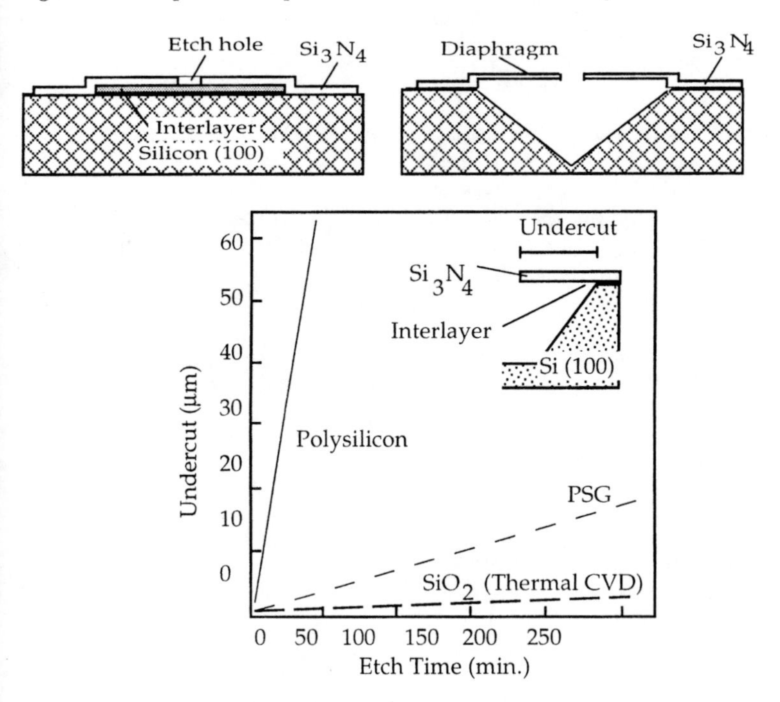

Figure 1.12 The undercut-etching of poly-silicon, phosphosilicate glass, and CVD SiO_2 [58].

In silicon bulk micromachining, usually the mask material is SiO_2 or Si_3N_4, and photolithography and heavy boron diffusion are used to define the microstructure. The anisotropic etchant readily etches silicon and is stopped by the mask material and heavily doped silicon regions. Microstructures, such as cantilever beams, microbridges, diaphragms, and torsional structures, are all fabricated using this method [38-42].

When using anisotropic etchants to fabricate microstructures, the corners and edges often become rounded because of the enhanced etching of their large surface areas. This problem can be circumvented by adding "compensation" areas to these weak spots. A simple compensation scheme, shown in figure 1.13, is shown to be effective and results in sharp edges [41]. The exact dimensions of the added areas depend on the etchant. In reference [41] a simple method to empirically determine them is explained.

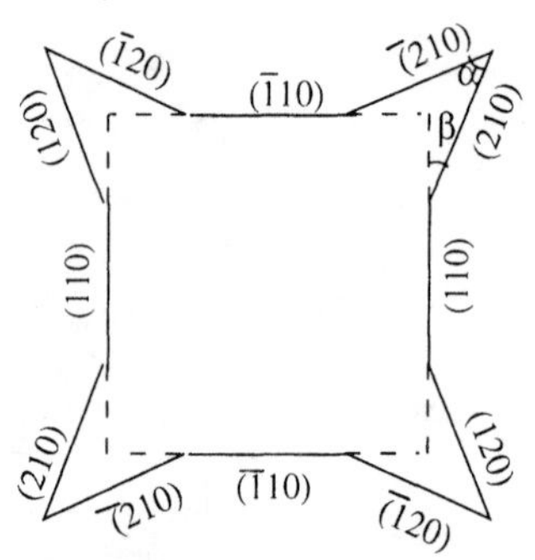

Figure 1.13 Compensation edges used to prevent "rounding" of the corners in silicon micromachining (a=53.13°, b=18.43°) [41].

In bulk micromachining, the microstructure is formed because of its lack of dissolution in the etchants. In some cases, silicon is heavily doped by boron to reduce its etching rate [15,17,35]. In some cases, heavy boron doping changes the mechanical properties of silicon, as discussed in the previous section. This may be undesirable in some applications, such as silicon diaphragms used in pressure sensors. Hence, in addition to heavy boron doping, other methods, such as p-n junction etch-stops, have also been developed, as discussed previously.

1.4.3 Surface Micromachining

Essential to surface micromachining is the concept of sacrificial layers that readily etch away, leaving behind etch-resistant layers. This method was first demonstrated using metallic films and is used to fabricate microstructures from low-pressure chemical vapor deposition (LPCVD) silicon nitrite, polysilicon, SiO_2 and polymers.

The important steps involved in surface micromachining are shown in figure 1.14. A sacrificial layer is deposited on the silicon substrate (figure 1.14a), which may first have been coated with an isolation layer (figure 1.14b).

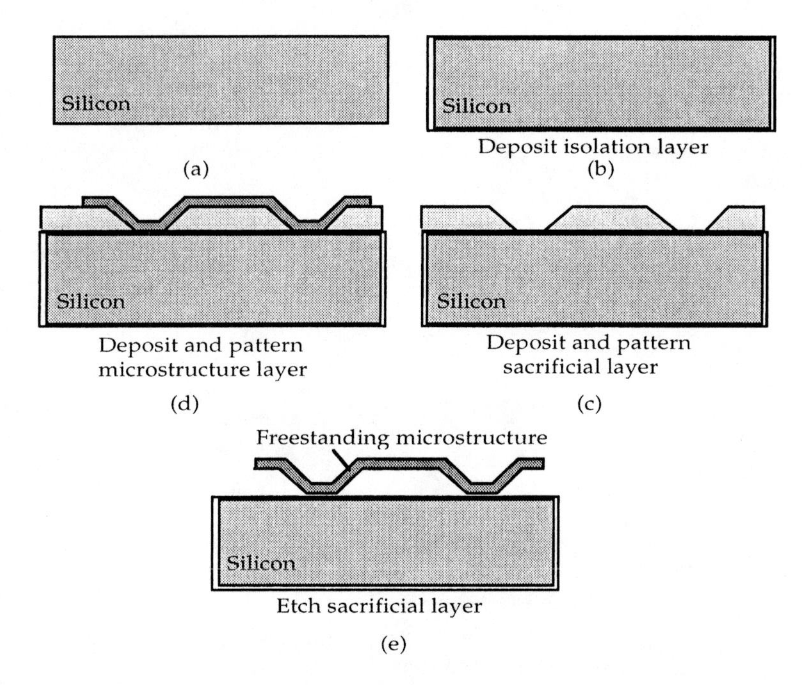

Figure 1.14 Important fabrication steps of surface micromachining.

Using regular photolithography and patterning, regions are etched in the sacrificial layer (figure 1.14c) and the microstructural thin film is deposited and patterned (figure 1.14d). Selective etching of the sacrificial layer leaves a nearly free-standing micromechanical structure (figure 1.14e). As long as the sacrificial layer can be etched without significant etching or attack of the microstructure, the technique is applicable.

Silicon nitride micromechanical beams can be fabricated by etching a polysilicon sacrificial layer in KOH, with the silicon substrate protected by an oxide or silicon nitride isolation layer. If a window is opened in the isolation region underneath the silicon nitride film, the KOH etches a cavity in the substrate after removing the polysilicon. Polysilicon microstructures, such as beams and plates, are made by etching an underlying oxide film using various forms of HF. CVD phosphosilicate glass (PSG) is HF etched much faster than thermal or undoped CVD oxides, making it attractive as the sacrificial layer.

1.5 Microfabrication of Specific Devices

Here, we cover fabrication of four specific electric microactuators: i) a variable capacitance micromotor, ii) a piezoelectrically actuated microcantilever beam, iii) flip-up free-space Fresnel microlenses, and iv) LIGA fabricated micromotors. Fabrication of these devices can be viewed as representative of most microactuators.

1.5.1 Variable Capacitance Micromotor (VCM)

VCM can be considered a typical example of micromechanisms which are free to execute 2-D and 3-D motions and which are detached from the substrate. Solid links, gear boxes and other micromechanisms can be fabricated using steps identical or similar to the fabrication steps shown in figure 1.15 [43].

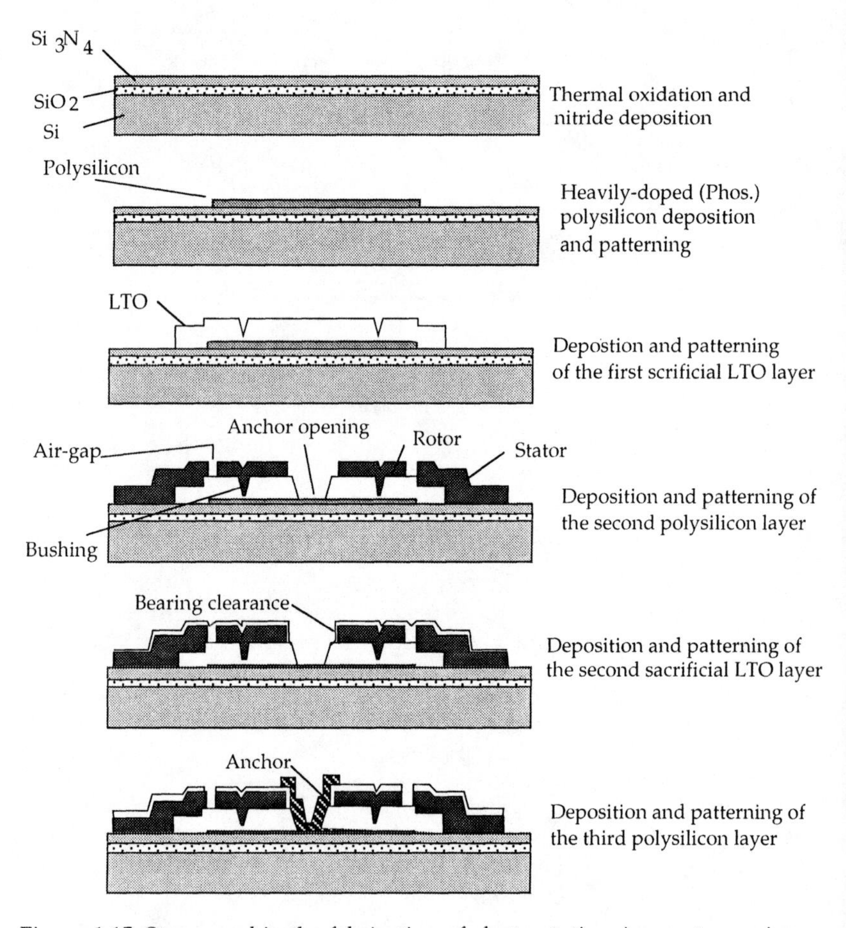

Figure 1.15 Steps used in the fabrication of electrostatic micromotors using low temperature oxide sacrificial layers and two layers of polysilicon [43].

a) Deposit 1 μm silicon-rich Si_3N_4 over 1 μm thermal oxide to isolate the substrate. b) Deposit a 0.35 μm-thick heavily-doped (phosphorus) polysilicon layer and pattern to form the substrate shield. c) Deposit the first low temperature oxide (LTO) sacrificial layer (2.3 μm) using CVD. Pattern to define the bushing molds and time etch the LTO in buffered oxide etch to a depth of 1.8 μm. The LTO layer is then patterned and etched to open windows down to the nitride layer for the stator anchors. d) Deposit a

2.5 μm-thick heavily phosphorous-doped polysilicon layer. Pattern it to define the rotor and stators. e) Deposit the second sacrificial 0.3 μm LTO layer. Pattern it to define the bearing clearance. Etch it to expose the electric shield below which the bearing is then deposited. f) Deposit a 1 μm-thick heavily-doped (phosphorus) polysilicon layer. Pattern it to form the bearing. g) Immerse the complete device in a 1:1 HF:deionized water solution to dissolve the sacrificial LTO and release the rotor. To reduce the final HF release time, a sufficient number of holes is patterned in the rotor to expedite the under-etching of polysilicon.

1.5.2 Microcantilever Beam with Piezoelectric Actuator

Figure 1.16 schematically shows the fabrication steps of fabricating a conducting tip [44,45]. These steps are:

a) After Si_3N_4 deposition over (100) silicon, lithography is used to pattern the nitride film and open a square window (3μm x3μm).

b) Anisotropic etching is used to etch the silicon and expose the <111> planes.

c) Metallization and patterning are used to deposit W over the etched pyramidal pit. At this stage a variety of methods, including electro-plating, can be used to build up W over the pit.

d) Finally, the silicon is completely etched to release the Si_3N_4 cantilever beam with the integrated tungsten tip.

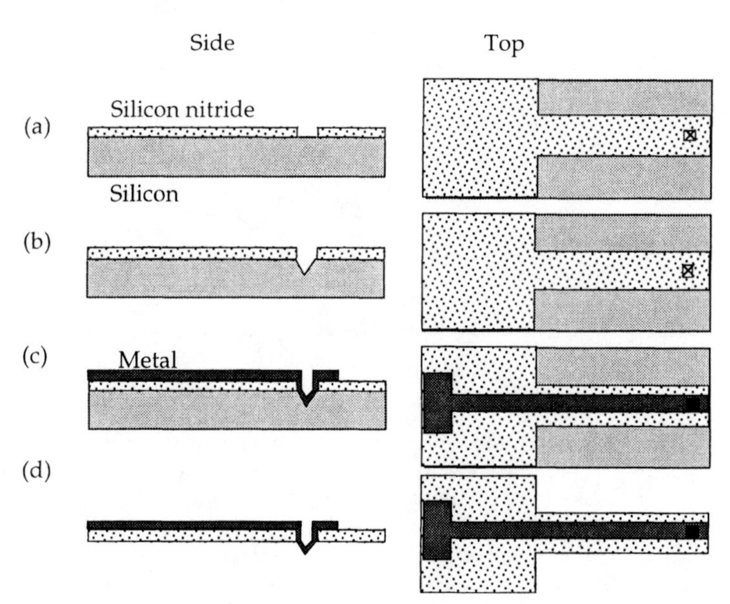

Figure 1.16 Steps used in the fabrication of a conducting tip supported by a Si_3N_4 layer [44].

To integrate piezoelectric strips with the probe or other structures (figure 1.17), ZnO or PZT piezoelectric layers are usually used. Deposition of piezoelectric layers is quite tricky because of the anisotropic nature of the layers and the fact that the actuator functions based on the proper choice of applied electric field direction with respect to the crystallographic axis.

C-axis-oriented, high-quality ZnO layers have been deposited successfully by many researchers in the past [5,44]. It has been found that it is important to have a highly-polished (minimum roughness) substrate to obtain good quality and well-oriented ZnO films. Deposition of ZnO using reactive sputtering of pure Zn in an O_2/Ar (80%/20%) plasma in a d.c. magnetron sputtering chamber has been reported. This method resulted in the deposition of ZnO films with their C-axis perpendicular to the substrate.

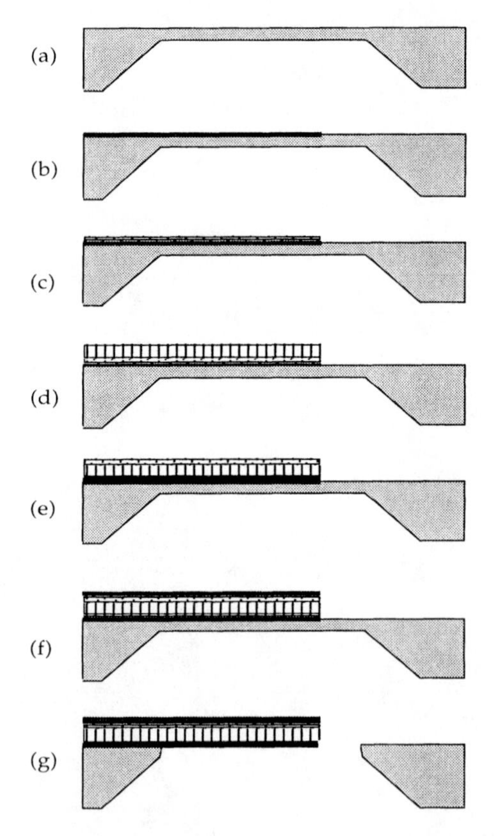

Figure 1.17 Steps used in the fabrication of a ZnO piezoelectric actuator probe. (a) Bulk micromachining to define the tip area. (b) and (c) Deposit and pattern the bottom electrode (Si_3N_4/Al/Si_3N_4 on Si). (d) Deposit ZnO and pattern. (e) Deposit and pattern Si_3N_4. (f) Deposit and pattern Al (the top electrode). (g) Etch the silicon to release the composite cantilever beam [44].

Most wet etches attack ZnO, and plasma etching is usually used to pattern dielectrics on ZnO. Usually, a thin layer of Si_3N_4 deposited over ZnO improves its electric characteristics by preventing charge injection into the piezoelectric ZnO layer. This method also improved the drift and low frequency characteristics of the ZnO for the same reason. Figure 1.17 shows some of the steps used in the fabrication of cantilever beams with ZnO piezoelectric actuators.

$PbZr_xTi_{1-x}O_3$ (PZT), with its very high piezoelectricity, is the material of choice in actuators. It can be deposited using different methods, including sol-gel, co-sputtering, etc. In the sol-gel process, the organometallic precursors $Pb(C_2H_4O_2)_2{:}3H_2O$, $Zr(OCH_2CH_2CH_3)_4$ and $Ti[OCH(CH_3)_2]_4$ are used. In sputtered film, three metallic targets (Pb, Zr and Ti) are used simultaneously in a magnetron. Subsequent annealing at temperatures up to 600 °C is needed to achieve the desired stoichiometry and structure. To orient the PZT film, a thin layer of lead titanate (<100> orientation) is usually sputtered first. A metallic barrier layer to prevent Pb/Si inter-diffusion is also employed. It has been found that a layer of Ta acts as an acceptable barrier layer and at the same time withstands annealing of the film. PZT films are poled to impart them with the desired piezoelectricity. The poling is usually achieved by applying a voltage across the PZT at elevated temperatures and as the PZT is cooled down to room temperature [45].

1.5.3 Flip-up Free-space Fresnel Microlenses

As mentioned in section 1.4.3 silicon nitride micromechanical beams can be fabricated by etching a polysilicon sacrificial layer in KOH, with the silicon substrate protected by an oxide or silicon nitride isolation layer. In figure 1.18, an example of this procedure is shown, where a Fresnel lens structure was fabricated using the above procedure. These steps are [46]:

i) A 2 µm-thick phosphosilicate glass (PSG) is deposited on the silicon substrate (PSG-1). This layer acts as a sacrificial layer and is removed later to free the structure above it.

ii) A polysilicon layer (poly) of 2 µm thickness is grown on PSG-1 (poly-1). This layer is patterned and contains the optical devices.

iii) Another layer of PSG (0.5µm thick) is deposited over poly-1 (PSG-2). This layer also acts as a sacrificial layer and, when dissolved, it provides a free space between the plate, which contains the optical devices, and the staples that are used to restrain the plate's movements.

iv) Finally, a second layer of poly is deposited over PSG-2 (poly-2). This layer is patterned and it produces the staples.

A limitation on the process is the attack by buffered HF or HF vapors on $POCl_3$-doped polysilicon films, especially when they are deposited on oxides containing phosphorus. LPCVD silicon nitride etches much more slowly in HF than in oxide films, especially when deposited with a silicon-rich composition, making it a desirable isolation film. A further limitation on polysilicon surface structures is that large-area structures, such as the Fresnel lens structure discussed above, tend to deflect and attach to the substrate after the final rinsing step. Nonetheless, this technology is very promising and will greatly enrich the free-space opto-mechanical system's capabilities.

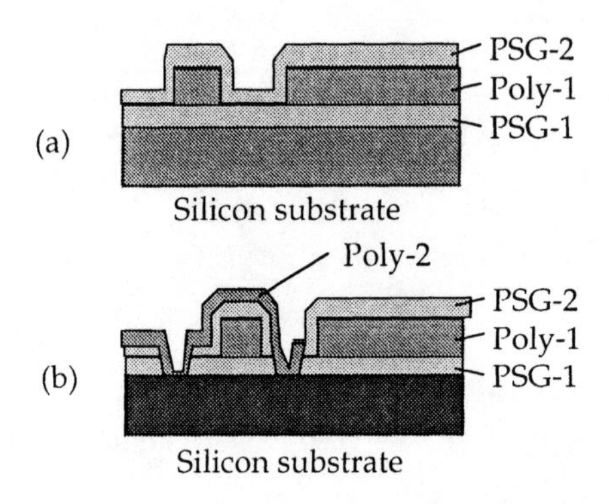

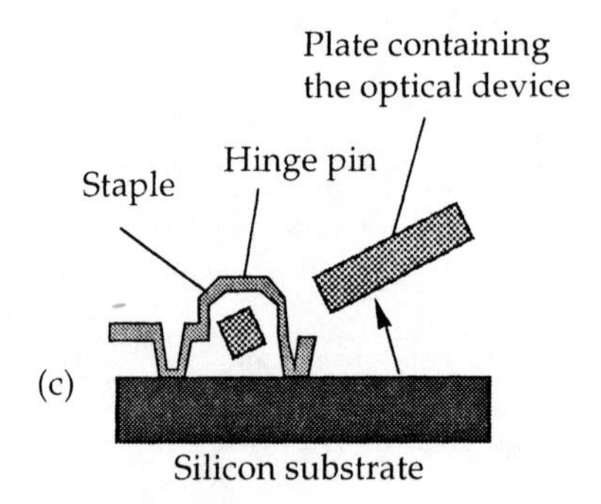

Figure 1.18 Fabrication steps used in fabricating free-space micro optical systems [46].

1.5.4 LIGA

Using the LIGA (a German acronym with an English translation of lithography, electroforming and molding) process, metallic and other structures with very high-aspect ratios (up to 1 mm in height and a few μm in width) can be easily fabricated [47,48]. These structures are subsequently used as a mandrel for injection molding. Or they can be used directly as the device structures.

In the LIGA fabrication process, a coherent synchrotron X-ray radiation source is used to perform lithography on a poly(methyl methacrylate) (PMMA) resist system with a thickness ranging from 1 μm to 1 mm. Figure 1.19 shows the typical steps used in the LIGA process. The x-ray mask consists of a Si_3N_4 diaphragm with Au metallization lines that prevent the x-ray from reaching the PMMA. The mask is usually fabricated using bulk micromachining of silicon.

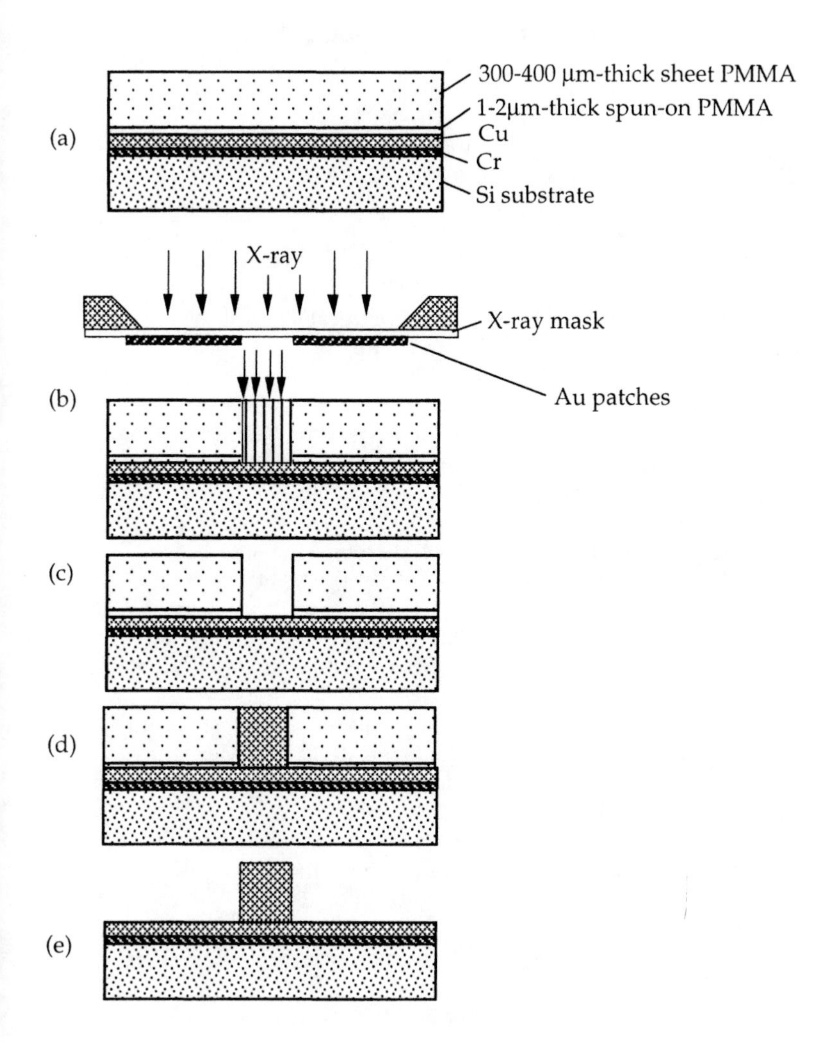

Figure 1.19 The typical LIGA process used in fabrication of microactuators [47,48].

1.6 References

1. M. Tabib-Azar, "Sensor Parameters and Characterization." In: VCH Handbook series, Volume I; Fundamentals. Edited by W. H. Ko and T. Grandke, pp. 18-42 (1990).
2. H. N. Norton, Handbook of Transducers. Prentice Hall, Englewood Cliffs, NJ (1989).
3. A. Garcia, and M. Tabib-Azar, “Sensing Means and Sensor Shells: A New Method of Comparative Study of Piezoelectric, Piezoresistive, Electrostatic, Magnetic, and Optical Sensors.” Sensors and Actuators A. Physical Vol. 48 (2), pp. 87-100 (1995).

4. M. Tabib-Azar, Integrated Optics, Microstructures and Sensors. Kluwer Academic Publishings, Boston, MA (1995).
5. K. L. Chopra, Thin Film Phenomena. R. E. Krieger Publishing Company, Malabar, Florida, p.p. 10-42 (1985).
6. A. P. Boresi, and O. M. Sidebottom, Advanced Mechanics of Materials. Fourth edition, John Wiley, New York, (1984).
7. J. D. Ferry, Viscoelastic Properties of Polymers. Third edtition, John Wiley & Sons, p.p. 11-14 (1980).
8. M. Tabib-Azar, "Optically Controlled Silicon Microactuators." Nanotechn. 1, p.p.81 (1990).
9. M. Tabib-Azar and J. S. Leane, "Direct Optical Control for a Silicon MicroActuator." Sensors and Actuators, Vol. A (21), p.p. 229-235 (1989).
10. J.J. Sniegowski, H. Guckel, and T.R. Christenson, "Performance Characteristics of Second Generation Polysilicon Resonating Beam Force Transducers." Technical Digest of IEEE Solid-State Sensor and Actuator Workshop, IEEE Publ. # 90CH2783-9, p.p. 9-12 (1990).
11. R. T. Howe, R. S. Muller, "Resonant-Microbridge Vapor Sensor." IEEE Trans. Electron Devices, Vol. ED-33 (4), p.p. 499-506 (1986).
12. X. Ding, W.H. Ko, and J. Mansour, "Residual and Mechanical Properties of Boron-doped p^+-Silicon Films." Sensors and Actuators, A21-A23, p.p. 866-871 (1990).
13. J. Y. Pan, P. Lin, F. Maseeh, and S.D. Senturia, "Verification of FEM Analysis of Load-Deflection Methods for Measuring Mechanical Properties of Thin Films." Technical Digest of IEEE Solid-State Sensor and Actuator Workshop, IEEE Publ. # 90CH2783-9, p.p. 70-73 (1990).
14. T.S.J. Lammerink, and W. Wlodarski, "Integrated Thermally Excited Resonant Diaphragm Pressure Sensor." Transducers '85, IEEE Publication # 85CH2127-9, p.p. 97-100
15. K. Petersen, "Dynamic Micromechanics on Silicon: Techniques and Devices." IEEE Transaction on Electron Devices, Vol. ED-25 (10), p.p. 1241-1250 (1978).
16. K. Petersen, "Silicon as Mechanical Material." Proc. IEEE, Vol. 70 (5), p.p. 420-457 (1982).
17. M. Tabib-Azar, K. Wong, and W. Ko, "Aging Phenomena in Heavily Doped (p+) Micromachined Silicon Cantilever Beams." Sensors and Actuators A Vol. 33, p.p. 199-206 (1992).
18. D. Sarid, Scanning Force Microscopy. Oxford University Press, New York, p. 16 (1991).
19. M. A. Neifeld, M. Tabib-Azar, Pin-Ju Hsiang, and Augusto Garcia-Valenzuela, "Silicon Smart Spatial Light Modulators for Optical Computing." Presented at the Optical Society of America 1992 Annual Meeting in Albuquerque, New Mexico.
20. K. Gabriel, F. Behi, R. Mahadevan and M. Mehregany, "In Situ Friction and Wear Measurements in Integrated Polysilicon Mechanisms." Sensors and Actuators, A21-A23, pp. 184-188 (1990).
21. Y.-C. Tai and R. S. Muller, "IC-Processed Electrostatic Synchronous Micromotors." Sensors and Actuators, Vol. 20, pp. 49-55 (1989).
22. S. K. Ghandhi, VLSI Fabrication Principles. John Wiley, New York,

p.p.427-429 (1983).
23. W.D. Westwood, "Physical Vapor Deposition." In: Microelectronic Materials and Processes. Editor: R.A. Levy, p.p. 133-195 (1986).
24. Y. Pauleau, "Interconnect Materials." In: Microelectronic Materials and Processes. Editor: R.A. Levy, Kluwer Academic Publishers, Boston MA, p.p. 642-644 (1986).
25. Page 510-514 of [22].
26. E. G. Spencer and P.H. Schmidt, "Ion-Beam Techniques for Device Fabrication." J. Vac. Sci. Technol. Vol. 8, S52 (1971).
27. J. D. Mackenzie, and D. R. Ulrich, "Sol-Gel Optics, Present Status and Future Trends." Proc. of SPIE Conf. on Sol-Gel Optics, Vol. 1328, p.p. 2-13 (1990).
28. B. D. Fabes, et al., "Laser Processing of Channel Waveguide Structures in Sol-Gel Coatings." Proc. of SPIE on Sol-Gel Optics, Vol. 1328, p.p. 319-328 (1990).
29. J. D. Mackenzie, Proc. of SPIE, Vol. 878, p. 128 (1988).
30. K. Heuberger and W. Lukosz, "Embossing Technique for Fabricating Surface Relief Gratings on Hard Oxide Waveguides." Applied Optics, Vol. 25 (9), p.p. 1499-1504 (1986).
31. D. W. Hewak. and J. W. Y. Lit, "Fabrication of Tapers and Lenslike Waveguides by a Microcontrolled Dip Coating Procedure." Applied Optics, Vol. 27 (21), p.p. 4562-4564 (1988).
32. W. Kern, "Chemical Vapor Deposition." In: Microelectronic Materials and Processes. Editor: R.A. Levy, p.p. 203-240 (1986).
33. L. Esaki, "A Bird's-Eye View on the Evolution of Semiconductor Superlattices and Quantum Wells." IEEE Journal of Quantum Electronics, Vol. QE-22 (9), p.p. 1611-1624 (1986).
34. M. Fogiel, Modern Microelectronic. Research and Education Association, New York, p.p. 405-407 (1981).
35. H. Seidel, "The Mechanism of Anisotropic Silicon Etching and Its Relevance for Micromachining." In Microsensors. Edited by R.S. Muller, R.T. Howe, S.D. Senturia, R.L. Smith, and R.M. White, IEEE Press Selected Reprint Series, p.p. 104-109 (1990).
36. B. Kloeck, et al., "Study of Electrochemical Etch-Stop for High-Precision Thickness Control of Silicon Membranes." IEEE Transactions on Electron Devices, Vol. 36(4), p.p. 663-669 (1989).
37. G.K. Herb, "Plasma Etching Technology-An Overview." In: Plasma Etching; An Introduction, Edited by D. M. Manos, and D.L. Flamm, Academic Press, Inc., p.p. 1-87 (1989).
38. R. T. Howe, "Surface Micromachining for Microsensors and Microactuators." J. Vac. Sci. Technol. B, Vol. 6(6), p.p. 1809-1813 (1988).
39. E. Obermeier, P. Kopstynski, and R. Niebl, "Characteristics of Polysilicon Layers and Their Application in Sensors." In Microsensors. Edited by R.S. Muller, R.T. Howe, S.D. Senturia, R.L. Smith, and R.M. White, IEEE Press Selected Reprint Series, p.p. 83-86 (1990).
40. S. Sugiyama, et al., "MicroDiaphragm Pressure Sensor." IEEE International Devices Meeting, p.p. 184-187 (1986).
41. X-P. Wu, and W.H. Ko, "Compensation Corner Undercutting in Anisotropic Etching of (100) Silicon." Sensors and Actuators, Vol. 18, p.p. 207-215 (1989).
42. J.H. Jerman, D. J. Clift, and S.R. Mallinson, "A Miniature Fabry-Perot

Interferometer with Corrugated Silicon Diaphragm Support." Technical Digest of IEEE Solid-State Sensor and Actuator Workshop, IEEE Publ. # 90CH2783-9, p.p. 140-144 (1990).

43. M. Mehregany, "Silicon Microactuators." In: Advances in Actuators. Edited by: A. P. Dorey and J. H. Moore, IOP Publication, Philadelphia, pp. 135-170 (1995).
44. S. Akamine, T. R. Albrecht, M. J. Zdwblick and C. F. Quate, "A Planar Process for Microfabrication of a Scanning Tunneling Microscope." Sensors and Actuators, A21-A23, pp. 964-970 (1990).
45. P. Muralt, et. al., "Fabrication and Characterization of PZT Thin-Film Vibrators for Micromotors." Sensors and Actuators A 48, pp. 157-165 (1995).
46. M. C. Wu, L. Y. Lin, and S.S. Lee, "Micromachined Free-Space Integrated Optics." Proc. of SPIE Conf. on Integrated Optics and Microstructures II, eds: M. Tabib-Azar, D. Polla, and K.K. Wong, Vol. 2291, p.p. 40-51 (1994).
47. H. Guckel, K. J. Skrobis, T.R. Christenson, J. Klein, "Micromechanics for Actuators via Deep X-Ray Lithography." SPIE, Vol. 2194, pp. 2-10 (1994).
48. H. Guckel. et. al., "Deep-X-Ray and UV Lithographies for Micromechanics." IEEE Solid-State Sensor and Actuator Workshop, Hilton Head Island, pp. 117-122 (1990).

Electrical Microactuators

2.1 Introduction

Electrical actuation in its electrostatic form relies on the coulombic attraction of oppositely charged material bodies. This mechanism has been known for a long time and coulombic actuators, in their simplest form, rely on the attraction of two charged plates or mechanical objects [1-6]. The force that can be produced by an electrostatic actuator is discussed in detail later in this chapter.

For the time being, we take the stored energy as an indication of the amount of work that the actuator can perform [4,5] and note that the stored energy is directly proportional to the capacitance of the actuator (C) that is, in turn, indirectly proportional to the distance between the energized plates. Thus, the distance between these plates is the important design parameter of electrostatic actuators. To produce reasonable forces with voltages on the order of a hundred volts or so, the distance between energized plates should be made as small as possible. It is also well-known that the breakdown of air depends on the separation between the charged plates and increases from $3x10^4$ V/cm for macroscopic (≈ 1cm) plate separations to $>10^6$ V/cm for microscopic (≈1-500 μm) plate separations [6]. In the microscopic regime, the stored energy in the electrostatic field ($=\frac{1}{2}\varepsilon_0 E^2$, where ε_0 is the permittivity of free space and E is the electric field), becomes comparable to that of the magnetostatic energy densities of 10^5-10^6 J/m^3.

Because of the above considerations, large-scale electrostatic machines are not very practical. On the other hand, micron-scale machines fabricated with IC fabrication methods [7] can readily take advantage of electrostatic actuation. In these structures, the separation between different plates having areas in the range of 10-1000 μm^2, with plate separations in the range 500 Å - 10 μm, can be easily designed and fabricated.

Other forms of electrostatic actuators that use piezoelectric layers [8] are also designed and readily fabricated. In these actuators, the piezoelectric layer is directly deposited over the deformable silicon or SiO_2 (or in some cases Si_3N_4) structures and makes an intimate contact with them. Also, the thickness of deposited piezoelectric layers can be carefully controlled to keep the voltage necessary for maximum deformation low and acceptable. The energy density stored in the piezoelectric actuators is also quite large in the 10^5-10^6 J/m^3 range in the lead zirconate titanate (PZT) materials discussed in section 2.3. In these materials, the dielectric constant can be quite large, which offsets their somewhat lower breakdown fields.

In the following sections we discuss the above and other actuation methods that directly convert electrostatic signals to mechanical work.

2.2 Electrostatic Microactuators

Electrostatic actuators utilize coulombic attraction between two bodies to induce displacement or to exert force [5,6]. Some of the various situations that may exist are discussed in the next few sections. In all these cases, the electrostatic force can be calculated from the stored energy (U_E) in the electrostatic field of the device. In general, this energy is given by $U_E = \frac{1}{2} CV^2$, where C is the capacitance of the device, and the force is given by the gradient of U_E:

$$F = -\nabla U_E \tag{2.1}$$

In component form, the above equation can be written as:

$$F_x = -\partial U_E / \partial x \,,\; F_y = -\partial U_E / \partial y \,,\; \text{and } F_z = -\partial U_E / \partial z \tag{2.2}$$

The next step is to determine how the capacitance of the structure (C) changes when the actuator is excited. Before we discuss a few different cases in the following sections, we note that the force will be generated in the direction of reducing the energy of the total system (U_T) and the battery (U_B) rather than that of the capacitor alone. This is an important point to keep in mind.

For example, in simple cases, the total system consists of a battery, or a source of the electric field, and a parallel-plate capacitor ($=\varepsilon A/x$, where A is the plate area, ε is the permittivity and x is the plate separation). As the value of the capacitor increases (i.e.: x decreases), the stored electrostatic energy in it increases, provided that the voltage across the capacitor is kept constant. Thus, based on the stored energy in the capacitor alone, the force will be in the direction of *reducing the capacitance* of the structure. Clearly this is not correct. The electrochemical energy stored in the battery as well as the total free energy of the system, on the other hand, will be reduced when the capacitance is increased. This can be seen by noting that the battery performs the necessary work to bring the plates of the capacitor closer to each other and in the process its stored electrochemical energy becomes smaller. Since the electric field in the capacitor changes as a function of time, a displacement current flows in the circuit. The total energy of the system, thus, is reduced because of radiation and resistive losses from the flow of the current through wires, etc. Taking these losses to be small, compared with stored energies in the battery and the capacitor, we note that $\partial U_B/\partial x = -\partial U_E/\partial x$. Thus, the energy of the source of the electric field, i.e. the battery, decreases as the plates get closer to each other. The force, therefore, will act to reduce the distance between the plates to reduce U_B.

Some electrostatic actuators exhibit force and displacement behavior which are quite non-linear, as discussed in the following sections. In these actuators, mechanical non-linearities coupled with the electrostatic force non-linearities render the analytical solution of the actuator behavior almost impossible. In these cases, numerical methods are needed to simulate their performance and to design them [9]. In this book we almost exclusively use analytical methods to illustrate the underlying concepts rather than the precise values of force and displacement. There are many efforts to produce computer-aided tools using finite element and finite difference methods to design microactuators [10].

2.2.1 Perpendicular Motion

In figure 2.1 the voltage across the capacitor is fixed, there are no y or z motions and the top plate can move only in the x-direction. The capacitance is given by $\varepsilon_{air}A/d$, where A is the capacitor area and "d" is the distance between the top and bottom plates. Thus, the force is given by:

$$F_x = \frac{1}{2} V^2 \varepsilon_{air} A / d^2 \tag{2.3}$$

indicating that the electrostatic force is directed in the x-direction and in the direction to reduce the distance between the top and bottom plates.

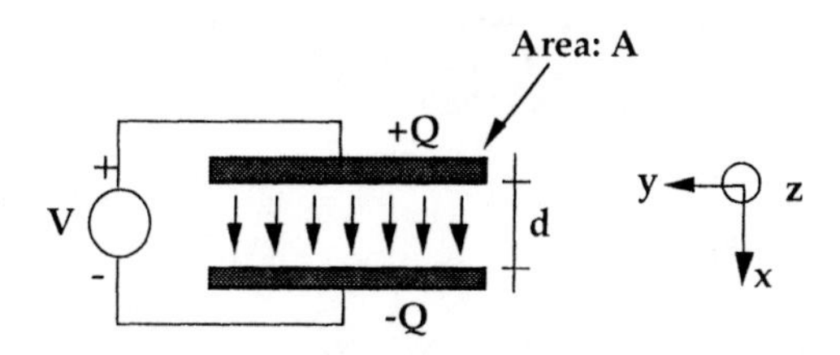

Figure 2.1 A schematic of a simple parallel plate configuration electrostatic actuator. When unrestricted, the electrostatic force displaces the top and bottom plates to reduce their distance "d".

In actual applications, the actuator is usually attached to a load against which it performs work. In most cases, the load can be modeled by a mass and spring system. A simple spring-actuator configuration is shown in the figure 2.2 inset. In this configuration, the electrostatic actuator performs work against the spring. Using the concept of load line discussed in Chapter 1, we can analyze the performance of the electrostatic actuator by first drawing the force-displacement lines of the parallel-plate electrostatic actuator with different voltages applied. Superimposed on these curves, we also draw the force-displacement line of the spring. Equation 2.3 along with Hooke's law give the force-displacement relationships for the actuator and for the spring, respectively. For the actuator-spring system shown in the figure 2.2 inset, these equations are:

$$F_{actuator} = -\frac{1}{2} V^2 \varepsilon_{air} A / x^2 \tag{2.4}$$

$$F_{spring} = k(d - x) \tag{2.5}$$

By setting $F_{actuator}=F_{spring}$, we find the equilibrium position of the top plate of the electrostatic actuator at any given voltage. This problem can also be solved graphically, as shown in figure 2.2, where we have calculated forces using a 100 μm X 100 μm top plate with separation 10μm (=d) and a spring constant of 5x10^{-8} N/μm (=k). The series of $F_{actuator}$-x curves shown in figure 2.2 were generated with applied voltages of 5 to 10 volts, with one-volt increments.

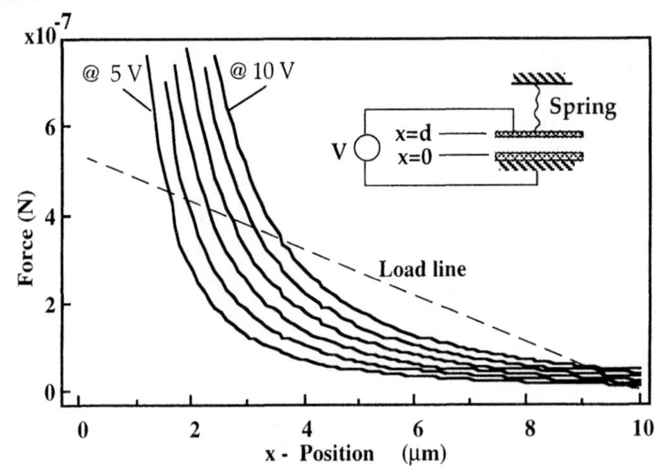

Figure 2.2 The force versus position of the actuator (solid lines) and the spring (broken line).

Although there are two intersections between the load line and each of the $F_{actuator}$-x curves, there is only one acceptable and stable solution as can be seen from the energy curves shown in figure 2.3. The energy of the combined actuator-and-spring system has two parts: one part due to the attractive electrostatic force generated in the actuator and the other component due to the spring. The energy of the combined system can thus be written as:

$$U_T = \frac{1}{2}k(d-x)^2 - \frac{1}{2}\frac{\varepsilon_{air}A}{x}V^2 \tag{2.6}$$

Figure 2.3 depicts U_T-x of the combined system for V=5 and 10 volts. Near $x=0$, the energy reaches a maximum, while near $x=d$ the energy has a minimum. At both these points, the force is zero (figure 2.4a), but as seen from the energy curve, only the latter is a stable equilibrium position.

The combined actuator-spring system can be modeled by a spring having an "effective" Hooke's constant (k_{eff}) given by:

$$k_{eff} = \frac{\partial^2 U_T}{\partial x^2} = k - \frac{\varepsilon_{air}AV^2}{x^3} \tag{2.7}$$

It is interesting to note that according to the above equation, the spring attached to the top plate exhibits an effective spring constant that is less than the Hooke's constant of the spring itself. The effective Hooke's constant (shown in figure 2.4b) depends on the magnitude of the voltage applied across the parallel plate capacitor. This can be seen from the curvature of the U_T-x curves (equation (2.7)) shown in figure 2.3.

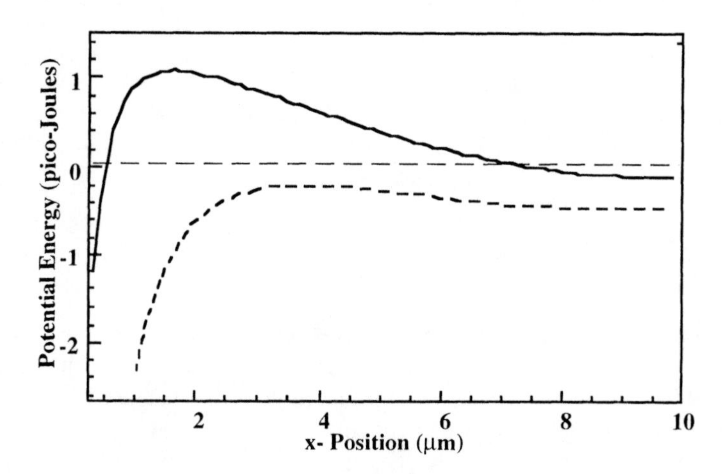

Figure 2.3 The potential energy of the combined actuator-spring system as a function of the separation between the top and bottom plate at 5 volts (solid line) and 10 volts (broken line).

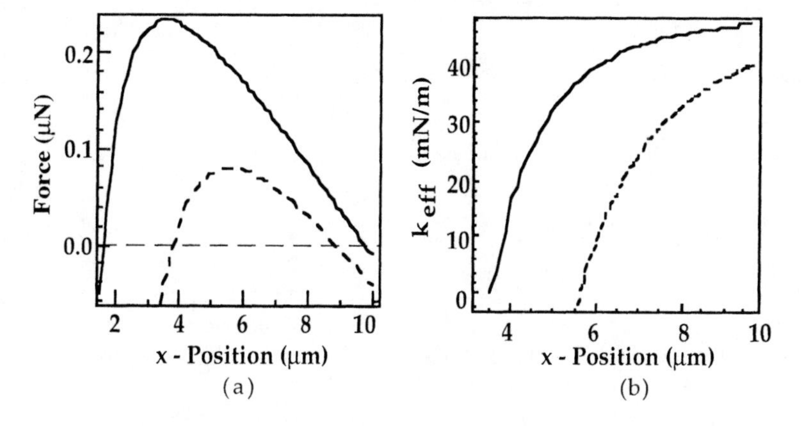

Figure 2.4 The position dependence of force (a) and the effective Hooke's constant (b) of the combined actuator-spring system at 5 volts (solid lines) and at 10 volts (broken lines).

2.2.2 Lateral Motion

When the parallel plates are displaced laterally as shown in figure 2.5, the electrostatic force in the y-direction is given by:

$$F_y = -\frac{\partial U_E}{\partial y} \tag{2.8}$$

To find an expression for F_y, we note that the capacitance of the parallel plate structure shown in figure 2.5 is directly proportional to the overlap areas between the top and bottom plates:

$$C=\varepsilon_{air}A/d = \varepsilon_{air}yz/d \tag{2.9}$$

where y and z are the amount of overlap between the top and bottom plates respectively, in the y- and x-directions. We further assume that the amount of overlap between these plates is fixed in the z direction. Noting that $U_E = \frac{1}{2} CV^2$, the two equations above can be solved to yield F_y:

$$F_y = -\frac{\partial U_E}{\partial y} = \frac{1}{2}\frac{\varepsilon_0 z}{d}V^2 \qquad y<\ell \tag{2.10}$$

It is interesting to note that F_y does not depend on y for $y<\ell$ or for $y> \ell$, and it reverses sign as y becomes equal to ℓ. For $y\approx \ell$, F_y is zero. A more exact analysis of this problem can be performed to take into account the fringing fields in order to find the exact dependence of F_y on y.

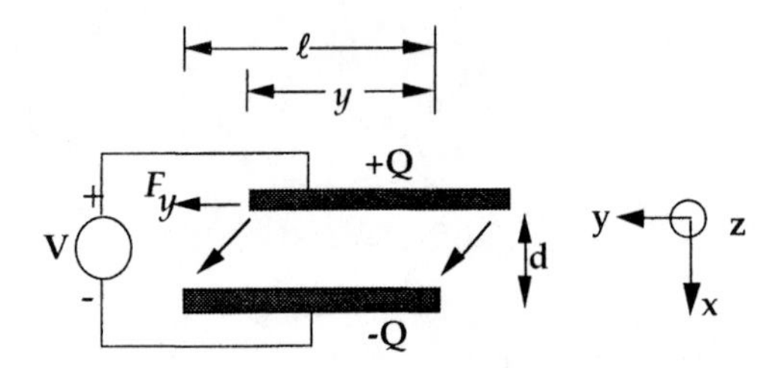

Figure 2.5 A schematic of a simple parallel plate electrostatic actuator with y-directed motion.

In electrostatic linear micromotors, the presence of x-directed force (i.e., more generally, out-of-plane forces) is quite undesirable since it tends to "clamp" the rotor (the moving plate) down to the stator (the stationary plate) [11]. A spacer and/or a dielectric layer, therefore, is needed to prevent the rotor from making an electrical contact with the stator and to restrict its motion in the y-direction.

2.2.3 Dielectric Slab

The parallel plates in the above structures can be completely fixed and actuation can be achieved by letting a dielectric slab move into the structure in response to the electrostatic field as shown in figure 2.6. The force that is exerted on the dielectric slab can also be easily derived from the electrostatic energy. As shown in figure 2.6, two capacitors in parallel form the actuator structure. These capacitors can be written as:

$$C_{air}=\varepsilon_0(\ell-y)z/d \tag{2.11}$$

$$C_d = \frac{\varepsilon_d y z}{d} \tag{2.12}$$

where ε_d is the dielectric constant of the slab with thickness $d' \approx d$, and all the other dimensions are defined in figure 2.6. The total capacitance of the structure is:

$$C_T = C_{air} + C_d \tag{2.13}$$

The energy stored in the electrostatic field is:

$$U_E = \frac{1}{2} V^2 \left[\frac{\varepsilon_0 \ell z}{d} + \frac{(\varepsilon_d - \varepsilon_0) y z}{d} \right] \tag{2.14}$$

The force is:

$$F_y = \frac{1}{2} V^2 \frac{(\varepsilon_d - \varepsilon_0) z}{d} \tag{2.15}$$

which is in the positive y-direction trying to pull the dielectric slab into the capacitor.

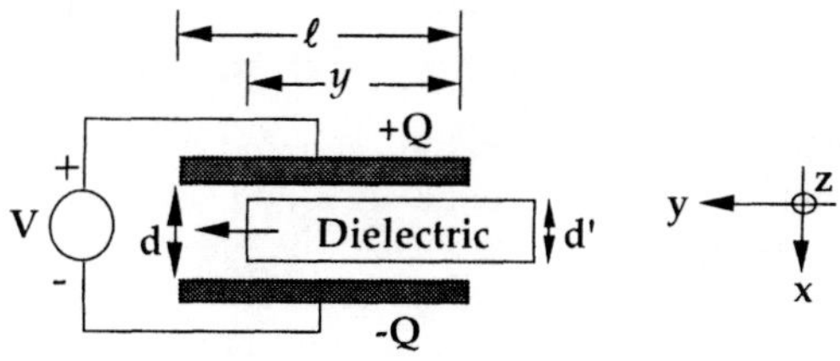

Figure 2.6 A schematic of a parallel-plate electrostatic actuator with a dielectric moving part.

It is interesting to note that in the above scheme, the x-directed forces cancel each other and the dielectric slab only experiences a y-directed force [12]. In actual devices, however, it is almost impossible to achieve equal distances between the rotor and the stators and, due to the $1/x^2$ dependence of the electrostatic forces (equation 2.4), the rotor experiences a perpendicular "clamping" force due to the stator that is slightly closer.

2.2.4 Combination Perpendicular and Parallel Motions

The dielectric slab in the above example can be replaced by a conducting slab more suitable for silicon micromachining. The conducting slab in silicon micromachined actuators is usually a heavily-doped polysilicon mechanical structure as discussed in section 2.2.8. Figure 2.7a schematically shows the actuator structure. Clearly we cannot assume $d' \approx d$ since this will result in electrically shorting the capacitor plates. Thus, there are three capacitors involved in these structures, denoted by C_1-C_3, as shown in figure 2.7a.

We assume that the moving slab does not have any x- or z-degree of freedom and it can move only in the y-direction. If not restricted in the x-direction, the conductor slab will find its resting place touching one of the plates. Using a procedure similar to the above, the force on the conductor is calculated and is given by:

$$F_y = \frac{1}{2}\frac{\varepsilon_0 z d'}{d(d - d')}V^2 \qquad (2.16)$$

This force does not depend on y, as expected (see equation (2.10) for the lateral force), and it is due to the fringing fields, as shown in figure 2.7a. Thus, to increase the force in this case, the only parameter that can be adjusted is the clearance between the conductor and the plates (d-d'). It is not very practical to make this very small because of the mask alignment requirement, which, at best, is around 1 μm. The force can be increased, however, by putting many of these actuators in parallel, as discussed in section 2.3.3. Alternatively, the plates can also be tilted, as shown in figure 2.7b, to combine the x-directed motion and the corresponding capacitance dependence with the y-directed motion. In this case the force depends on y.

As discussed before, the stability of the position of the rotor with respect to the stators is important in electrostatic actuators that utilize the above actuation methods. In the symmetric structures shown in figure 2.7, the clamping forces generated by the stators in the x-direction cancel out, provided that the distance between the rotor and the stators is kept equal. In actual devices, this symmetry condition is quite difficult to achieve and, due to the $1/x^2$ dependence of the clamping forces on the stator-rotor distance, a slight imbalance results in the attraction of the rotor to the stator that is slightly closer. Various active and passive control methods can be used to address this issue [13,14]. Active control methods are quite difficult to implement due to the small sizes of these actuators which require very sensitive and small sensors, etc.

There are efforts to develop active methods with active close-loop control methods [13,14]. Passive control methods, however, can be readily used to adjust the rotor-stator distance and prevent it from becoming unstable . One such proposed method uses the rotor-stator capacitance (C_{rs}) as the control signal. As the rotor distance becomes smaller, the C_{rs} increases and the rotor voltage, or charges, can be reduced accordingly to reduce the electrostatic force on the rotor and to stabilize its distance. The idea explored in [13] is to incorporate the C_{rs} in an LC oscillator tank with a resonant frequency much larger than the mechanical natural frequency (ω_{mrs}) of the rotor-stator system. Then, the LC tank oscillator is excited with a source signal having a frequency larger than the center frequency of the tank. Under this condition, the electrostatic force exerted on the rotor will be the rms value of the ac electrostatic force generated in the C_{rs}. As the C_{rs} becomes larger, the LC tank oscillation amplitude drops, reducing the rms electrostatic force on the rotor. This method is quite fast and it does not require complicated electronics or sophisticated sensors.

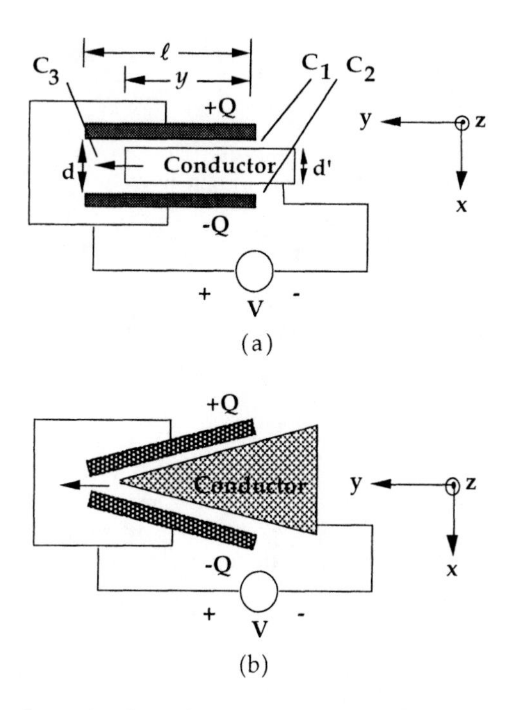

Figure 2.7 A schematic of an electrostatic actuator capable of generating large forces.

2.2.5 General Formulation

Forces experienced by dielectrics in the presence of electromagnetic forces are expressed in terms of Maxwell stress tensor discussed in many standard textbooks in electromagnetism [4]. The volume force (F_v) on a dielectric with mass density g, charge density ρ, dielectric constant ε in the presence of an electric field E is given by:

$$F_v = \rho E - \frac{1}{2}E^2\nabla\varepsilon + \frac{1}{2}\nabla\left(E^2\frac{d\varepsilon}{dg}g\right) \tag{2.17}$$

The first term is the electrostatic volume force and the second term is non-zero when the dielectric is inhomogeneous. The third term is known as the electrostriction term and it gives a volume force on a dielectric in an inhomogeneous field.

The volume force is related to the stress tensor **T** (with components $T_{\alpha\beta}$) through the following equation:

$$F_{v\alpha} = \frac{\partial T_{\alpha\beta}}{\partial x_\beta} \tag{2.18}$$

It can be shown that equation 2.17 can be written in the stress tensor form as:

$$T_{\alpha\beta} = \kappa\varepsilon_0 \left[E_\alpha E_\beta - \frac{\delta_{\alpha\beta}}{2}\left(1 - \frac{g}{\kappa}\frac{d\kappa}{dg}\right) E_\gamma E_\gamma \right] \tag{2.19}$$

where $\delta_{\alpha\beta}$ is one if $\alpha=\beta$ and zero otherwise. The αth component of the force transmitted across the surface element dS_β, is given by:

$$dF_\alpha = \sum_{\beta=1}^{3} T_{\alpha\beta} dS_\beta \tag{2.20}$$

and F_α is related to the αth component of the volume force $F_{v\alpha}$ through a volume integral:

$$F_\alpha = \int F_{v\alpha} dv \tag{2.21}$$

Equations 2.19-2.21 along with the Clausius-Mossotti equation [15], which relates the dielectric constant to the mass density, are sufficient in solving most of the problems in electrical actuators in the electrostatic regime.

The above general equations can be used to find the force that will be applied to a liquid dielectric in the configuration shown in figure 2.8. Using a similar arrangement, liquid pumps can be constructed, as discussed in section 2.2.9.

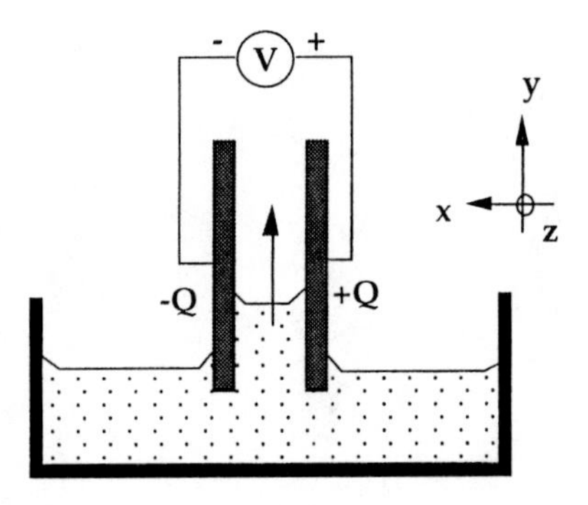

Figure 2.8 A schematic of a parallel plate electrostatic actuator applying a volume force on a liquid dielectric.

In the following sections we discuss application of electrostatic forces in linear micromotors (section 2.2.6), rotational micromotors (section 2.2.7), resonant structures (section 2.2.8) and electro-hydrodynamic pumps (section 2.2.9).

2.2.6 Linear Electrostatic Micromotors

Many linear electrostatic motor configurations are reported in the literature [12,13,16-18]. These can be divided into two categories of structures that utilize rotors, those that are free (actually quasi-free) to move in a plane and those that have limited movement and have to execute an oscillatory (back and forth) motion. This latter category is treated in section 2.2.8, where we cover resonant structure, and here we treat the former.

Figure 2.9a schematically shows a linear electrostatic motor where a moving part (rotor) is situated over strips of metals deposited over the substrate (stator). By applying voltage pulses between different poles on the rotor and the stator, one can make the rotor move forward or backward. The structure of this actuator is very similar to the superconducting magnetic levitation actuator (Meissner effect) discussed in section 3.7 of Chapter 3.

The actuation mechanism employed in the linear electrostatic motor depicted in figure 2.9a is discussed in section 2.2.2, where parallel plates of the capacitor are allowed to move laterally and parallel to each other [12,17,18]. In these schemes, there is clearly also a perpendicular electrostatic force present that tends to clamp the plates to each other (as discussed in section 2.2.1). In linear motors, hence, either a dielectric is deposited over the stator to prevent perpendicular motion and short circuiting of the stator-rotor poles, or a space is used to keep the rotor some distance away from the stator.

To reduce the clamping force, a couple of different methods are introduced in the literature and implemented. The obvious solution to this problem is to include another stator plate right above the rotor to cancel out the clamping force generated by the bottom stator by symmetrically applying an opposing clamping force by the top stator. This method is very effective, but, unfortunately, requires additional fabrication steps.

In reference [12] a variation of the structure shown in figure 2.9a was used to fabricate a linear electrostatic actuator that in turn was used to measure the mechanical properties of polymers. In its design, the rotor was a slotted polymer film that was suspended between two stators with electrodes. Upon excitation of the opposite stator electrodes, the polyimide film was pulled in electrostatically. The rotor pitch of 18 μm and stator electrode width of 9 μm with a 1 μm air gap on both sides of the rotor was used. The polyimide rotor (1x1 mm^2 area and ε_r=3) thickness was 10 μm, and application of 100 V excitation to the stators generated an electrostatic force of 38 μN. Rotor displacement of as large as 12 μm at 2x10^{-2} GPA residual stress with polyimide Young's modulus of 3x10^9 Pa was predicted.

A second approach to reducing the clamping force in electrostatic linear motors is to use multi-phase excitation of the rotor-stator poles [13,16]. This approach takes advantage of the dipole-dipole interaction between the rotor and the stator.

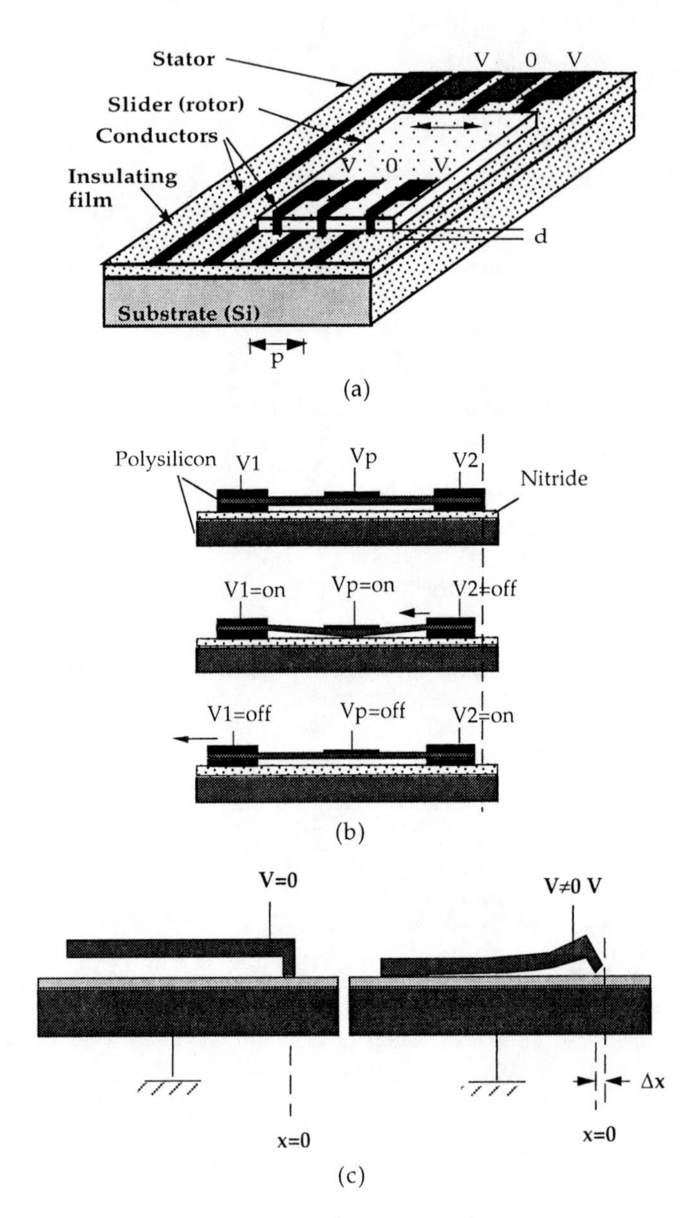

Figure 2.9 (a) A schematic of a linear electrostatic micromotor [16]. (b) A schematic of a high-force and high-precision linear electrostatic stepper motor [18]. (c) A scratch-drive electrostatic linear actuator [19].

By exciting poles that are directly situated above each other with identical voltages, the clamping force is reduced. In this method, an important design parameter is the ratio of the stator and rotor electrode pitch (p) and the rotor-stator distance (d). At the electrode pitch to distance ratio of 1.5 [16], it has been demonstrated that the useful (parallel) force could reach 40% of the clamping force. A linear motor of 1 mm^2 translator area with d=2.5 μm and p=1.5d that produced 8 μm travel at 6 V and generated 1.44 μN (0.2 $\mu N/V^2$) was fabricated.

Using a structure similar to figure 2.9a, an electrostatic linear actuator that produced 10^{-3} N at 200 volts with 10 μm travel was fabricated [17].

Figure 2.9b shows another linear electrostatic microactuator that was capable of producing large forces (40 μN) at 40 V clamping voltage (*V1* and *V2*) and 25 V actuator voltage (*Vp*) [18]. It could also move with a speed of 100 μm/s at the excitation frequency of 1200 Hz and it had a maximum displacement of 43 μm under these conditions. The actuator simply consisted of a 200 μm -long, 0.5 μm -thick and 100 μm -wide polycrystalline plate with four legs (only two are shown in the figure). The nitride thickness was around 0.5 μm and the plate/nitride air-gap was 2.4 μm. The pull-down voltage was around 15 volts.

Instead of using electrostatic clamping on the feet of the above actuator, one can also take advantage of the friction between the feet and the substrate to construct a linear actuator called a scratch-drive actuator, schematically shown in figure 2.9c. This actuator is becoming very popular because of its very long travel distance, high speed and possible operation at low excitation. It can also step motion with very high resolution of less than 25 nm with proper design. Scratch-drive actuators are used to construct *XYZ* stages that could step with 27 nm resolution with 90 V excitation [19].

2.2.7 Rotary Electrostatic Micromotors

Electrostatic Curie Wheel. A ferro-electric material, such as polycrystalline $BaTiO_3$, can be used to build an electrostatic Curie wheel. The dielectric constant of $BaTiO_3$ depends on temperature, as shown in figure 2.10a. It is relatively constant up to the Curie temperature, beyond which it starts decreasing drastically.

An electrostatic motor can be implemented by taking advantage of the dielectric behavior of $BaTiO_3$, as shown in figure 2.10b. The conducting wheel is coated with a thin layer of polycrystalline $BaTiO_3$. A light source is used to locally heat the $BaTiO_3$ layer adjacent to a small plate to reduce its dielectric constant. An electric field is applied between this plate and the conducting wheel. As the $BaTiO_3$ film heats up to temperatures above its Curie temperature (adjustable by doping between 50-180 °C), its dielectric constant diminishes. This is a situation very similar to the force experienced by the dielectric slab inserted into a parallel plate capacitor discussed in section 2.2.3, and we can use the formula developed there to calculate the force, torque and angular velocity.

With a modest voltage of 10 volts, it is possible to generate a force of 0.18 μN, according to equation (2.15), at the surface of the wheel underneath the plate. This corresponds to a torque of 9×10^{-10} N.m. In these calculations we have used plate-to-wheel spacing of 1 μm, wheel and plate width of 0.1 mm, plate length of 0.1 mm, wheel radius of 0.5 mm, a modest change in the dielectric constant of $100\varepsilon_0$ (upon heating) and ferroelectric thickness of 10 μm. Other means can also be used to heat the ferroelectric coating material. A series of heaters embedded in the wheel or heating caused by passage of a current through the $BaTiO_3$ layer, among other methods, can be used.

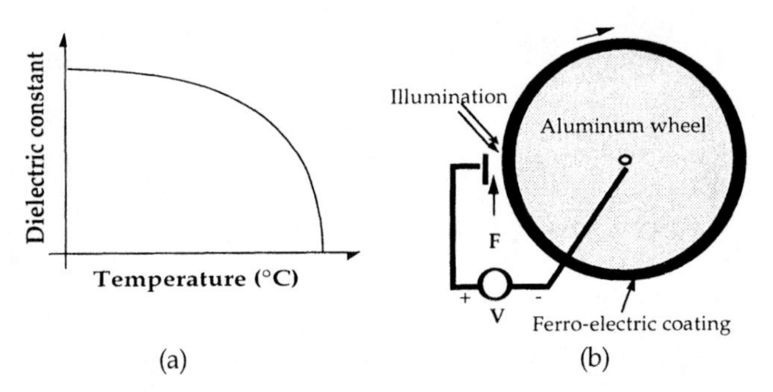

Figure 2.10 a) The dielectric constant of polycrystalline $BaTiO_3$ as a function of temperature. b) A schematic of an electrostatic Curie wheel.

Microfabricated Variable-Capacitance Electrostatic Motors. A variable-capacitance micromotor (VCM) is shown in figure 2.11. It consists of a rotor with poles that form a variable capacitor with the stator poles situated on the stator loop [20-30].

The rotor is grounded and the stator electrodes are excited with voltage pulses having appropriate phases to move the rotor. To reduce the electrostatic clamping of the rotor to the substrate, a conducting layer (heavily-doped polysilicon layer) is situated underneath the rotor that shields the electrostatic forces generated by the substrate charges. To reduce the electrostatic clamping of the rotor to the substrate, an electrostatic shield layer (heavily-doped polysilicon layer) that shields the electrostatic forces generated by the substrate charges is situated underneath the rotor.

The micromotor was fabricated by various groups, using the surface micromachining methods described in Chapter 1 and section 2.11. Its performance is limited by wear and tear occurring at the central pin and between the rotor and the substrate [28].

There are many different variations on the basic configuration shown in the above figure. These include a salient-pole, center-pin, top-drive motor with the stator poles protruding over the rotor, and a salient-pole, center-pin, side-drive motor. The top-drive configuration had many operational as well as fabrication difficulties which made it problematic [28]. To name a few, these problems included an out-of-plane force component that made the rotor unstable, a clamping force that increased the wear of the rotor bushing, and the requirement of protruding stator poles that introduced some fabrication difficulties.

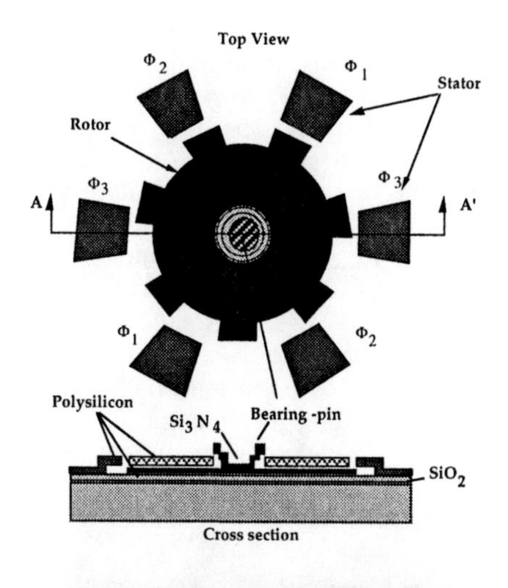

Figure 2.11 (a) A schematic of a variable capacitance electrostatic micromotor. (b) A SEM picture of a typical micromotor. The diameter of this micromotor is 500 μm [28,29.a].

The side-drive motor was the most successful configuration. Its rotor-stator gap varied between 1-3 μm and various stator-to-rotor pole ratios of 3-1, 3-2, and 2-1 were implemented, which required two-phase excitation for the latter and three-phase excitation for the others [28]. The excitation voltages between 30 V - 120 V, depending on the stator-to-rotor air-gap, were used [28].

With center pin clearance of 0.3 μm and stator-rotor airgap of 1.5 μm, rotor diameter of 100 μm, 15,000 rpm with a few pN.m motive torque (it can be shown that the torque T(θ) is given by: $T(\theta) = \frac{1}{2}V^2\frac{\partial C(\theta)}{\partial \theta}$, where θ is the angle of rotation and C is the stator-rotor capacitance) at excitation voltages 40 V have been demonstrated [22].

Operation of these micromotors in air, silicone oil and de-ionized water have been demonstrated. For a detailed review of wear and friction issues in these micromotors, the reader should refer to [23-30].

Micromotors with gas-microbearing [30.a] and lubricants [30.b] with reduced friction and wear have also been designed and fabricated. A larger version of VCM, which exhibited rotational speeds as high as 3800 rpm in air and 4400 rpm in vacuum, and which generated a motive torque of 10^{-4} gf cm at excitation voltages of 300-600 V, has also been fabricated and reported [31].

Microfabricated Electrostatic Wobble Micromotors. A wobble motor is schematically shown in figure 2.12 and it consists of a rotor that wobbles around the center bearing pin [32-34]. The rotation of the rotor is affected by sequentially exciting the stator electrodes in the direction of the rotation. A normal force develops between the excited stator electrode and the electrically-grounded rotor pulling the rotor closer to the excited electrode. As the electrical excitation travels around the stator loop, the rotor rolls and rotates on the bearing. The rotating action comes about due to the smaller size of the rotor's inner diameter compared with the outer diameter of the bearing pin.

Given excitation speed of v_e (electrodes/second), number of electrodes in the stator loop of n_e, bearing circumference of p_b and rotor inner perimeter of p_{ir}, the angular velocity of the rotor is given by $2\pi\ (p_{ir}-p_b)\ v_e/(n_e\ p_{ir})$. Typical rotational speeds of 1000 rpm, at excitation voltage of 100 V for wobble micrometer gear ratios (p_{ir}/p_b) of 70-200, have been reported [32].

A somewhat larger wobble micromotor with conical rotor was designed and fabricated. This micromotor had a stator diameter of 2 mm and 60 stator electrodes. It operated with an excitation voltage in the range 80-120 V [33].

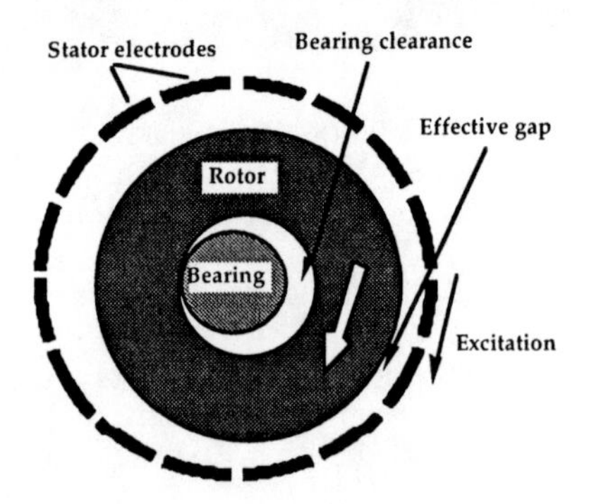

Figure 2.12 A schematic of the wobble micromotor.

A wobble motor with cylindrical rotor was reported [34]. This micromotor could produce a motive torque of 1.2 dyne-cm, and rotation speed of 1000 rpm at 200 volts excitation voltage. It had four stator poles, with a stator loop diameter of 571.5 μm and a rotor diameter of 495.3 μm. Its rotor length was 0.1 cm and its stator had an anodic dielectric passivation of 30.5 μm in thickness.

Static-Induction Motor. Both of the micromotors discussed above belong to the VCM family. Variable-capacitance micromotors have some inherent disadvantages. The stator-rotor gap must be relatively small in order to obtain large variations in capacitance, posing some fabrication problems. VCM is a synchronous motor and it requires rotor position feedback for even minimal performance. Most importantly, however, the rotor in these micromotors is subject to a transverse electrostatic force which leads to undesirable bushing friction and wear.

An induction motor (IM), on the other hand, is an asynchronous motor, so rotor position feedback is not needed. The rotor can be a smooth conductor or a fluid one, in contrast to a VCM, which requires rotors with physically salient poles. The main drawback of IM is the strong dependence of its performance on the conductivity of its rotor [35].

Figure 2.13 schematically shows a rotary electrostatic IM where an annular disk rotor is sandwiched between two plates that contain the stator electrodes. The rotor has uniform permittivity and conductivity, and it rotates between two plates that contain the stator electrodes, as shown in the figure. The stator electrodes are excited sequentially with a voltage pulse that generates azimuthally-traveling potential waves that have equal temporal and spatial frequencies at the top and bottom stator plates.

In simple terms, the IM's operation can be explained as follows. As the potential waves travel around the stator, they induce free image charges on the rotor. These image charges travel in the direction of the potential waves, but lag behind due to charge relaxation in the rotor and stator-rotor gap. The resulting displacement between the image charges and the potential wave results in a torque acting on the rotor. If the charge relaxation time constant is very fast, the image charges do not lag behind, and if the charge relaxation time is very slow, very few charges are induced in the rotor. Thus, the torque on the rotor diminishes under both of these conditions.

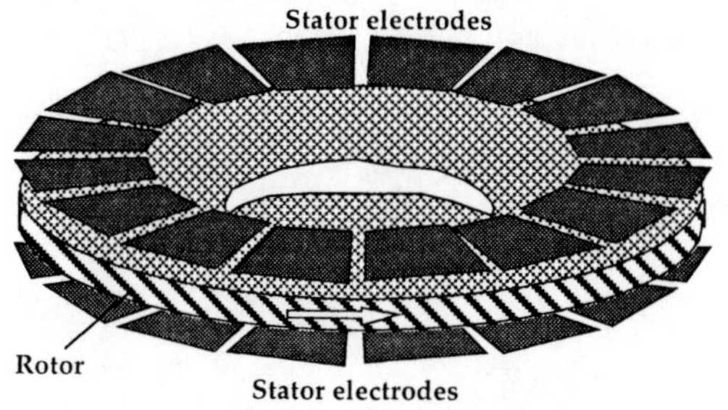

Figure 2.13 A schematic of an electrostatic induction motor.

An exact analytical solution and an analysis of IM is given in [35]. It shows that it is possible to generate torques in the nN-m range at rotational velocities of 10,000 r/s, using a rotor of 50μm outer radius and 35 μm inner radius, and a voltage amplitude of 100 volts at 50 kHz. The rotor thickness was 1 μm, its resistivity was 10^6 Ω cm, and its stators had 4 electrodes. The operation principle of IM is almost identical to that of the electro-hydrodynamic pumps discussed in section 2.5. A variety of IM structures can be devised using piezoelectric materials such as $LiNbO_3$, quartz or PZT. In these materials, surface acoustic waves are accompanied by a traveling potential wave that can be used to construct induction motors or fluid pumps.

2.2.8 Resonant Structures: Comb-Drives

According to equation 2.10, two parameters can be changed to increase the electrostatic force generated by a sliding parallel-plate capacitance configuration: the plates' distance and their area. In planar technologies, the distance between two plates is easily defined by photolithography and very small distances are avoided to prevent possible short circuit as the structure deforms. The plate area, however, is defined by the thickness of the heavily-doped polysilicon layer, and it is clearly not feasible to increase this thickness above 50-100 μm because of deposition limitations. Many capacitors, however, can be combined in parallel to increase their effective overlap areas and, hence, to increase the force generated by them. Such a structure, schematically shown in figure 2.14, has gained considerable popularity in recent years because of its modularity and versatility [36-42]. This device is also a resonant structure capable of storing mechanical energy in its vibrational mode. Structures with resonances in the range 10 to 75 kHz and mechanical Q of 100 in air were produced [22.a].

The comb-drive consists of two sets of capacitor banks, each containing a number of capacitors in parallel. Either, these two capacitor banks can either be excited sequentially to produce a push-pull effect, or one of them can be used as the sense arm to sense the magnitude and frequency of the actuation by monitoring its capacitance as a function of time.

Comb-drive actuators are used in a variety of applications. They are incorporated in *xy*-microstages [40] with typical displacements of 3.61×10^{-2} $\mu m/V^2$, and with force of 46.9 pN/V^2 with typical operating voltages in the 0-15 V range (nine-finger comb). They are also incorporated in optical switches and choppers [41-43], with 0.2 mN force at 50 V excitation and 10 μm displacement [43].

There are many variations in the geometry of the comb-drives reported in the literature. These include the parallel configuration (figure 2.14) as well as slanted, circular (spiral) [38], striated and T-bar [39] configurations. High aspect ratio, 3-D actuators that were capable of producing 0.2 mN force at 60 V have also been reported [44].

Comb-drives are also used as oscillators in microbalances, gyroscopes, and accelerometers. In these applications, high-Q as well as stable oscillator operation are of great importance. Various methods to improve Q and adjust the resonance frequency are also demonstrated [45].

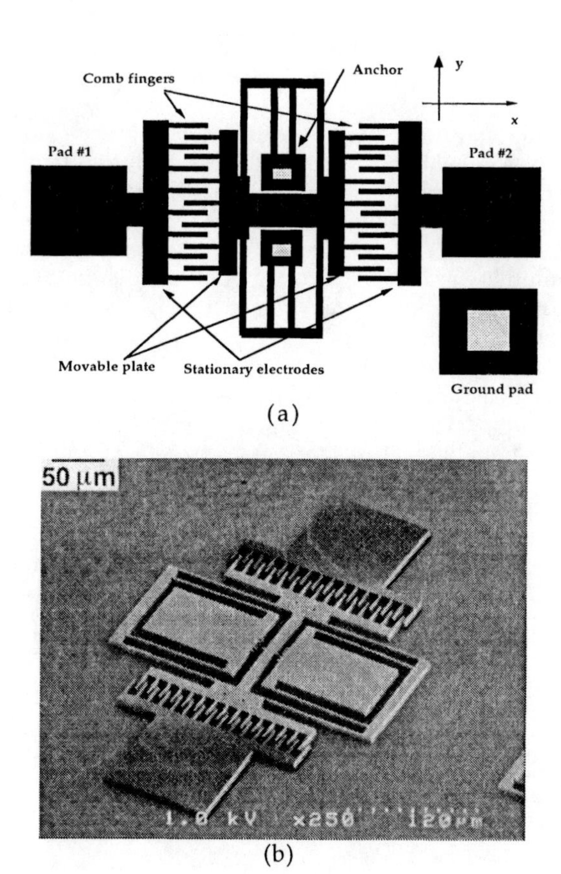

Figure 2.14 (a) A schematic of a comb-drive electrostatic microactuator.
(b) A microfabricated comb-drive [46].

2.2.9 Electro-Hydrodynamic Actuators

Electro-hydrodynamic pumps and actuators (EHD) were first proposed and built on a large scale in the 1960s. Their principle of operation is the coupling between a non-homogeneous electric field and an ionic or a polarizable fluid or gas. The force density acting on a dielectric fluid in the presence of a inhomogenous electric field **E** was given in section 2.2.5 and is very similar to the force generated in the electrostatic induction motor discussed before [4,47]:

$$F = q_f E + P \cdot \nabla E - \frac{1}{2} E^2 \nabla \varepsilon + \nabla \left[\frac{1}{2} \rho \frac{\delta \varepsilon}{\delta \rho} E^2 \right] \qquad (2.22)$$

where ε is the permittivity of the fluid, P is the polarization vector, ρ is the mass density and q_f is the free space-charge density.

The first term on the right hand side of the above equation corresponds to the Coulomb force, the second term is the Kelvin polarization force, the third term is the dielectric or Korteweg-Helmholz force. The last term is the electrostrictive force that is only relevant for compressible media.

The polarization force is zero at d.c. fields and at a.c. fields it is much smaller than the Coulomb forces. Thus, the first and third term in the above equation are the most important contribution to the force field. EHD pumps, therefore, require either a fluid with induced gradient in permittivity or free (or induced) space charges. Temperature gradients can be easily used to induce a permittivity gradient. Space charges can either be injected into the fluid by bare electrodes or they can be capacitively induced in the fluid by insulated electrodes or by electrodes with large surface-charge barriers.

Thus, depending on the main process that is used, one can realize different types of EHD pumps, schematically shown in figure 2.15, which include the EHD induction (both thermal and charge) pump and the unipolar or bipolar EHD injection pump.

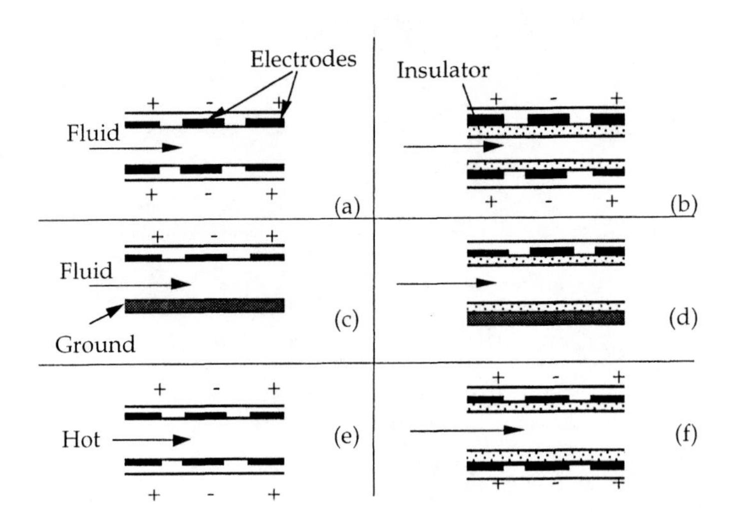

Figure 2.15 Various channel configurations used in electro-hydrodynamic pumps.

Figure 2.16 shows the ethanol flow rate of an EHD pump reported in [48]. The device was fabricated using silicon bulk micromachining and had a 3 mm x 3 mm active area and rectangular hole sizes of 140 μm. It achieved a maximum pressure of 2480 Pa at 700 V driving voltage. Its flow rate of ml/min was three orders of magnitude larger than that of piezoelectric or thermally-excited micropumps.

Using the above EHD pump, several organic solvents, such as methanol, ethanol, acetone, etc., could be pumped with different yields. De-ionized water could be pumped, but an electrolysis that resulted in an accumulation of gases between the grids occurred, decreasing the pumping rate as a function of time [48].

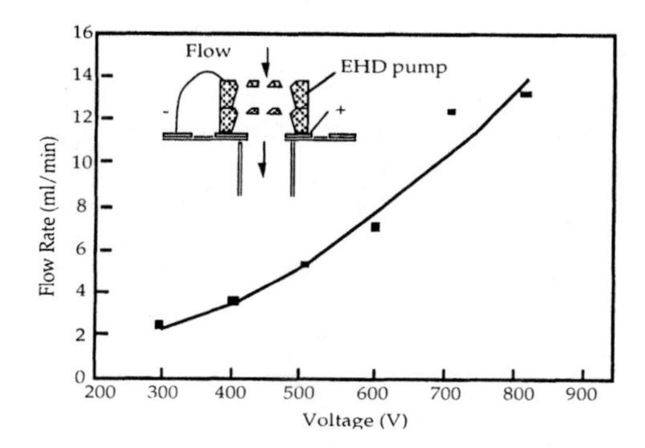

Figure 2.16 The ethanol flow rate versus applied voltage of the electro-hydrodynamic pump that is shown in the inset [48].

2.3 Piezoelectric Actuators

Piezoelectric actuators take advantage of the coupling that exists between mechanical deformation and electrical polarization in certain materials known as piezoelectrics [8, 49-55]. These materials produce charge on their surfaces when they are mechanically strained or, conversely, they become strained upon application of an electric field. Piezoelectric materials can be divided into single crystals, ceramics and polymers. Important examples of single crystal piezoelectrics are lithium sulfate, rochelle salt, quartz, tormaline, and $LiNbO_3$. Lead zirconate titanate (PZT), barium titanate, and ZnO are important ceramic piezoelectrics, and polyvinylidene fluoride is a notable example of a piezoelectric polymer.

Piezoelectric materials are usually of two kinds: materials with built-in net electric dipole moment and materials that need poling to become piezoelectric. Ceramic and polymeric piezoelectric materials usually need poling while some of the crystalline piezoelectrics do not. In these materials, domains with net electric dipoles arrange in such a manner as to cancel each other's electric field (such a cancellation reduces the free-energy of the systems.) In the case of polyvinylidene fluoride, stretching as well as electric poling can be used to make it piezoelectrically active. In the case of PZT, an electric field is applied at an elevated temperature to lower its coercive field (see figure 2.17), and the material is cooled down. It is not necessary to have the electric field on during the cool down stage since the domains do not spontaneously randomize.

Polymers and ceramics can be easily deposited over silicon (or associated materials such as Si_3N_4 and SiO_2) and subsequently patterned. Vacuum-evaporated hexagonal CdS, hexagonal and cubic ZnS and reactively sputtered as well as evaporated hexagonal ZnO have been studied in the past. Recent deposition methods include metal-organic chemical vapor deposition, rf-magnetron sputtering, sol-gel deposition, ionized cluster beam deposition, electron cyclotron resonance sputtering and laser ablation techniques. Different formulations of $Pb(Zr_xTi_{1-x})O_3$ (PZT) have received continued attention due to their large coupling coefficients. ZnO has received attention as a material for surface acoustic wave devices (SAW) and waveguides because of its strong piezolectricity, large refractive index (from visible to infrared) and transparency. Some of these deposition methods were discussed at the end of Chapter 1.

In piezoelectric materials, the strain-stress relationship is modified by the presence of the applied electric field. Thus, one can write [51]:

$$S = s\,T + d\,E \qquad (2.23)$$

where S is the strain (relative elongation), T is the stress (force per unit area), s is the coefficient connecting S and T, and d is the coupling coefficient between the electric field (E) and the strain. S, T, and E are vectors and s and d are tensors.

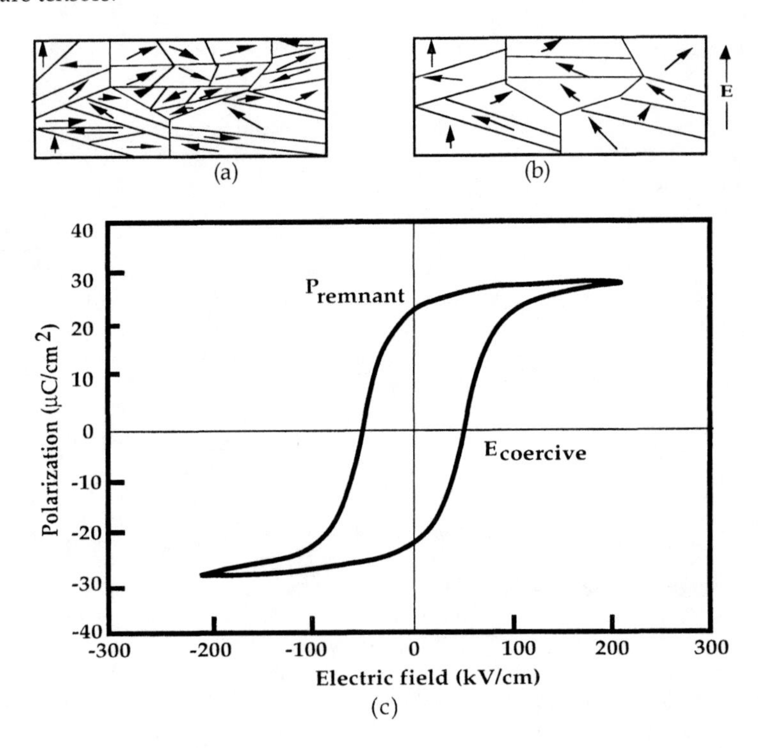

Figure 2.17 A schematic of domains in poly-crystalline ferroelectric and piezoelectric materials: (a) as prepared PZT and (b) poled PZT. (c) The hysteresis curve for a typical PZT sample showing the effect of poling [8,52].

Likewise, for the inverse process of generating an electric field by stressing the material, we have the following relationship:

$$E = -g\,T + (\varepsilon^T)^{-1} D \tag{2.24}$$

where g is the coefficient relating the stress to the electric field, D is the displacement vector, and ε^T is the tensor containing dielectric coefficients at zero tension. It is customary to use the following mnemonic rule to reduce the double index notation and to write the above equations in the component and matrix forms [51,53].

1	6	5	2	4	3
xx	xy	xz	yy	yz	zz

Displacements have single indices associated with their direction, denoted by 1 for x-direction, 2 for y-direction and 3 for z-direction. Also, the preferred axis of the material is designated as the z or 3 axis. Using crystal symmetries, various components of **s** and ***d*** tensors are related to each other or found to be zero. These symmetry groups are quite involved and beyond the scope of this book. In table 2.1 we list the important piezoelectric coefficients in selected materials. For example, for a free standing PZT (T=0) with a field applied in the z-direction, as shown in figure 2.18a, the displacement in the z-direction is given by:

$$s_3 = d_{33} E_3 \tag{2.25}$$

Noting that s_3 is given by $\Delta z/z$, with z being the PZT length in z-direction and $E_3 = V/z$, we find $\Delta z = d_{33}V$. It is interesting to note that for the electric field applied in the z-direction, the change in PZT length is independent of its length. For a voltage of 100 V, we use d_{33} from table 2.1 and find Δz of 289 Å for PZT4. Likewise, we can calculate the change in the x-direction due the applied field in the z-direction ($s_1 = \Delta x/x$) using a d_{31} ($=d_{13}$) coefficient.

$$s_1 = d_{31} E_3 \tag{2.26}$$

$$\Delta x = x\, d_{31} V/z \tag{2.27}$$

where x is the length of the PZT in the x-direction.

From table 2.1 we note that d_{31} is-1.23 Å/V for PZT4. Taking z and x both to be 1 μm, we calculate Δx of -123 Å.

Another important piezoelectric parameter is the coupling coefficient (K^2) that is defined as [51]:

$$K^2 = \frac{W_{mechanical}}{W_{electrical}} \tag{2.28}$$

where $W_{mechanical}$ is the energy stored mechanically and $W_{electrical}$ is the total energy stored electrically. Coupling coefficient is a measure of how efficiently an electrical signal is converted to a mechanical signal (or vice versa in the receiving mode). K^2 in typical piezoelectric materials are given in table 2.1.

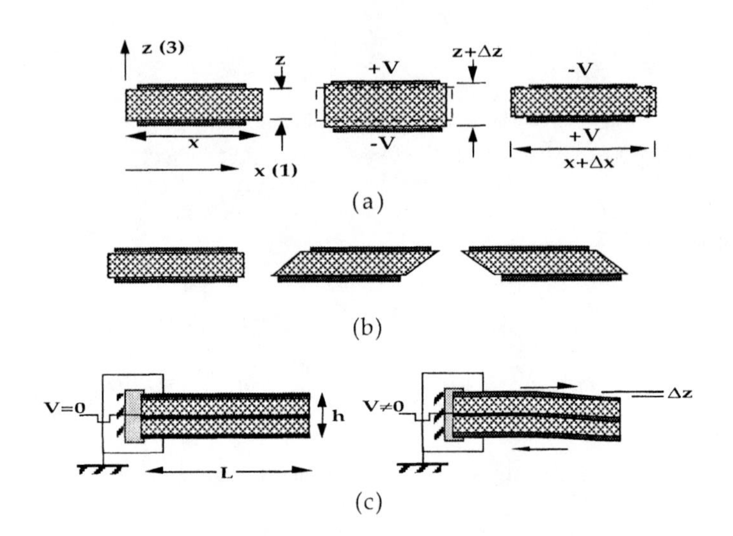

Figure 2.18 Various excitation modes of piezoelectric devices. (a) Thickness expansion (TE) mode, (b) thickness shear (TS) mode, and (c) bimorph structure composed of two TE mode layers [52].

Table 2.1 Electrical properties of selected piezoelectrics at 30 °C [8,49-55].

Material	$\varepsilon_{33}/\varepsilon_0$	d_{33} Å/V	d_{13} Å/V	d_{15}, d_{32} Å/V	k_{33}	T_Θ °C	Q	g_{33}, g_{31} mV m/N
PZT4^	1300	2.89	-1.23	4.96	0.7	328	500	26.1
PZT5A^	1700	3.47	-1.71	5.84	0.7	365	75	24.8
PZT5H^	3400	5.93	-2.74	7.41	0.7	193	65	19.7
PZT7D^	----	2.25	-1.00	---	0.5	325	500	---
PZT8^	1000	2.25	-.97	3.30	0.6	300	1000	25.4
PVDF	12	-0.35	0.28	--, .04	0.20	170	10	-339, 216
ZnO	10.9	0.12	-0.05	-0.08	0.48	----	---	---
AlN	10.7	0.05	---	----	0.31	---	---	---
Quartz*	4.6	0.02	---	---	0.09	---	$>10^6$	---
$LiNbO_3$	30	0.06	-.01	.68	0.17	1210	---	---
$LiTaO_3$	45	0.08	-.02	.26	0.19	660	---	---

^ Bulk poled PZT's. Values in thin film PZT's are somewhat lower by as much as 60%.

*For 0° X-cut: Quartz has many different crystal cuts chosen for their temperature coefficients and other properties.

In actual applications the temperature effects should be taken into account because, especially in ceramic piezoelectric materials such as PZT, the piezoelectric coefficients depend on the temperature. The applied fields should also be kept below the depolarizing field of the material, also a function of temperature. (In poling piezolectrics, an electric field is usually applied at elevated temperature. Apparently, if a strong field, called a depolarizing field, in a direction other than the poling direction is applied, the piezoelectric property is modified.) In the case of poled piezoelectrics, such as PZT's, it is found that the piezoelectric properties, such as the coupling coefficients, vary as a function of time. An empirical equation is found to describe this time dependence and it is given by: $k(t)=k(0)(1+\Delta \log t)$ where Δ is the relative variation of the coupling constant per time decade and $k(0)$ is the coupling constant after the poling.

Typical values of Δ is -1.7 % for PZT4 and -2.3 % for PZT8. PZT7D has the lowest Δ of - 0.006%.

Some typical piezoelectric transducer displacement modes are shown in figure 2.18. Not shown in this figure are face shear (FS), transverse length expansion (LE_t), parallel length expansion (LE_p) and planar expansion (PE_t) modes. The bimorph configuration shown in figure 2.18c is of great importance in piezoelectric microactuators. The two layers involved in this structure can both be piezoelectric layers excited in TE mode (figure 2.18a), or one of these layers can be a non-piezoelectric material such as SiO_2, Si_3N_4 or Si. When both layers are piezoelectric, it can be shown that the deflection at the tip of the bimorph cantilever beam (neglecting the metal layers) is given by [53]:

$$\Delta z = 3d_{31}V\frac{L^2}{h^2} \tag{2.29}$$

where L is the beam length, h is its width and V is the applied voltage. For a beam 600 μm length, and thickness of 10 μm made of two layers of PZT4 material, Δz is 4.43 μm at 10 volts. Figure 2.19 shows the force-Δz curve for this actuator. It can be shown that the force needed to deflect the beam Δz amount is given by

$$F = \frac{\Delta z Y b h^3}{4L^3} \tag{2.30}$$

where b is the width of the beam and Y is the Young's modulus.

The Young's modulus for PZT4 is 7.5×10^{10} N/m^2, resulting in a force of $\approx$ 1.9 μN for b=5 μm. This is the force responsible for the deflection of the beam. Thus, if the deformation is restricted, the force can be used to perform work against an external load, such as a spring. The equilibrium position of the beam in the presence of such a load is clearly determined by the force balance equation. Treating the bimorph actuator as a spring with an effective Hooke's constant of $k_{eff}=Ybh^3/4L^3$, we note that the force balance equation can be written as:

$$F_{actuator}=F_{spring} \tag{2.31}$$

F_{spring} is given by $k_s(z)$ for a spring with spring constant k_s, and $F_{actuator}$ is given by $k_{eff}(\Delta z-z)$ (for $z<\Delta z$). These two forces are equal when $z= \Delta z k_{eff}/(k_s+k_{eff})$, where Δz is the equilibrium position of the cantilever beam tip when there is no load ($F_{spring}=0$). The above equation can also be graphically solved as shown in figure 2.19.

Table 2.2 Mechanical properties of selected materials at 30 °C [8,49-55].

Material	Y 10^{10}N/m^2	ρ g/cm^3	c km/s Longitudinal	c km/s Shear
PZT4	7.5	7.6	3.3	-----
PZT5A	6.1	7.5	4.60	1.75
PZT5H	7	7.5	4.60	1.75
PZT7D	9.2	7.6	2.9	-----
PZT8	8.7	7.6	3.4	-----
PVF	2x10^{-5}	1.76	≈0.8	-----
ZnO	12		6.37	2.73
Quartz	9.7 ∥z 7.6 ⊥z	-----	5.74	3.33 5.1
SiO_2	7	2.2	5.1	2.8
Si_3N_4*	14	3.1	-------	-----
Al	7	2.7	6.42	3.04
Au	8	19.3	3.24	1.20

* PECVD Si_3N_4.

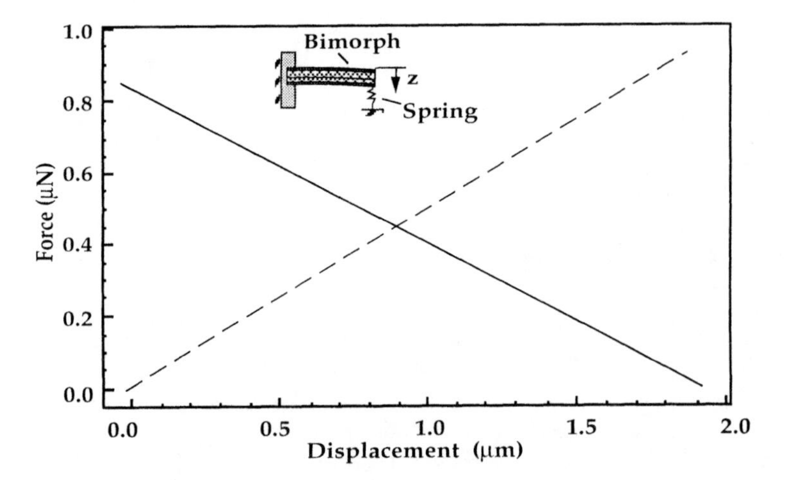

Figure 2.19 The force versus displacement characteristics of the bimorph piezoelectric (PZT4) actuator shown in figure 2.16c. The effective Hooke's constant of the bimorph was 0.43 N/m and that of the spring was 0.5 N/m. The equilibrium point is at z=0.89 μm.

2.3.1 Linear Micromotors

Microworms were developed to facilitate alignments of fiber optics. One consists of a piezoelectric strip with four legs. The legs are electrically isolated from each other. Four switches are used to connect each of the legs to the power supply, as schematically shown in figure 2.20. When a leg is connected to the power supply, it sticks to the surface due to electrostatic attraction. The other aspect of the microworm is its piezoelectric body that contracts or expands when a voltage is applied across it. Thus, a typical operation cycle consists of clamping two front legs down, then applying a voltage to the piezoelectric strip to contract it, and then clamping the back legs down, followed by releasing the front legs and disconnecting the power supply to the piezoelectric strip. By repeating the cycle, the microworm crawls along the surface. The simplest forms of microworms can move only on a straight line in a forward/backward direction. It is possible to segment the piezoelectric strip into two sections with two independent voltage excitations and achieve a curvilinear movement of the microworm as well.

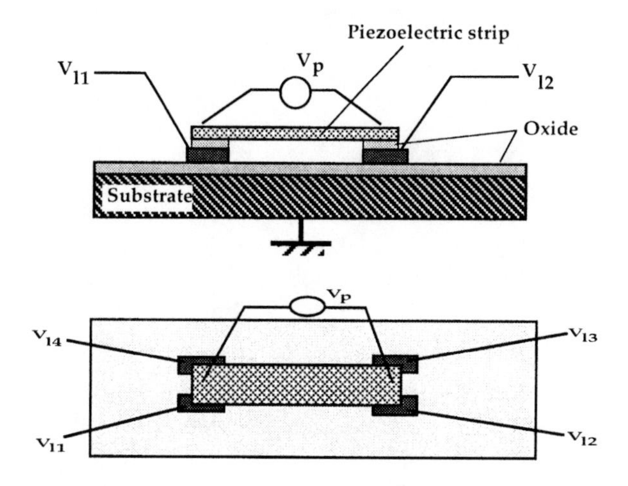

Figure 2.20 A schematic of a microworm linear motor.

Typical clamping voltages of 50-100 V, drive voltages of 10-50 V and speed of a few μm/s are easy to achieve. This actuator, in a slightly different form, is commercially available under the trade name Inchworm, and it is used in some scanning tunneling electron microscopes as coarse positioners [53].

2.3.2 Rotational Micromotors

There is an interest in developing piezoelectric rotational micromotors because, compared to magnetic micromotors, they can easily be scaled down and, compared to electrostatic motors, they have larger energy density [56-61]. Figure 2.21 schematically shows the structure and the operation principle of this motor. As the diaphragm bends upward, it squeezes the tilted elastic legs of the rotor [60]. This causes the rotor to move because of the tilting of the legs and the friction between the legs and the diaphragm. When the diaphragm bends downward, the friction is reduced, the legs relax and the rotor becomes stationary. Operated at the resonance frequency of the diaphragm, this motor can generate forces and torques in excess of a 30 nN-m, and speeds in excess of 200 rpm.

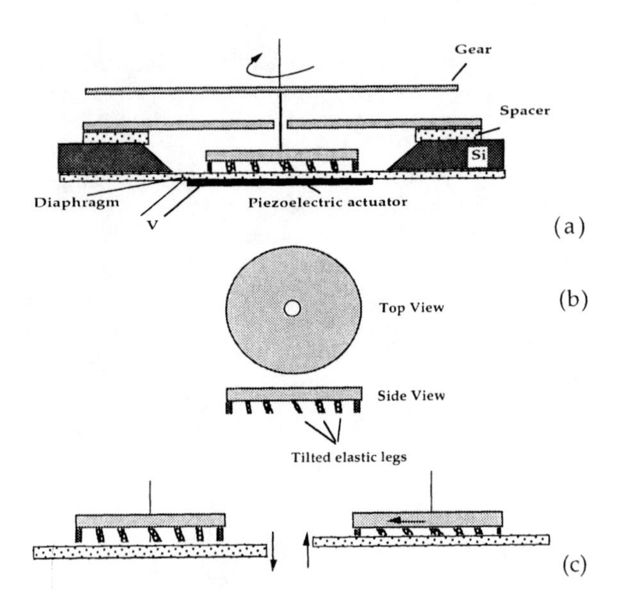

Figure 2.21 a) A schematic of a rotational piezoelectric micromotor. b) The top and side views of the rotor. c) The rectification of the oscillatory motion of the diaphragm by the tilted elastic legs of the rotor [60].

There are other rotational piezoelectric micromotor architectures as well [62]. In one approach, a cylindrical piezoelectric tube is excited by four electrodes to execute a bending motion that travels around the circumference of the cylinder as shown in figure 2.22.

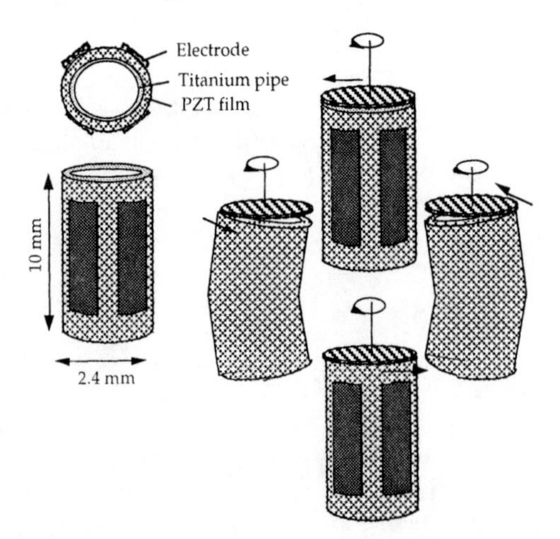

Figure 2.22 A schematic of the structure of the ultrasonic motor [61].

The rotor of the above motor is mechanically coupled through friction to the piezoelectric cylinder. The pre-loading of the rotor as well as the coupling friction are important design parameters of this motor. Rotational speeds in excess of 300 rpm at 20-30 V excitation signal are reported.

2.3.3 Piezoelectric Resonant Structures

Piezoelectrically-driven resonant structures such as diaphragms, microbridges, and cantilever beams are quite common. A microbalance is an example of a resonant structure that is readily used in thickness monitors, gas sensors and other applications [62-63].

In microactuators, piezoelectrically-driven mechanical resonant structures are used in: i) atomic force microscopes, ii) gas sensors and microbalances, iii) motor drivers, and iv) micropumps and switches.

Cantilever beams, diaphragms, and microbridges are usually used as the resonant structure, and a layer of piezoelectric material, co-fabricated with them, is used to excite them. In atomic force microscopes, the piezo-electrically-driven cantilever beam is used to scan the surface of the sample, and the interaction between the cantilever beam tip and the sample modifies the oscillation amplitude of the beam and shifts its resonance frequency [53,54]. A second strip of piezoelectric can be used to monitor the oscillation amplitudes.

In gas sensors and microbalances, a diaphragm is usually used, and the mass loading is measured by monitoring the shift in the resonance of the structure. Most often, the resonator is coated with a gas-sensitive film that also provides it with selectivity. Gas concentrations as low as one part per billion can be easily detected using this method [62,63]. These devices are also used as thickness monitoring devices in evaporation and deposition systems.

Oscillatory motion generated by resonant structures can also be used to generate linear motion in solids, liquids and gases. We saw an example of such a device in section 2.3.2. Micropumps that use piezoelectrically-driven diaphragms to pump liquids and gases are discussed in Chapter 6.

2.4 Electro-Rheological Actuators

Electro-rheological fluids (ER) consist of a dielectric carrier fluid that contains semiconducting particles with dimensions in the range from 0.04 μm to 50 μm [64]. A sudden application of an electric field of sufficient intensity polarizes these particles, causing them to arrange themselves into chains, as schematically shown in figure 2.23. This process results in the formation of particle bridges across the fluid between the positive and negative electrodes. Using an optical microscope, the formation of these particle chains and their subsequent breakdown upon the interruption of the electric field can be readily observed.

There have not been any reports regarding incorporation of ER fluids in microactuators. With the emergence of microfluidic systems, however, ER fluids may find interesting and enabling applications in these actuators. In the macroscopic domain, it is readily shown that ER actuators can generate large forces comparable to those of hydraulics, which have the best power-to-weight and force-to-mass ratios [64]. ER actuators are also shown to have a much simpler structure and a lower weight when compared to hydraulics.

The typical electric fields needed to cause the ER fluids to show significant shear stress or solid-like behavior are around 2×10^4 V/cm. Simple experiments have shown that a significant increase in resistance-to-flow occurs as the electric field strength increases from zero to 4×10^4 V/cm. The maximum field strength is limited to fields above which dielectric breakdown occurs.

Clearly, ER fluids have many parameters that need to be carefully chosen for a given application. As the ratio of the carrier fluid to solid particles decreases, the zero-field viscosity of the fluid becomes large and the current conduction through the ER fluid also increases. On the other hand, larger ratios of particles-to-carrier fluids is desirable in order to have large changes in the shear resistance of the ER as a function of the applied field. To reduce the electrical conduction of the ER fluids, the amount of water in the ER fluid has to be reduced. Water is also undesirable because it causes deterioration of the ER fluid as a function of time. The particle design is also very important. The interface between the particles and the carrier fluid needs to be carefully passivated. Particles are usually composed of SiO_2 plus additives such as NaCl and KCl, etc. Surfactants are usually used to modify the surface interaction of the silica particles and the carrier fluid.

Application of ER fluids in microactuators can be in different forms. We envision that in gear boxes, schematically shown in figure 2.23b, ER fluids may find applications. In flow mode, they can be used as electrically controlled valves.

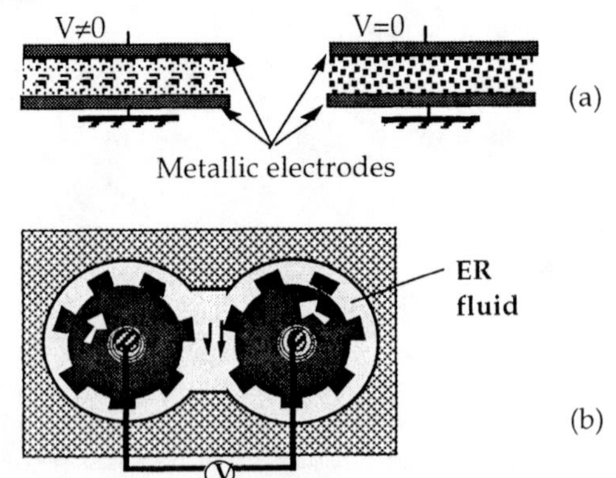

Figure 2.23 (a) Parallel plates immersed in an electro-rheological fluid. A sudden application of a sufficiently large electric field causes them to arrange themselves into chains increasing the shear resistance of the device to shear flow, and it results in solid-like behavior of the fluid. (b) A schematic of a microactuator gear box with ER coupling fluid.

Due to the large forces that can be generated by ER fluids, large force microactuators may utilize ER fluids in the near future. Coupling between the visco-elastic properties and an electric field in liquids can also be induced if the liquid contains charged or polarizable ions.

2.5 Applications in Optics

Electrostatic microactuators, so far being able to exert only small forces, are excellent candidates for manipulating light. In the following sections, we discuss application of electrostatic microactuators in free-space optical devices (section 2.5.1) [67-84] and in guided-wave or integrated optics (section 2.5.2) [85-103].

2.5.1 Free-Space Optics

One of the most important applications of microelectromechanical systems and actuators is perhaps in the area of free-space optics; after all, it does not take much force to bend a beam of light. Applications in this area include spatial light modulators (both in reflection and transmission mode), optical switches and choppers, optical scanners, spatial filters, variable apertures, lenses (graded index, graded profile, Fresnel, etc.) and adaptive optic devices, such as self-focusing lenses and adjustable mirrors. These components are necessary in optical communication, computing and other application areas, such as product identification, etc.

Free-space optical devices tend to have large sizes, and, due to the vibration and temperature variations in an optical system, these components need to be adjusted frequently. By using micromachining fabrication methods, these devices can be made as small as necessary (actually some optical devices must have large dimensions to have acceptable performance, and their dimensions should also be kept much larger than the wavelength of light to avoid diffractions, etc.) and can be made very inexpensively. Also, by co-fabricating the optical devices on the same substrate, the necessity to align them frequently can also be eliminated. These optical components can be integrated with optical detectors and electronics, which, along with proper actuators, can be used to make the optical components adaptable.

There are many examples of optical devices designed and fabricated using micromachining technologies [67,68]. Here, we cover free-space optical devices that are integrated with actuators.

Spatial Light Modulators. These modulators have a variety of applications in spatial light filtering and adaptive reduction of phase distortion in optical communication and optical computing. Figure 2.24 shows a simple example of a spatial light modulator implemented using micromachining. Aluminum mirrors 50 μm in diameter and 4750 Å in thickness that were supported along their axis by two hinges 1 mm wide and 1000 Å thick. The mirror-electrode airgap was 5 μm and 20 V excitation signal was required. The tilt angle was 10 ° and the response time was 10 μs [69].

The mirrors were attached to the frame by cantilever beams that acted as torsional springs by suspending the mirrors over the substrate, as shown in figure 2.24 [67]. In these devices, polysilicon pads situated underneath the mirrors acted as electrodes that, when excited, exerted an electrostatic force on the suspended mirrors. The mirrors were 30 x 30 μm, situated 2 μm above the substrate; and the torsional beams were 15 μm long, 0.4 μm wide and 2 μm thick. The tilt angle, when the mirror was actuated, was 7.6 degrees, and the actuation occurred at 30 V. When the voltage was reduced to below 20 V, the mirrors were released. A 32 x 32 array of mirrors was designed, fabricated and successfully tested [67].

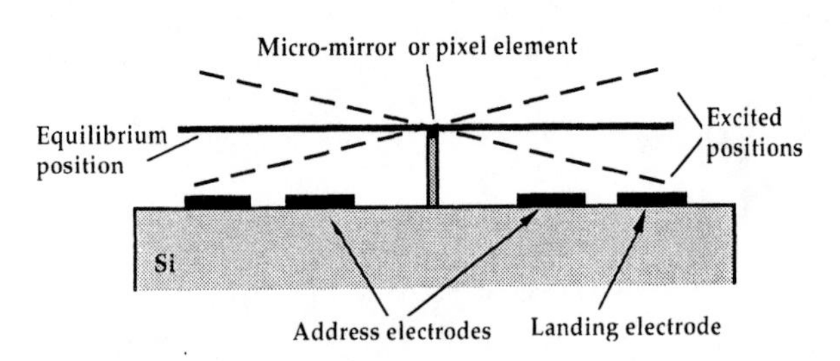

Figure 2.24 A schematic cross-section of a torsion-beam digital micromirror device [67].

Single crystal silicon micromirrors were also fabricated and reported recently [68]. The mirrors were 300 x 240 μm single-crystal silicon covered with sputtered aluminum and suspended with torsional beams (7 height: 1 width aspect ratio for stiffness) over a substrate and actuated using a comb-drive situated on the edge of the mirror. The mirror deflection was 2 μm at 40 V excitation.

400 μm x 400 μm and 30 μm-thick single-crystalline silicon mirror arrays were also fabricated and tested. These mirrors tilted 10° when excited by 30 V and had an operation frequency range of 100 Hz to 20 kHz. The mirror-substrate airgap was 5 μm [69].

100 μm x 100 μm mirrors with non-planar actuation electrodes to reduce the actuation voltage were also fabricated and tested [70]. They also used torsional beams (25 μm long, 7 μm wide and 0.4 μm thick) for suspension and support. The mirrors were gold coated (2 μm) while the torsional beams were nickel (04 μm thick). The torsional angle was 15° at 35 V at 5 μm mirror-electrode airgap.

A mirror that could be deformed continuously by 9 actuator elements located underneath its surface was designed and fabricated using micromachining. The electrostatic actuators could be excited in the range of 0-110 V to deform the mirror locally by 0-0.8 μm. Since the mirror continuously covered these actuators, its deformation was smooth. The mirror dimension was 560 x 560 x 1.5 μm, the actuator dimension was 200 x 200 x 2.0 μm, and the actuator gap was 2.0 μm [71].

Optical Choppers. Conventional mechanical optical choppers are quite large and have a maximum chopping rate of only a few kHz. Electro-optic and acousto-optic modulators can operate up to a few hundred MHz, but require coherent light. Polarizing filters and Kerr cells can also be used, but they require high operating voltages, in excess of a few hundred volts, and their off-to-on transmission ratio depends on the wavelength of the light. Microfabricated optical choppers have the potentials of high-performance, no wavelength dependence, low cost, very light weight and small dimensions, which enable them to be directly attached to an optical fiber.

Figure 2.25a schematically shows an optical chopper that is actuated electrostatically with comb-drives [72]. Its shutter plate (120x195 μm) has an opening that, when aligned with the substrate opening, lets the light through. The shutter blade as well as the connecting beam and the comb-drive are polysilicon.

To improve the opacity of the shutter plate to infrared, the above device had to be coated with a metallic layer. The operation frequency of the chopper was from 1 to 31 kHz, depending on whether it was operated in a vacuum or not. Figure 2.25b shows the deflection amplitude as a function of the actuation voltage in air and in a vacuum [72].

Another example of an optical chopper using a VCM was reported recently [73]. The optical path passes through the rotor and the substrate underneath the rotor so that as the rotor rotates, it turns the light on and off. The aperture size was 200 x 300 μm^2 with minimum driving voltages of 35-60 V with more than 7000 rpm angular frequency.

In another approach [74], a laterally-driven cantilever beam was used to move a shutter blade (30 μm in width) over an opening in the substrate (30 μm in diameter). As the cantilever beam was electrostatically driven sideways, its shutter blade moved over the substrate opening, turning the light on and off. Switching time of 50 μs at excitation voltage of around 75 volts was obtained. The Q-factor of the laterally-driven cantilever beam was 33 and its resonance frequency was 9.5 kHz.

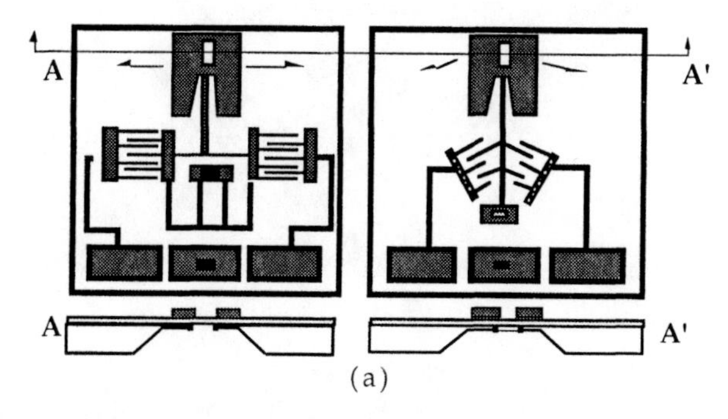

Figure 2.25 a) A schematic of optical choppers with linear and angular comb electrostatic actuators [72].

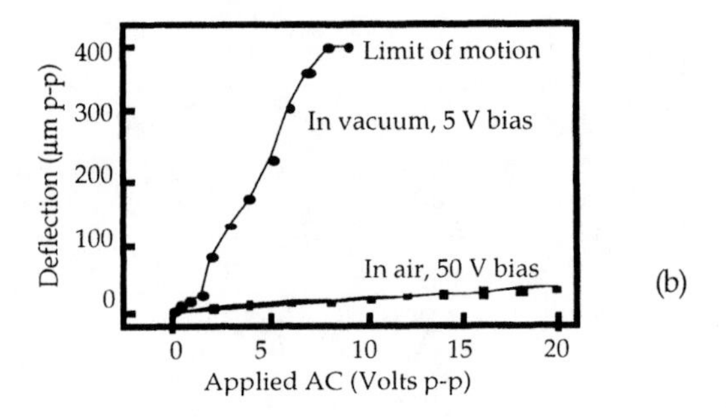

Figure 2.25 b) The deflection of these choppers in air and in a vacuum [72].

Optical Scanners. Scanners are another optical component that can significantly benefit from micromachining technologies. Scanners are used in a variety of optical systems, including laser radar, optical communication systems, object identification and sizing (by triangulation), distance measurements and holographic data storage. There are many different architectures that can be used to scan a beam. These range from using lenses in a binary optic [66] device to using mirrors attached to micromotors [75], to mirrors attached to electrostatically- or piezoelectrically-actuated cantilever beams [76,77], to electrostatically-actuated micromirror devices [78], to vertical and slanted micromirrors [79], to 2-D piezoelectrically-driven micromirrors [80].

Figure 2.26 schematically shows an example of a piezoelectrically-actuated scanner [80]. Mirror areas of 12 mm^2 as well as 6 mm^2 were designed and fabricated, requiring 2-4 p-p V (< 1 W) to scan 10 to 25 degrees. Operation frequencies in the range 100 Hz to 2 kHz were obtained. The beam dimensions were 6.8 mm in length, 3.4 mm in width and 10 μm in thickness.

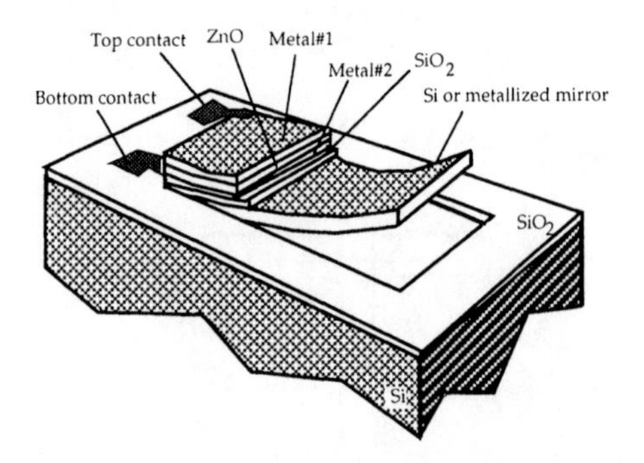

Figure 2.26 A schematic of a piezoelectrically-actuated silicon optical scanner [80].

300 μm x 400 μm flip-up mirrors electrostatically actuated with comb-drives were also reported. These scanners had two different configurations, reaching 12° and 28° scan angles with corresponding operation frequencies of 2.1 kHz and 3.1 kHz and 0.68 m-deg. jitter. The scan angle- excitation voltage was linear, with 6.8° at 50 volts [78].

Another scanner used a rectangular micromirror attached to a rectangular PZT piezoelectric actuator with a cantilever beam (leaf spring). It exhibited 4.7° scan angle at 4.5 V. Its overall dimension was 2 mm x 4 mm x 0.4 mm, its actuator was 653 μm x 227 μm x 22 μm with 2 μm PZT; its leaf-spring was 1564 μm x 13 μm x 5 μm; and its mirror was 977 μm x 1484 μm x 22 μm. It exhibited both translational and rotational resonance behaviors at, respectively, 110 Hz and 214 Hz [80].

Scanning mirrors using magnetic actuation have also been recently reported. They used an inductor coil to actuate the micromirrors in a magnetic field generated by a permanent magnet. Scan angles in excess of 60 degrees were obtained. These scanners were used in holographic data storage and retrieval, and demonstrated successful operation where hundreds of holograms were stored and retrieved [82]. We discuss magnetic actuators in Chapter 3.

Adaptive Optics. The human eye is probably the best example of an adaptive system. Muscles attached to its lens enable it to focus on objects from very far distances to as close as 10 cm. Its dynamic range is also very large, enabling it to function under intense light down to a very dim environment and, in some cases, it can even act as a photon counter. In satellite optical communications or remote imaging applications, the light beam passes over regions in the atmosphere with varying air density, clouds and other time-dependent phenomena, giving rise to a time dependence of a local refractive index. Thus, the wavefront of the light beam becomes distorted and changes as a function of time. Optical systems used in these applications need to be adaptive to achieve optimal performance. Arrays of micromirrors have been used in the past to correct the wavefront distortion, increasing the range and resolution of remote imaging systems and improving the signal-to-noise ratio in optical satellite communications.

Figure 2.27 schematically shows an example of an optical adaptive system. When the wavefront correction is achieved, the signal becomes maximum [71,83,84]. Thus, micromirrors are adjusted individually to achieve maximum signal detection. A variety of control schemes and strategies can be used in these applications, including classical control, neural network-based adaptive non-linear control, and other artificial intelligence-based machine control and learning methods.

One of the best applications of micromachining is probably in the area of optics. In addition, an adaptive optic system can benefit a lot from structures that can be mass-produced using micromachining methods. Presently, adaptive mirrors are micromachined and assembled by hand and, by using micromachining techniques, precision, weight and cost can be improved tremendously.

Deformable mirrors, as discussed in the spatial light modulator section, can be directly used in adaptive optics by providing feed-back, as schematically shown in figure 2.27 [76,83,84]. One major requirement in adaptive mirror systems is that the mirrors should be excitable continuously with an analog signal although there are also ways of constructing digital systems as well.

Lenses. Lenses are one of the most important components of optical systems. Lenses, as well as binary optic lenses, are designed and fabricated using micromachining [67]. Here we discuss some of the lenses that are integrated with actuators and are adaptive.

Micro-Fresnel lenses integrated with a self-assembled *XYZ* stage actuated with scratch drive actuators were described in [76]. These microlenses could be displaced vertically above the substrate with a travel range of 300 μm. Another interesting adaptive lens is described in [83], where two nearly parallel glass plates form the top and bottom of a cavity that is filled with a high-refractive index liquid. A stack of piezoelectric plates is arranged to deform the top glass plate, exerting a force on it upon excitation. As the top glass plate is pulled out or pushed in, a lens which has a curvature that is controlled by the amount of the deformation of the top glass plate is formed. The lens could be driven with 7 ms voltage pulses (150 Hz).

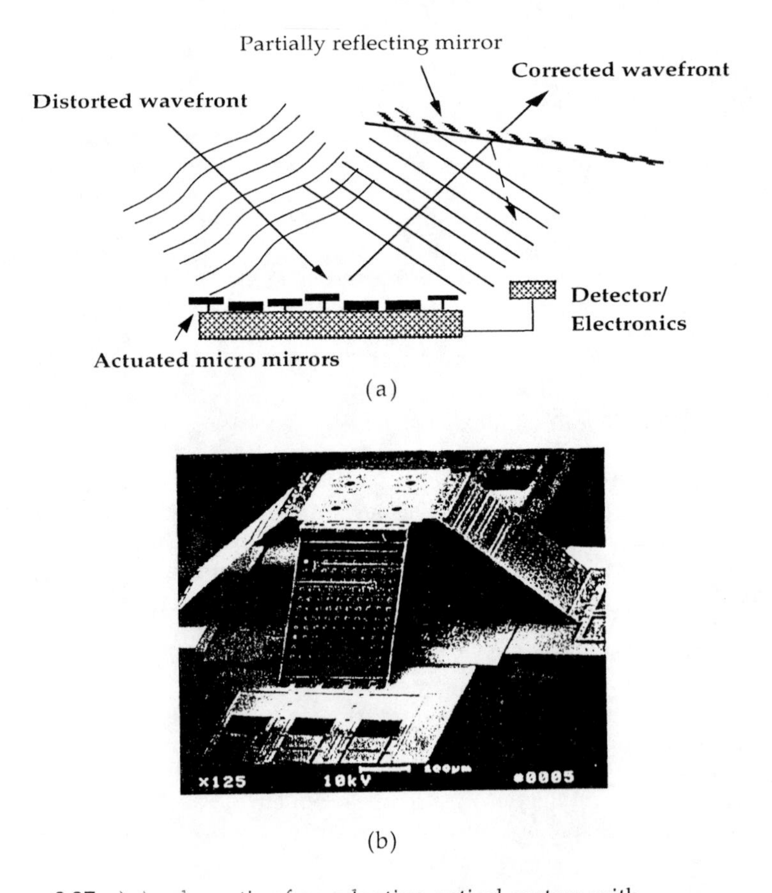

Figure 2.27 a) A schematic of an adaptive optical system with electrostatically actuated micromirrors.
b) SEM of self-assembled micro-*XYZ* stage with integrated micro-Fresnel lens [76].

2.5.2 Integrated Optics

Beginning with the fabrication of the first microscale optical waveguide, progress in integrated optics has been driven, in large part, by developments in the area of microstructures. In the last 8-10 years, the technology of microstructures has progressed extremely rapidly, enabling the development of a wide variety of novel integrated optic devices for sensing, actuation [85-89] and communication [90]. Utilization of microstructures in free-space optics has also received much attention lately and has resulted in a variety of optical devices, including scanners, lenses and switches, as discussed above. Here, we concentrate on guided wave optics and focus our attention on the following questions:

i) How integrated optics benefits from microstructures
ii) How useful devices (mainly sensors and actuators) can be devised using integrated optics and microstructures

Integrated optics has benefited enormously from the development and fabrication of microstructures [90-103], which, in general, are of a physical scale, typically 1-500 μm, which is appropriate for the control of optical radiation. In the microoptics arena, the low power output of most microactuation mechanisms is not a disadvantage since little energy needs to be expended in order to change the path of an optical beam. Because microstructures are fabricated using batch processes, these devices are potentially low cost and the monolithic nature of these devices enhances reliability.

The use of microstructures to convert a mechanical displacement to a change in optical intensity has served as the basis for sensors as well as switches, since many physical parameters can be converted to a displacement or strain [85,90]. Integrated optic sensors, when linked by optical fibers, can provide ready isolation from high voltages and can provide a high degree of immunity to electromagnetic interference, as well. In many instances, the physical parameter is sensed by detecting displacements, in the range of 10^{-2} to 10 μm, of the cantilever beam, microbridge or diaphragm. This range, on the low side (10^{-2} to 1 μm), falls favorably in the realm of evanescent-field sensing, and, on the high side (1 to 100 μm), renders itself to propagating wave sensing.

Microactuators have benefited from the application of microstructures and integrated optics as well. Integrated optic position sensors can improve the reliability of closed-loop actuator systems that must operate in a hostile environment. In addition, a number of groups have worked on developing optically-driven actuators in order to produce actuator systems that are fully optically linked.

In the following sections we discuss the above developments in detail and we provide a map of progress in these areas.

Innovative Integrated Optics Microstructures. Integrated optics circuits require optical waveguides which connect various passive and active optical devices. Almost universally, these optical devices are appropriate microstructures with suitable refractive indices and other optical parameters. In this section we discuss microstructures that enhance the functionality or yield of the already-existing IO circuits either by improving manufacturability or by providing them with a means to achieve functionalities that are otherwise not possible. The following examples illustrate these features. Utilization of simple V- or U-grooves enable insertion of fiber optics into the IO circuits with relative ease and they simplify alignment procedures. They also increase the temperature stability and reliability of the IO circuits. On the other hand, incorporation of a cantilever beam in the path of an IO waveguide introduces the possibility of switching the light in a waveguide made from a non-electro-optic material such as SiO_2 and Si_3N_4 [85,89,90].

Waveguides are probably the most common integrated optic elements. They are fabricated using a variety of methods [85], including standard photo-lithography and wet and dry etching, direct patterning during growth and embossing.

Optical Modulators and Switches. The micromachined optical switches offer higher isolation, lower power consumption and the possibility of a bi-stable operation.

A simple optical switch using a cantilever beam actuator on the path of an optical waveguide is shown in figure 2.28. The multimode polyimide waveguides were deposited and patterned post fabrication, using bulk micromachining of the cantilever beam (600μm x 50μm x 5μm). This simple un-optimized device operated with 50 volts and achieved isolation of 20 dB [85].

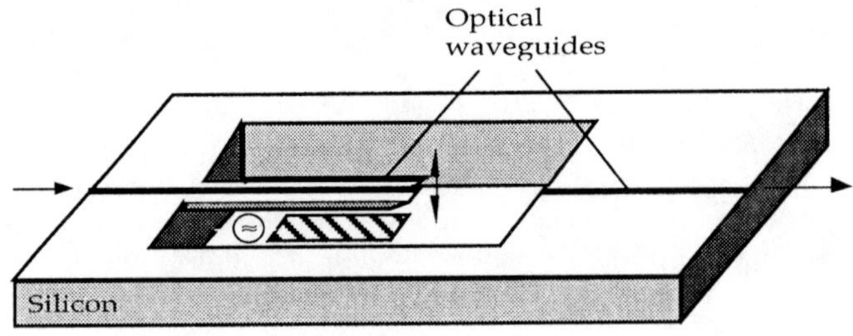

Figure 2.28 A silicon microcantilever beam and integrated optical waveguides. (The cantilever beam is 600 x 50 x 5 μm^3, fabricated by p^+, boron diffusion and EPW anisotropic etchant).

In micromechanical optical switches, vibration and acceleration resistance are important considerations, and the resonant frequency of the structure should be carefully chosen to render it insensitive to environmental vibrations or noise. Actually, these devices can act as a very sensitive acoustic pick-up microphone if not designed properly.

Figure 2.29 shows an example of a 1 X 2 optical fiber switch where the micromachined structure is used to laterally move the input fiber optics between two output fiber optics [94]. The actuation is accomplished by a local heater, and an anisotropically etched trench in silicon aligns the send and receive optical fibers. The design goal was to achieve coupling loss of -0.4 dB while the device showed -5.8 dB. The isolation in the off state was better than -66.5 dB. This switch had a maximum operation frequency of 1.3 Hz and required around 495 mW for switching.

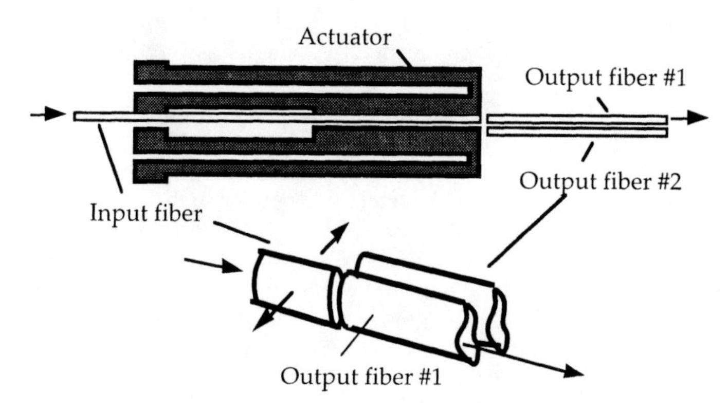

Figure 2.29 A 1x2 micromachined optical-fiber switch with a bimorph heater [94].

A 2x2 micromachined optical switch, shown in figure 2.30, has also been reported [95]. Insertion loss of this switch was less than 3.1 dB and its cross-talk was less than -40 dB. It operated with an applied voltage of 100 volts (practically no conduction current). Its switching speed was around 0.5 ms.

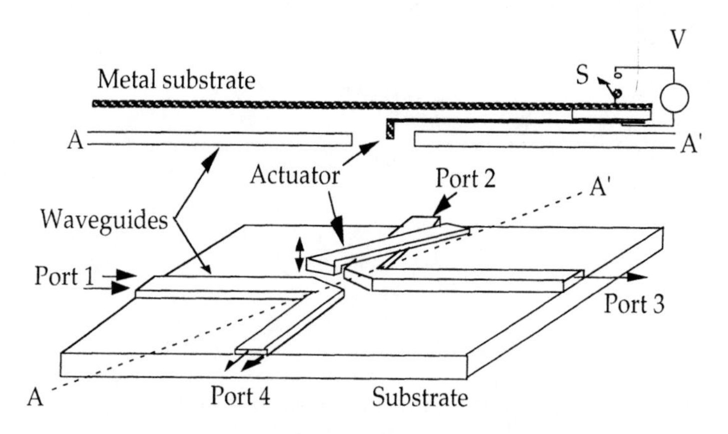

Figure 2.30 An electrostatically-driven 2x2 micromachined optical switch [95].

An optical switch using a flip-up mirror driven by a scratch drive actuator has also been reported. It could withstand vibration acceleration of up to 89 g's at 200 Hz to 10 kHz. It could be turned on at 15 ms and turned off at 6 ms [76]. It was used to construct a 2x2 fiber optic switch. Its scratch drive actuators could move at a speed of 2500 mm/sec at 50 kHz and 100 V excitation signal.

A 10 ms 2x2 fiber optic bypass switch that uses magnetic actuation method to move the mirror in and out of the fiber-optic plane in a configuration very similar to figure 2.30 has also been fabricated and reported [77]. High-aspect ratio silicon micromirrors driven with a comb-drive were also designed and fabricated. In these devices, lateral mirror movement of 34 μm was achieved by a driving voltage of 30 V. The actuation could be affected at 978 Hz driving frequency.

Optical Modulators and Switches Using Evanescent Interactions. The preceding section described optical modulators and switches that had a microstructure intersecting an optical waveguide path to achieve the desired switching or modulation. It is also possible to construct modulators and switches by using microstructures that only interact with the guided light through evanescent fields. It can be shown that when an optical wave is confined within a dielectric guide, there is an exponentially-decaying field just outside the guide's boundaries [85,90]. The decay length depends on the index of refraction of the surrounding media (cladding) as well as on the particular mode in which the light is propagating. Such exponentially-decaying fields are referred to as evanescent fields. On a typical optical waveguide, the decay length of the evanescent fields is on the order of the wavelength. Consider a three-layer waveguide where one of the cladding layers is air (or a vacuum). If an external object is brought within the reach of the evanescent field ($\sim\lambda$) of the waveguide core, new boundary conditions are imposed on the wave propagating down the waveguide. The guiding structure becomes a four-layer waveguide.

Various modulators and switches that take advantage of the above interaction have been reported [90,92]. Figures 2.31 and 2.32 show two different structures that take advantage of evanescent field coupling between the guided light and a microstructure situated above the waveguide. In both these structures, a micromachined structure is located over a waveguide. Upon actuation, the microstructure deforms and touches the waveguide. Through the evanescent field interaction, the deformed microstructure modifies the effective refractive index of the waveguide. Since the TM and TE waves are affected differently, the polarization state of the guided wave also changes.

Electrostatic actuation of the metallic membrane, causing it to contact the waveguide, alters the propagation constant of the guided mode [9]. These switches have an actuation speed of 1 μs and a scaleable actuation voltage of 5-45 volts. A 6.5 extinction ratio was obtained at 1.3 μm wavelength with the routing switch shown in figure 2.32b. These devices use the snap action properties of the membrane and they achieve analog modulation by adjusting the applied voltage level after the membrane is pulled down. This varies the area of contact between the metallic membrane and the waveguide. In devices reported in [91,92], the modulation is achieved by varying the airgap between the cantilever beam and the waveguide.

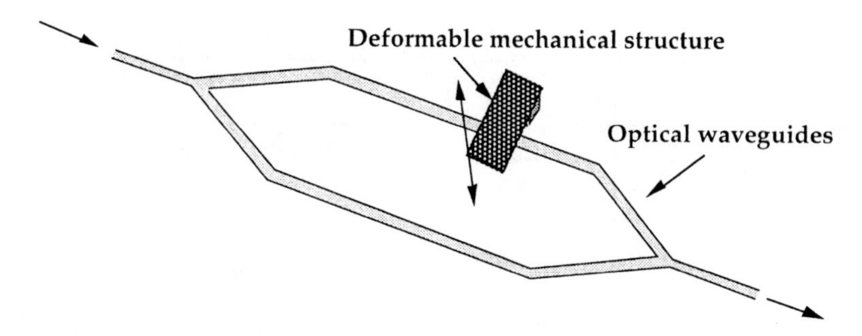

Figure 2.31 A cantilever-type micromechanical waveguide switch [89,90].

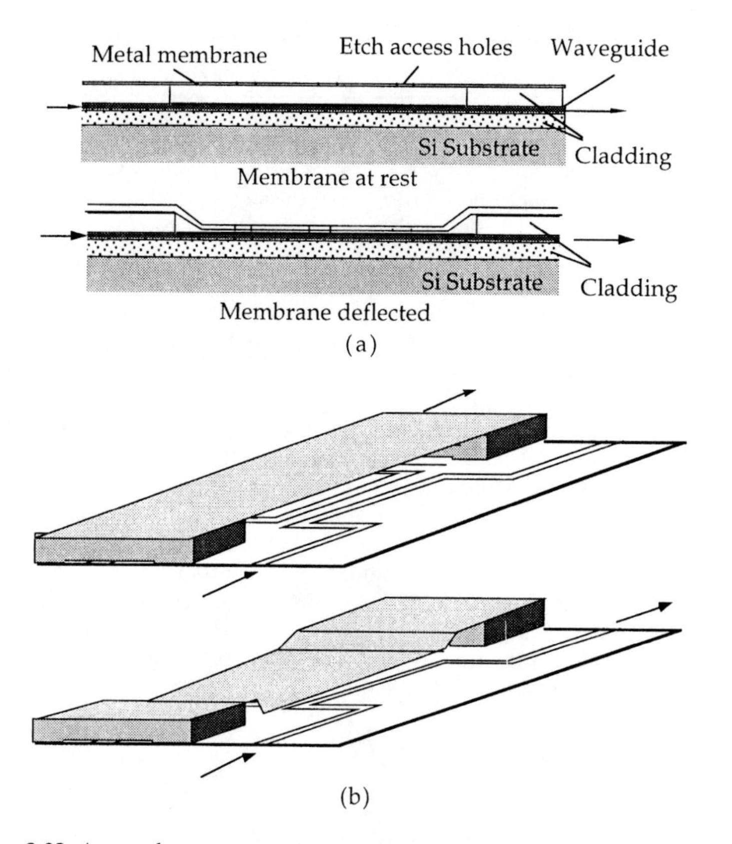

Figure 2.32 A membrane-type micromechanical 1x2 waveguide switch.
(a) Side view, (b) perspective [91].

Table 2.3 Modulation-depth and bandwidth of various types of modulators.

Type	Materials/ Structure	Bandwidth	Modulation Depth (dB)	Operation Voltage (V
Mach-Zehnder EO	$LiNbO_3$	40 GHz	-3	3.6 @ 1 mA
Mechanical anti-reflection switch, Fabry-Perot	Si_3N_4 Diaphragm	5 MHz	-20	50 @ 1 mA
Deformable grating	Si_3N_4 Beams	6.1 MHz	-16	3
All-Si	Single crystal Si	200 MHz	-0.5	50 @ 10 mA
Si TO	All-Si or SiO_2 or Si_3N_4	700 kHz	-3.9	0.25
Wannier-Stark EA	InP	18 GHz	-20	1 @ 3 mA
Quantum-Confined Stark effect EA	InGaAsP MQW	26 GHz	-20	2 @ 3 mA
Liquid crystal EO		18 GHz		

TO: Thermo-optic EO: Electro-optic
EA: Electro-absorption MQW: Multiple quantum well
Modulation Depth: measure of (off signal)/(on signal).

2.6 Applications in Acoustics and Ultrasonics

Both piezoelectric and electrostatic actuation methods are used to generate sound waves over a large frequency range. Common loudspeakers are the electromagnetic-mechanical devices discussed in Chapter 3 and, due to their coils and permanent magnets, they can be bulky and not readily scaleable to smaller dimensions. On the other hand, both the piezoelectric and electrostatic actuation methods can be easily scaled down and, in conjunction with silicon micromachining, can produce audio and ultrasonic output devices capable of operating in a variety of liquid and gaseous environments. Efficiency of sound generation, in audio and ultrasonic devices depends on the actuation mechanism as well as the impedance matching between the device and its environment. Using metals, polysilicon, single crystalline silicon, oxides and nitrides, as well as, polymeric membrane materials, micromachined devices can be readily tailored to have the desirable acoustic impedance.

Moreover, phased array structures can be constructed by addressable arrays of ultrasonic output devices that would be capable of sending the sound wave in different directions by properly phasing the input to each device on the array. Or, alternatively, in the input mode, these phased arrays can be used to determine the direction of the received sound waves.

Figure 2.33 schematically shows the structure of an ultrasonic transducer that was designed and fabricated in an array format that contained in excess of a few thousand elements (area ≈ 1 cm^2 and individual device diameter of 60 μm). The array exhibited a dynamic range in excess of 110 dB at 2.3 MHz in air. In water, a single pair of transducers was able to operate from 2 MHz to 20 MHz [104].

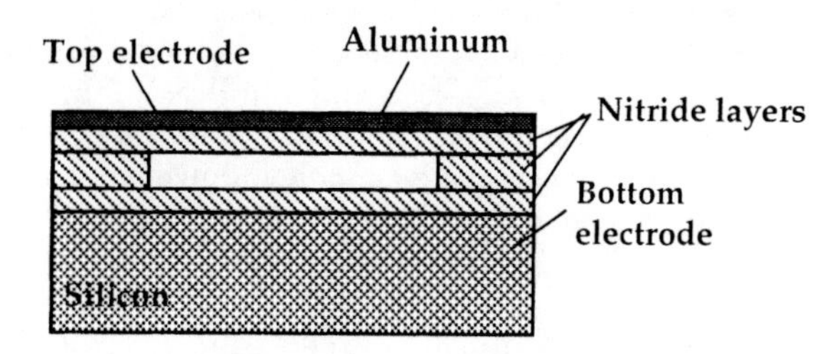

Figure 2.33 A schematic of a single-element micromachined ultrasonic transducer [104].

An ultrasonic microcutter has also been designed and fabricated. It consists of a silicon cantilever beam with a tapered tip to act as a cutter and a strip of piezoelectric film deposited over the cantilever beam to vibrate it laterally. Large ultrasonic displacements, around 40 μm, at operation frequencies in the range 50-150 kHz were obtained [105].

2.7 Applications in Fluidics

Fluidics refers to devices that control and manipulate the flow of fluids. Taking a more general definition of fluids, gases are also included. A typical fluidic device may contain reservoirs of different fluids and channels to guide these different fluids into mixing chambers. There may be regions to provide the possibility of combustion or oxidation/reaction of these fluids, and there may be moving parts that can be powered by these reactions or chambers with embedded sensors to analyze the fluids and their byproducts. Spark plugs to initiate reactions and heaters/coolers to control the reaction and fluid temperatures will also be included. Optical analysis schemes, such as spectroscopy, along with other molecular and atomic microscopes, can readily be integrated with other micromachined structures to construct such an elaborate fluidic machine. Electron guns, mass spectrometers, flow cytometers, micro electrophoresis, stripping voltametry and micro blood-chemistry analyzers fabricated using micromachining are proposed in the literature [106-118].

Different components that may be needed in a fluidic system are: micropumps, vacuum pumps, valves, chambers, a variety of sensors (including gas, ion, species, pressure, temperature, force and displacement), rotors and pistons. Some of these are discussed in Chapters 4 and 6.

2.8 Applications in Relays

Solid-state relays based on semiconductor electronic devices are quite fast (≈0.01 μs) and they can handle large voltages and currents (up to 1000 Amps at 1000 Volts). The only shortcoming in these devices is that their off-to-on resistance ratio is not very high. Typical on-resistance of 0.1 Ω and off-resistance of 10^7 Ω is encountered in these devices resulting in an off-to-on ratio of 10^8. With electromagnetic relays, this ratio can be increased to 10^{10}, with even better voltage and current handling capabilities. These relays have much better input/output isolation and their power consumption can be much lower than the solid-state switches at high voltages and currents. It is also possible to have a latch mechanism to enable them to stay on or off without control power. Their main shortcomings are that they are bulky (compared to solid-state switches), the mechanical vibration/acceleration can cause them to turn on/off, and they are slow (≈ μs-ms). They are also expensive since they are usually composed of an electromagnet and a mechanical part that need to be assembled together to form the device.

Microwave switches are an important component of microwave communication systems. These switches are used to direct the signal from the antenna to the front-end amplifier in the receiving mode or to direct the output of the power amplifier to the antenna in the transmission mode. Clearly, the switch should be able to handle very large powers and it should have minimum insertion loss since, in the receiving mode, the signal is very weak and any noise generated by the switch or its insertion loss are undesirable.

Silicon micromachining can be used to fabricate miniature relays with superior characteristics, such as very small dimensions (≈100 μm), latching capabilities, excellent immunity to vibration (up to a few hundred g's over a wide spectrum), low power consumption (μWatts), excellent off-to-on resistance ratio (10^{10}), and very low cost (using batch fabrication). Owing to their small dimensions, arrays of these switches are designed and fabricated.

In Chapters 3 and 4 we discuss examples of magnetic and thermally actuated relays. Figure 2.34 schematically shows the structure of an electrostatic relay that was designed and fabricated using silicon micromachining. These devices had gold-gold contact points and operated with actuating voltages between 30-300 V with an off-resistance in excess of 10^{12} Ω. Their on-resistance was on the order of 0.5 Ω and it is predicted that it can be further reduced to around 10^{-2} Ω by using multiple contact points and better cantilever beam design. The beam length was 50 μm, its width 10 μm and its thickness was around 1 μm and was situated 5 μm above the gate electrode. It performed over 10^9 cycles without deterioration in its performance.

Electrostatic relays that operated up to GHz range have also been reported [119]. In another electrostatic relay, contact resistances in the range of 12 Ω - 80 Ω, 1 μW switching power and switching time constant of 2.6 μs were observed [120].

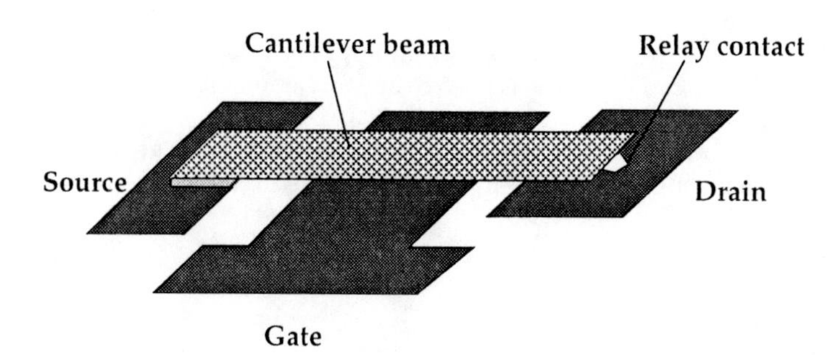

Figure 2.34 A schematic of an electrostatic switch fabricated using silicon micromachining [120].

2.9 Applications in Scanning Probes and Microrobotics

Scanning tunneling microscopes and atomic force microscopes are important examples of scanning probes [121-133]. Mechanical actuation is required, in one form or another, in the positioning and various operations of these probes. For example, in the scanning tunneling microscopes, the tip-to-wafer distance is adjusted to keep the tunneling current constant at a given voltage. Or, in the spectroscopy mode, the tip-to-surface distance is kept constant and current versus voltage measurements are performed. In atomic force microscopy, there is a mode of operation that vibrates the tip over the surface and images the surface by measuring the change in the resonant frequency of the cantilever beam.

In all these applications, either the tip-to-surface distance should be adjusted or the cantilever beam should be adjusted. Integrating piezoelectric layers with a cantilever beam or a mechanical structure (microbridges, diaphragms, etc.) offers the following advantages: i) since the piezoelectric film makes intimate contact with the rest of the structure, the device is more reliable; ii) since the film thickness can be controlled quite accurately by the deposition method, it can be made quite thin, resulting in the lower voltages needed for actuation; iii) devices can be batch-fabricated inexpensively; and iv) eventually electronics and sensors can be integrated with microactuators for in-situ signal processing and active control.

The next section discusses microrobotics, and local probes can be viewed as very simple examples of microrobots with only 3-5 degrees of freedom. In building microstructures or in diagnostic and high-resolution imaging applications they can be used in parallel to simultaneously cover large areas.

Local probes include scanning tunneling microscopes, atomic force microscopes, high spatial resolution thermal probes, and capacitance or charge-sensitive probes. In most of these probes there is a mechanism to adjust the position of the probe tip over the sample surface. Various methods, including cylindrical piezoelectric motors, thin film piezoelectric actuators and electrostatic actuators, have been used in the past. An important trend in all of these applications has been to integrate the actuator with the probe to reduce cost, to improve reliability and long- and short-term drifts, and to increase the overall functionality of the probe.

Figure 2.35 shows an example of a piezoelectrically-activated cantilever beam used in scanning tunneling microscopy (STM). These integrated probes have been fabricated and characterized in the past [125,126].

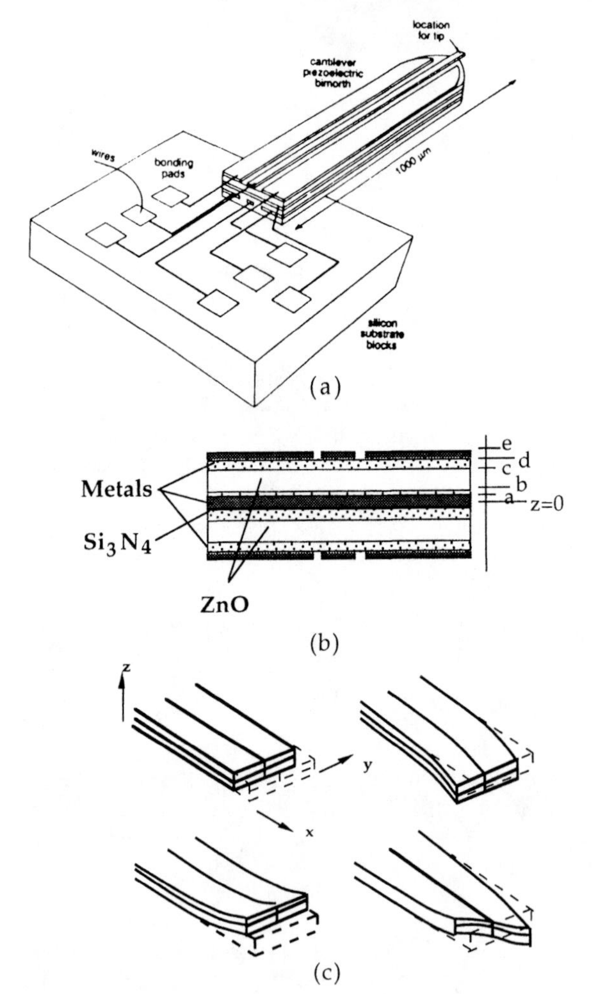

Figure 2.35 (a) An STM tip with piezoelectric actuators used to both laterally and vertically move the tip. (b) The beam cross section and (c) different modes of actuation [126].

The integrated tip was composed of two layers of magnetron sputtered ZnO, three layers of aluminum metallization, and four layers of Si_3N_4 (deposited using low pressure and plasma-enhanced chemical vapor deposition methods). The ZnO layer had its c-axis perpendicular to the plane of the tip. It was found that the smoothness of the substrate surface was very crucial in obtaining high-quality piezoelectric ZnO layers. The ZnO, dielectric and metal layers were 3.0, 0.2 and 0.5 μm thick. According to figure 2.35b, hence, a=0.25 μm, b=0.45 μm, c=3.45 μm, d=3.65 μm and e=4.15 μm. The elongation in the x-direction is given by [126]:

$$\Delta x = \frac{L\, d_{31}\, Y_z\, V\,(c-b)}{[Y_a(a+e-d) + Y_b(b-a+d-c) + Y_z(c-b)](d-a)} \qquad (2.32)$$

where a, b, c, d and e are the different layer thicknesses shown in figure 2.35b, and the coordinate system is defined in figure 21.9c, where Y_z is the Young modulus of the ZnO, and Y_a and Y_b are those of the metals and dielectrics respectively. V is the applied voltage and L is the beam length. Y's are given in table 2.2, L=1000 μm and d_{31} is given in table 2.1. The displacement in the z-direction is given by [126]:

$$\Delta z = \frac{3\,L^2\, d_{31}\, Y_z\, V\,(c^2-b^2)}{4\,[Y_a(a^3+e^3-d^3) + Y_b(b^3-a^3+d^3-c^3) + Y_z(c^3-b^3)](d-a)} \qquad (2.33)$$

For the above probe, it was shown that $\Delta x/V$ was 14 Å/V and $\Delta z/V$ was 2500 Å/V. The effective Hooke's constant of this probe in the z-direction was 3 N/m.

Figure 2.36 shows a schematic of a wafer-scale deposition facility [133]. Scanning probes are enclosed in microchambers with their own supply lines, as shown in figures 2.36 and 2.37. Micropumps and valves are all fabricated on silicon, using the MEMS technology demonstrated over the past 5-10 years. Parallel local probes deposit patches of semiconductors, metals, and insulators over the substrate. These probes, combined with the linear motion of the substrate underneath them, have four degrees of freedom.

Microrobotics. Microrobots with an overall size of a large grasshopper (≈ 1 cm^3) are being developed to perform tasks in tight, hard-to-reach and hazardous places. These mechanical bugs, one schematically shown in figure 2.38a, could perform surgery within the human body some day, repair underground electrical cables, or repair fiber optic junctions in remote locations. Microactuators are clearly a necessary component of these devices.

Reliable, efficient devices with very small power consumptions are needed to enable microrobots to navigate, as well as to cut tissues or work on other tasks. An average robot may require in excess of 10 actuators to perform such mundane tasks as moving around, avoiding obstacles and lifting simple objects. Thus, power consumption is an important parameter if these microrobots are to be self-sufficient and stand alone. Piezoelectric microactuators may prove to be quite suitable in microrobotics. In addition to actuators, sensors, on-board electronics, power sources and receiver/transmitters will also be contained inside these robots.

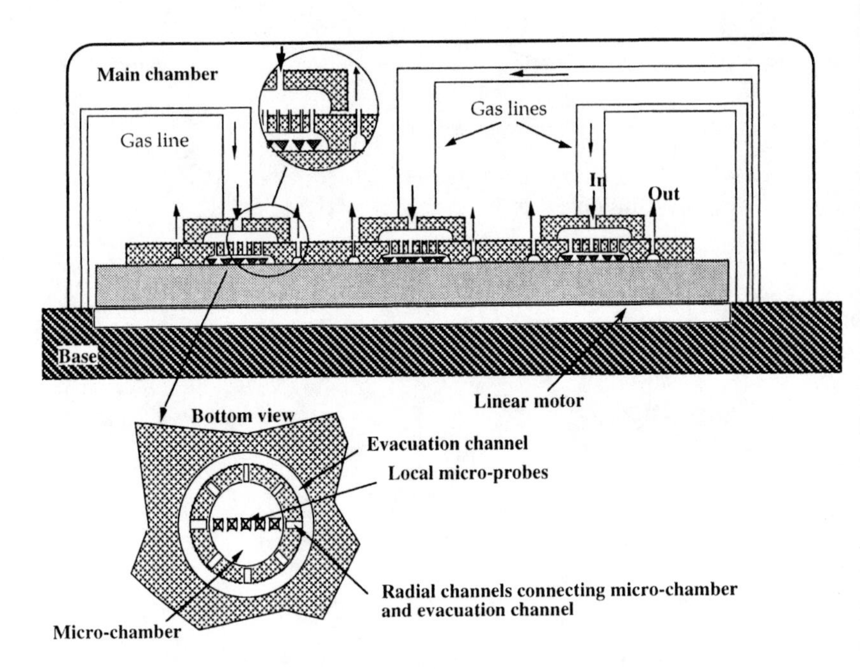

Figure 2.36 An overall view of the fabrication facility [133].

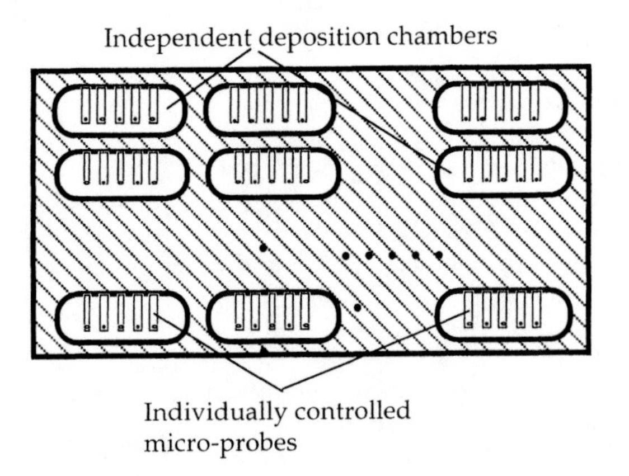

Figure 2.37 The different microchambers enclosing groups of parallel local probes can be used to deposit different materials in parallel.

Figure 2.38b shows a photograph of a microinspection machine which consisted of a piezoelectric stack actuator to move the microrobot in the forward and backward direction packaged in an ultra-thin (60 μm thick) aluminum casing, an eddy current sensor, cooling fins, and clamps. It moves forward when the stack piezoelectric actuator expands quickly (in the direction of the motion) and retrieves slowly, enabling the clamps to stick to the inner surface of the tubing in the expansion part of the actuator cycle due to friction. In the backward motion, the actuator expands slowly and retrieves quickly. The body of this microinspection system was 5.5 mm in diameter [134].

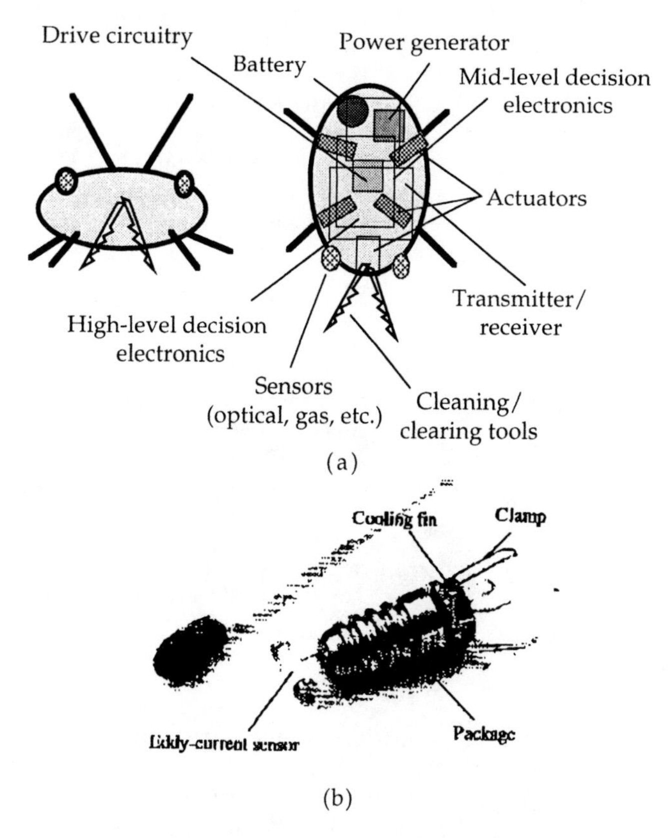

Figure 2.38 (a) A schematic of a 1 cm^3 self-sufficient microrobot.
(b) A photograph of a microinspection machine [134].

2.10 Applications in Surgery

In certain surgeries, such as brain tumor removal or other operations on the brain to electrocute hyper-active centers in Parkinson patients and in ophthalmic microsurgery, precision positioning and cutting is needed [135-137]. The tools that are used in these situations consist at present of handformed stainless steel with shaft diameters of approximately 900-1000 μm and typical tool tip displacements from 200-300 μm with forces felt by the surgeon ranging from 6-150 g. Micromachined devices can be used to replace these tools and improve performance by 20-30 %, while reducing cost.

Some of the microactuators that have already been discussed in this chapter are readily capable of achieving these displacements and forces. There are many groups in the US alone that have started the design and fabrication of microsurgery devices, using micromachining methods. Micromachining can be used either to fabricate tips and shafts to be insert d in these already existing microsurgery tools or to design and fabricate completely novel tools with actuated tips, electronics to control the position and force, and sensors to detect temperature, light intensity, and other parameters at the tip of the tool. These novel tools can also include a mechanical magnification mechanism so that the resistance that the tool tip experiences in cutting a tissue can be magnified. This magnified tissue resistance will be felt by the surgeon performing the task and it will prevent him/her from cutting through the tissue inadvertently. The technology to realize all of these functionalities exists, and it is only a matter of time before microsurgery tools become commercially available.

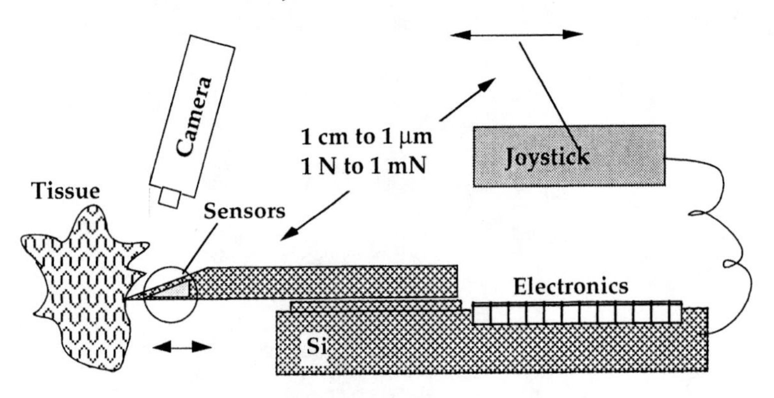

Figure 2.39 A schematic of a surgery knife attached to a precision microactuator with on-board electronics for sensing and knife motion control. The electronics plus the active joystick also enable force amplification to prevent the operator from tearing apart the tissue.

2.11 References

1. P. Benjamin, A History of Electricity, John Wiley, New York, pp. 506-507 (1898).
2. J. Sparks (ed.), The Works of Benjamin Franklin." Vol. 5, Whittemore, Niles, and Hall, Boston, p. 301 (1856).
3. Oleg D. Jefimenko, Electrostatic Motors. Electret Scientific Company, Star City (1973).
4. W.K.H. Panofsky, and M. Phillips, Classical Electricity and Magnetism. Addison-Wesley Publishing Co., Inc., Reading, MA, p. 106 (1955).
5. H. H. Woodson, and J. R. Melcher, Electromechanical Dynamics, Part I: Discrete Systems. John Wiley & Sons, Inc., New York, pp. 60-88 (1968).
6. W. S. N. Trimmer and K. J. Gabriel, "Design Considerations for Practical Electrostatic MicroMotor." Sensors and Actuators, Vol. 11,

pp. 189-206 (1987).
7. K. Petersen, "Silicon as Mechanical Material." Proc. IEEE, Vol. 70 (5), pp. 420-457 (1982).
8. T. Ikeda, Fundamentals of Piezoelectricity. Oxford University Press, New York, pp. 213-226 (1984).
9. H. Fujita and T. Ikoma, "Numerical Determination of the Electromechanical Field for a Micro Servosystem." Sensors and Actuators, A21-A23, pp. 215-218 (1990).
10. S. D. Senturia, "The future of Microsensors and Microactuator Design." Sensors and Actuators A 56, pp. 125-127 (1996).
11. R. H. Price, J. E. Wood and S. C. Jacobson, "Modeling Considerations for Electrostatic Forces in Electrostatic Microactuators." Sensors and Actuators, Vol. 20, pp. 107-114 (1989).
12. R. Mahadevan, M. Mehreghany, K. J. Gabriel, "Application of Electric Microactuators to Silicon Micromechanics." Sensors and Actuators, A21-A23, pp. 219-225 (1990).
13. S. Kumar, D. Cho, W. Carr, "A Proposal for Electrically Levitating Micromotors." Sensors and Actuators A24, pp. 141-149 (1990).
14. T. Niino, T. Higuchi and S. Egawa, Proc. IEEE Industry Applications Conf., Orlando, FL, pp. 1318-1325 (1995).
15. H. Frohlich, Theory of Dielectrics. 2nd Edition, Oxford University Press, New York, p. 26 (1986).
16. S. Hoen, P. Merchant, G. Koke and J. Williams, "Electrostatic Surface Drives: Theoretical Considerations and Fabrication." Transducers '97, pp. 41-44 (1997).
17. N. Triole, D. Hauden, P. Blind, M. Froelicher and L. Guadriot, "Three-Dimensional Silicon Electrostatic Linear Microactuator." Sensors and Actuators Vol. A 48, pp. 145-150 (1995).
18. N. Tas, J. Wissink, L. Sander, T. Lammerink and M. Elwenspoek, "The Shuffle Motor: A High-Force, High-Precision Linear Electrostatic Stepper Motor." Transducers '97, pp. 777-780 (1997).
19. L. Dellman et. al., "Fabrication Process of High Aspect Ratio Elastic Structures for Piezoelectric Motor Application." Transducers '97, pp. 641-644 (1997).
20. (a) L.-S. Fan, Y. -C. Tai and R. S. Muller, "IC-Processed Electrostatic Micromotors." 1988 IEEE Electron Devices Meet., San Francisco, CA, U.S.A, Dec. 11-14, pp. 666-669 (1988).
(b) M. Mehregany, S.F. Bart, L.S. Tavrow, J. H. Lang, S. D Senturia, and M. F. Schlecht, "A Study of Three Microfabricated Variable-Capacitance Motors." Sensors and Actuators, A21-A23, pp. 173-179 (1990).
21. R. S. Muller, "Microdynamics." Sensors and Actuators, Vol. A21-A23, pp. 1-8 (1990).
22. Y-C. Tai and R. S. Muller, "IC-Processed Electrostatic Synchronous Micromotors." Sensors and Actuators, Vol. 20, pp. 49-55 (1989).
23. K. J. Gabriel, F. Behi, R. Mahadevan, and M. Mehregany, "In-situ Friction and Wear Measurements in Integrated Polysilicon Mechanisms." Sensors and Actuators, A21-A23, pp. 184-188 (1990).
24. Yu-C. Tai, and R. S. Muller, "Frictional Study of IC-processed Micromotors." Sensors and Actuators, A21-A23, pp. 180-183 (1990).
25. E. J. Garcia, and J. J. Sniegowski, "Surface Micromachined

Microengine." Sensors and Actuators, A 48, pp. 203-214 (1995).
26. S. F. Bart, T. A. Lober, R. T. Howe, J. H. Lang and M. F. Schlecht, "Design Considerations for Micromachined Electric Actuators." Sensors and Actuators, Vol. 14, pp. 269-292 (1988).
27. L. S. Tavrow, S. F. Bart and J. H. Lang, "Operational Characteristics of Microfabricated Electric Motors." Sensors and Actuators A, 35, pp. 33-44 (1992).
28. M. Mehregany, " Silicon Microactuators." In: Advances in Actuators. Edited by A. P. Dorey, and J. H. Moore, Institute of Physics Publishing, Techno House, Redcliffe Way, Bristol BS1 6NX, UK (1995).
29. (a) K. C. Stark, Mechanical Coupling of Polysilicon Surface Micromachined Mechanisms. Ph.D. Dissertation, Case Western Reserve University (1997).
(b) X.-Q. Sun, Z.-J. Li, and L.-T Liu, "The On-Chip Detection of Micromotor Rotational Speed." Sensors and Actuator A 48, pp. 81-84 (1995).
30. a) J. -B. Huang, Q.-Y. Tong and P.-S. Mao, "Gas-Lubricated Microbearing for Microactuators." Sensors and Actuators A, 35, pp. 69-75 (1992).
b) H. Zarrad, et. al., "Optimization of Lubricants for Silica Micromotors." Sensors and Actuators, A 46-47, pp. 598-600 (1995).
31. A. Fujimoto, M. Sakata, M. Hirano and H. Goto, "Miniature Electrostatic Motor." Sensors and Actuators A, 24, pp. 43-46 (1990).
32. W. Trimmer and R. Jebens, "Harmonic Electrostatic Motors." Sensors and Actuators, 20, pp. 17-24 (198).
33. M. Sakata, Y. Hatazawa, A. Omodaka. T. Kudoh, and H. Fujita, "An Electrostatic Top Motor and its Characteristics." Sensors and Actuators, A21-A23, pp. 168-172 (1990).
34. S. C. Jacobson, R. H. Price, J. E. Wood, T. H. Rytting and M. Rafaelof, "A Design Overview of an Eccentric-Motion Electrostatic Microactuator (the Wobble Motor)." Sensors and Actuators, Vol. 20, pp. 1-16 (1989).
35. S. F. Bart and J. H. Lang, "An Analysis of Electroquasistatic Induction Micromotors." Sensors and Actuators, Vol. 20, pp. 97-106 (1989).
36. M. A. Schmidt and R. T. Howe, "Resonant Structures for Integrated Sensors." Tech. Digest, IEEE Solid-State Sensor Workshop, Hilton Head Island, SC, USA, pp. 94-97 (1986).
37. R. T. Howe, "Resonant Microsensors." Tech. Digest, 4th Int. Conf. Solid-State Sensors and Actuators (Transducers '87), Tokyo, Japan, pp. 843-848 (1987).
38. W. C. Tang, T.U- Chong, H. Nguyen and R. T. Howe, "Laterally Driven Polysilicon Resonant Microstructures." Sensors and Actuators, Vol. 20, pp. 25-32 (1989).
39. R. A. Brennen, M. G. Lim, A. P. Pisano, and A. T. Chou, "Large Displacement Linear Actuator," IEEE Workshop (1990).
40. V. P. Jaecklin, C. Linder, N. F. de Rooj and J.-M. Moret, "Comb Actuators for xy-Microstage." Sensors and Actuators A, Vol. 39, pp. 83-89 (1993).
41. M. T. Ching, R. A. Brennen, and R. M. White, "Microfabricated Optical Chopper." Optical Engineering, Vol. 33 (11), pp. 3634-3648 (1994).
42. M.-H. Kiang, D. A. Francis, C. J. Chang-Hasnain, O. Solgaard, K. Y. Lau and R. S. Muller, "Actuated Polysilicon Micromirrors for Raster-Scanning Displays." Transducers '97, pp. 323-326 (1997).

43. M. Kohl, J. Gottert and J. Mohr, "Verification of the Micromechanical Characteristics of Electrostatic Linear Actuators." Sensors and Actuators A 53, pp. 416-422 (1996).
44. L. S. Fan, S. J. Woodman and L. Crawforth, "Integrated Multilayer High Aspect Ratio Milliactuators." Sensors and Actuators A 48, pp. 221-227 (1995).
45. K. Wang, A.-C. Wong, W.-T. Hsu, and C. T.-C Nguyen, "Frequency Trimming and Q-Factor Enhancement of Micromachined Resonators via Localized Filament Annealing." Transducers '97, pp. 109-112 (1997).
46. K. B. Lee and Y.-H. Cho, "Frequency Tuning of a Laterally Driven Micromotor Using an Electrostatic Comb Array of Linearly Varied Length." Transducers '97, pp. 113-116 (1997).
47. S. F. Bart, L. S. Tavrow, M. Mehregany and J. H. Lang, "Microfabricated Electrohydrodynamic Pumps." Sensors and Actuators, A21-A23, pp. 193-197 (1990).
48. A. Richter, A. Plettner, K. A. Hofmann and H. Sandmaier, "A Micromachined Electrohydrodynamic (EHD) Pump." Sensors and Actuators A, Vol. 29, pp. 159-168 (1991).
49. R. Holland and E. P. Eer Nisse, Design of Resonant Piezoelectric Devices. The M.I.T. Press, Cambridge MA, pp. 42-95 (1968).
50. J. P. Shields, Basic Piezoelectricity. Howard S. Sams & Co. In. New York, pp. 27-40 (19966).
51. K. F. Etzold, "Ferroelectric and Piezoelectric Materials." In: Electrical Engineering Handbook. Edited by R. C. Dorf, CRC Press Boca Raton, Florida, pp. 1087-1097 (1993).
52. J. M. Giannotto, "Poled Ferroelectric Ceramic Devices." In: Electronics Engineer's Handbook. Edited by D. G. Fink, McGraw-Hill, New York, p. 7-58 (1975).
53. C. J. Chen, Introduction to Scanning Tunneling Microscopy. Oxford University Press, New York, pp. 213-235 (1993).
54. D. Sarid, Scanning Force Microscopy. Oxford University Press, New York, p. 16 (1991).
55. G. T. Davis, "Piezoelectric and Pyroelectric Polymers. In: Polymers for Electronic and Photonic Application. Edited by C. P. Wong, Academic Press, Inc., San Diego, pp. 435-461 (1993).
56. R. M. Moroney, R. M. White and R. T. Howe, "Ultrasonic Micromotors: Physics and Applications." IEEE Proceeding of Ultrasonics, Pub# CH832, pp. 182-187 (1990).
57. T. Morita, M. Kuosawa and T. Higuchi, "A Cylindrical Micro Ultrasonic Motor Fabricated by Improved Hydrothermal Method." Transducers '97, pp. 49-52.
58. A. M. Flynn, et. al., " Piezoelectric Micromotors for Microrobots." Proc. IEEE Ultrasonic Symp., pp. 1163-1172 (1990).
59. G. A. Racine, R. Luthier and N. F. de Rooj, "Hybrid Ultrasonic Micromachined Motors." Proc. IEEE MEMS, pp. 128-132 (1993).
60. P. Muralt et al., "Fabrication and Characterization of PZT Thin-Film Vibrators for Micromotors." Sensors and Actuators A Vol. 48, pp. 157-165 (1995).
61. T. Morita, M. Kurosawa, T. Higuchi, "An Ultrasonic Micromotor Using a Bending Cylindrical Transducer Based on PZT Thin Film." Sensors and Actuators A 50, pp. 75-80 (1995).

62. W. Gopel, "Ultimate Limits in the Miniaturization of Chemical Sensors." Sensors and Actuators A Vol. 56, pp. 83-102 (1996).
63. J. N. Zemel, "Future Directions for Thermal Information Sensors." Sensors and Actuators A 56, pp. 57-62 (1996).
64. A. P. Dorey, and J. H. Moore, Advances in Actuators. Institute of Physics Publishing, Techno House, Redcliffe Way, Bristol BS1 6NX, UK (1995).
65. M. E. Motamedi, "Micro-Opto-Electro-Mechanical Systems." Optical Eng. Vol. 33 (11), pp. 3505-3517 (1994).
66. M. E. Motamedi, et. al., "Development of Micro-Electro-Mechanical Optical Scanner." Opt. Eng. Vol. 36 (5), pp. 1346-1353 (1997).
67. V. P. Jaecklin, C. Linder and N. F. de Rooij, "Line-Addressable Torsional Micromirrors for Light Modulator Arrays." Sensors and Actuators A, 41-42, pp. 324-329 (1994).
68. Z. J. Yao and N. C. MacDonald, " Single Crystal Silicon Supported Thin Film Micromirrors for Optical Applications." Opt. Eng. 36(5), pp. 1408-1413 (1997).
69. W. Dotzel, T. Gessner, R. Hahn and C. Kaufmann, "Silicon Mirrors and Micromirror arrays for Spatial Laser Beam Modulation." Transducers '97, pp. 81-84 (1997).
70. B. Wagner, K. Reimer, A. Maciossek and U. Hofmann, "Infrared Micromirror Arrays with Large Pixel Size and Large Deflection Angle." Transducers '97, pp. 75-78 (1997).
71. T. G. Bifano, et. al., "Continuos-Membrane Surface-Micromachined Silicon Deformable Mirror." Opt. Eng., Vol. 36 (5), pp. 1354-1360 (1997).
72. M. T. Ching, R. A. Brennen and R. M. White, "Microfabricated Optical Chopper." In Miniature and Micro-Optics and Micromechanics Proc., SPIE 1992, pp. 40-46 (1993).
73. Th. Kraus, M. Batltzer, E. Obermeier, "A Micro Shutter for Applications in Optical and Thermal Detectors." Transducers '97, pp. 67-70 (1997).
74. G. Perragaux, P. Weiss, B. Kloek, H. Vuilliomenet and J. -P Thiebaud, "High-Speed Micro-Electromechanical Light Modulation Arrays." Transducers '97, pp. 71-74 (1997).
75. S. W. Smith, M. Mehregany, F. L. Merat, and D.A. Smith, "All-Silicon Waveguide and Bulk-Etched Alignment Structure on (110) Silicon for Integrated Micro-Opto-Mechanical Systems." Proceedings of International Integrated Optics and Microstructures III SPIE Conference Vol. # 2686, pp. 17-28 (1996).
76. S.-S. Lee, E. Motamedi and M. C. Wu, "Surface-Micromachined Free-Space Fiber Optic Switches with Integrated Microactuators for Optical Fiber Communication Systems." Transducers '97, pp. 85-88 (1997).
77. R. A. Miller, Y.-C. Tai, G. Xu, J. Bartha and F. Lin, "An Electromagnetic MEMS 2x2 Fiber Optic Bypass Switch." Transducers '97, pp. 89-92 (1997)
78. J. R. Reid, V. M. Bright and J. H. Comtois, "Automated Assembly of Flip-up Micromirrors." Transducers '97, pp. 347-350 (1997).
79. L. Fan, M. C. Wu, K. D. Choquette and M. H. Crawford, "Self-Assembled Microactuator XYZ Stages for Optical Scanning and Alignment." Transducers '97, pp. 319-322 (1997).

80. M. E. Motamedi, et. al., "Development of Micro-electro-mechanical Optical Scanner." Opt. Eng., Vol. 36 (5), pp. 1346-1353 (1997).
81. a) T. Kawabata, M. Ikeda, H. Goto, M. Matsumoto and T. Yada, "The 2-Dimensional Micro Scanner Integlated with PZT Thin Film Actuator." Transducers '97, pp. 339-342 (1997).
b) M. -H. Kiang, J. T. Nee, K. Y. Lau, and R. S. Muller, "Surface-Micromachined Diffraction Gratings for Scanning Spectroscopy Applications." Transducers '97, pp. 343-345 (1997).
82. R. A. Miller and Y.-C Tai, "Micromachined Electromagnetic Scanning Mirrors." Opt. Eng., Vol. 36 (5), pp. 1399-1407 (1997).
83. T. Kanedo, T. Ohmi, N. Ohya, N. Kawahara, "A New, Compact and Quick-response Dynamic Focusing Lens." Transducers '97, pp. 63-66 (1997).
84. G. Vdovin, S. Middelhoek and P. M. Sarro, "Technology and Application of Micromachined Silicon Adaptive Mirrors." Opt. Eng. Vol. 36(5), pp. 1382-1390 (1997).
85. M. Tabib-Azar, Integrated Optics and Microstructure Sensors. Kluwer Academic Publishings, Boston (1995).
86. M. Tabib-Azar, D. Polla, and K. Wang (Editors), Integrated Optics and Microstructures. Proceedings of International SPIE Conference (Pub. # 1793) (1992).
87. M. Tabib-Azar and D. Polla (Editors), Integrated Optics and Microstructures II. Proceedings of International SPIE Conference (Pub. # 2291) (1994).
88. M. Tabib-Azar (Editor), Integrated Optics and Microstructures III. Proceedings of International SPIE Conference (Pub. # 2686) (1996).
89. M. Tabib-Azar and G. Beheim, "Modern Trends in Microstructures and Integrated Optics for Communication, Sensing and Actuation." Optical Eng., Vol. 36 (5), (1997).
90. W. Lukosz and P. Pliska, "Electrostatically Actuated Integrated Optical Nanomechanical Devices." Proceeding of Integrated Optics and Microstructures, SPIE Vol. 1793, pp. 214-234 (1992).
91. G. A. Magel, "Integrated Optic Devices Using Micromachined Metal Membranes." Proceedings of International Integrated Optics and Microstructures III SPIE Conference Vol. 2686, pp. 54-63 (1996).
92. R. M. Boysel, et al, "Integration of Deformable Mirror Devices with Optical Fibers and Waveguides." SPIE Vol. 1793, pp. 34-41(1992).
93. S. Martellucci, A. Chester and M. Bertolotti, Advances in Integrated Optics. Plenum (1994).
94. L. A. Field, et al, "Micromachined 1x2 Optical-Fiber Switch." Sensors and Actuators A53, pp. 311-315 (1996).
95. K. Hogari, and T. Matsumoto, "Electrostatically Driven Micromechanical 2x2 Optical Switch." Applied Optics Vol. 30 (10), pp. 1253-1257 (1991).
96. R. Dangel, and W. Lukosz, "Electromechanically Actuated Integrated-Optical Mach-Zehnder Interferometer." Technical Digest series Volume 6, Integrated Photonics Research, April 29- May 2, Boston, MA, pp. 182-540 (1996).
97. R. A. Soref, J. P. Lorenzo, "Silicon Guided-Wave Optics." Solid-State Technology November , pp. 95-98 (1988).
98. M. A. Duguay, Y. Kokuban, T. L. Koch and L. Pfeiffer, "Antiresonant

reflecting optical waveguides in SiO_2-Si multilayer structures." Appl. Phys. Lett. 49, pp. 13-15 (1986).

99. A. Nathan, K. Benaissa, "Silicon Integrated Optic Devices and Micromechanical Sensors Based on ARROW." Proceedings of International Integrated Optics and Microstructures III SPIE Conference (Pub. # 2686), pp. 2-16 (1996).
100. K. H. Rollke, and W. Sohler, "Metal-Clad Waveguide as a Cutoff Polarizer for Integrated Optics." IEEE Journal of Quantum Electr. Vol. QE-13 (4) pp. 141-145 (1977).
101. K. Fischer, J. Muller, R. Hoffmann, F. Wasse, and D. Salle, "Elasto-optical Properties of SiON Layers in an Integrated Optical Interferometer Used as a Pressure Sensor." J. Lightwave Tech., Vol. 12 (1), pp. 163-169 (1994).
102. A. Garcia, and M. Tabib-Azar, "Sensing Means and Sensor Shells: A New Method of Comparative Study of Piezoelectric, Piezoresistive, Electrostatic, Magnetic, and Optical Sensors." Sensors and Actuators A. Physical Vol. 48 (2), pp. 87-100 (1995).
103. O. Parriaux, "Integrated Optics Sensors." In: Advances in Integrated Optics. Eds: S. Martellucci, A.N. Chester, and M. Bertolotti, Plenum Press, New York, pp. 227-242 (1994).
104. X. C. Jin, I. Ladabaum and B. T. Khuri-Yakub, "The Microfabrication of Capacitive Ultrasonic Transducers." Transducers '97, pp. 437-440 (1997).
105. A. Lal, R. M. White, "Silicon Micromachined Ultrasonic Micro-Cutter." Proceedings of IEEE Ultrasonics Symposium, pp. 1907-1911 (1994).
106. S. Shoji, S. S. Nakagawa and M. Esashi, "Micropump and Sample-Injector for Integrated Chemical Analyzing Systems." Sensors and Actuators, A21-A23, pp. 189-192 (1990).
107. R. Zengerle, J. Ulrich, S. Kluge, M. Richter and A. Richter, "A Bidirectional Silicon Micropump." Sensors and Actuators A, Vol. 50, pp. 81-86 (1995).
108. T. Gerlach, H. Wurmus, "Working Principles and Performance of the Dynamic Micropump." Sensors and Actuators A, Vol. 50, pp. 135-140 (1995).
109. M. Esashi, "Integrated Micro Flow Control Systems." Sensors and Actuators, A21-A23, pp. 161-167 (1990).
110. A. Olsson, G. Stemme and Erik Stemme, "A Valve-Less Planar Fluid Pump with two Pump Chambers." Sensors and Actuators A, Vol. 46-47, pp. 549-556 (1995).
111. L. Kuhn, E. Bassous and R. Lane, "Silicon Charge Electrode Array for Ink Jet Printing." IEEE Transaction on Electron Devices, Vol. ED-25 (10), pp. 1257-1260 (1978).
112. F. J. Kamphoefner, "Ink Jet Printing." IEEE Transaction on Electron Devices, Vol. ED-19 (4), pp. 584-593 (1972).
113. K. E. Petersen, "Fabrication of an Integrated, Planar Silicon Ink-Jet Structure." IEEE Transaction on Electron Devices, Vol. ED-26 (12), pp. 1918-1920 (1979).
114. R. D. Carnahan and S. L. Hou, "Ink Jet Technology." IEEE Transactions on Industry Applications, Vol. IA-13 (1) pp. 95-105 (1977).

115. R. G. Sweet, "High Frequency Recording with Electrostatically Deflected Ink Jets." The Review of Scientific Instruments, Vol. 36 (2), pp. 131-136 (1965).
116. E. Stemme and G. Stemme, "A Valveless Diffuser/Nozzle-Based Fluid Pump." Sensors and Actuators A, Vol. 39, pp. 159-167 (1993).
117. K. Petersen, "From Microsensors to Microinstruments." Sensors and Actuators A Vol. 56, pp. 143-149 (1996).
118. J. Fluitman, "Microsystem Technology: Objectives." Sensors and Actuators A Vol. 56, pp. 151-166 (1996).
119. S. Majumder, et. al., "Measurement and Modeling of Surface Micromachined, Electrostatically Actuated Microswitches." Transducers '97, pp. 1145-1148 (1997).
120. a) B. Romanowicz, Ph. Lerch, Ph. Renaud, E. Fullin and Y. de Coulon, "Simulation of Integrated Electromagnetic Device Systems." Transducers '97, pp. 1051-1054 (1997).
b) T. Seki, M. Sakata, T. Nakajima and M. Matsumoto, "Thermal Buckling Actuator for Micro Relays." Transducers '97, pp. 1153-1156 (1997).
121. H. J. Mamin, L. S. Fan, S. Hoen and D. Rugar, "Tip-Based Storage Using Micromechanical Cantilevers." Sensors and Actuators A, Vol. 48, pp. 215-219 (1995).
122. M. I. Lutwyche and Y. Wada, "Manufacture of Micromachined Scanning Tunneling Microscopes for Observation of the Tip Apex in a Transmission Electron Microscope." Sensors and Actuators A, Vol. 48, pp. 127-136 (1995).
123. P.-F. Indermuhle, V. P. Jaecklin, J. Brugger, C. Linder, N. F. de Rooij and M. Binggeli, "AFM Imaging with an xy-Micropositioner with Integrated Tip." Sensors and Actuators A 46-47, pp. 562-565 (1995).
124. J. Brugger, R. A. Buser and N. F. de Rooij, "Silicon Cantilevers and Tips for Scanning Force Microscopy." Sensors and Actuators A, Vol. 34, pp. 193-200 (1992).
125. T. R. Albrecht, S. Akamine, M. J. Zdeblick and C. F. Quate, "Microfabrication of Integrated Scanning Tunneling Microscope." J. Vac. Sci. Technol. A Vol. 8 (1), pp. 317-318 (1990).
126. S. Akamine, T. R. Albrecht, M. J. Zdeblick and C. F. Quate, "A Planar Process for Microfabrication of a Scanning Tunneling Microscope." Sensors and Actuators, A21-A23, pp. 964-970 (1990).
127. K. Matsumoto, M. Ishii, J-i Shirakashi, K. Segawa, Y. Oka, B. J. Vartanian and J. S. Harris, "Comparison of Experimental and Theoretical Results of Room Temperature Operated Single Electron Transistor made by STM/AFM Nano-Oxidation Process." Proc. of IEDM '95, pp. 363-366.
128. D. Samara, J. R. Williamson, C. K. Shih and S. K. Banerjee, "Scanning Tunneling Microscopy Induced Chemical-Vapor Deposition of Semiconductor Quantum Dots." J. Vac. Sci. Technol. B Vol. 14 (2), pp. 1344-1348 (1996).
129. G. J. Berry, J. A. Cairns and J. Thomson, "The Production of Fine Metal Tracks from a New Range of Organometallic Compounds." Sensors and Actuators A Vol. 51, pp. 47-50 (1995).

130. H. J. Mamin, L. S. Fan, S. Hoen and D. Rugar, "Tip-Based Data Storage Using Micromechanical Cantilevers." Sensors and Actuators A Vol. 48, pp. 215-219 (1995).
131. R. C. Barrett and C. F. Quate, "Charge Storage in a Nitride-Oxide-Silicon Medium by Scanning Capacitance Microscopy." J. Appl. Phys. Vol. 70 (5), pp. 2725-2733 (1991).
132. T. R. Albrecht, S. Akamine, T. E. Carver and C. F. Quate, "Microfabrication of Cantilever Styli for the Atomic Force Microscope." J. Vac. Sci. Technol. A Vol. 8 (4), 3386-3396 (1990).
133. Private communication with M. Tabib-Azar and Morton Litt.
134. T. Hattori, "Achievements of Japanese Micromachined Projects." Transducers '97, pp. 25-28 (1997).
135. S. Charles and R. Williams, "Micromachined Structures in Ophthalmic Microsurgery." Sensors and Actuators, A21-A23, pp. 263-266 (1990).
136. S. Shoji, M. Esashi and T. Matsuo, "Fabrication of an Integrated Microphone Head for Fault Analysis of MOS Integrated Circuits." Sensors and Actuators, Vol. 14, pp. 125-132 (1988).
137. G. Lim, K. Park, M. Sugihara, K. Minami and M. Esashi, "Future of Active Catheters." Sensors and Actuators A Vol. 56, pp. 113-121 (1996).

Magnetic Microactuators

3.1 Introduction

Magnetic actuators are perhaps among the oldest types of actuators. They are used in everyday appliances in the form of relays, electromotors and automatic valves. Magnetic actuation methods offer the possibility of generating repulsive forces in addition to attractive forces. This is in contrast to electrostatic schemes which offer only attractive forces. In most cases, the force versus displacement relationship in these devices is also much better behaved than in coulombic electrostatic actuators.

Implementation of magnetic actuation schemes at the "micro" scale, however, is quite challenging because most of these devices require coils and deposition of magnetic materials [1-20]. Coils with large inductances are difficult to implement in planar forms since many turns are required, resulting in large areas and large series resistances [8,9]. However, the critical current densities (defined as the limiting current density before excessive electro-migration sets in) in thin film conductors is larger than bulk conductors (typically $10^5 A/m^2$) by at least three to four orders of magnitude [3,10]. Hence, planar microcoils can generate larger magnetic fields proportional to their higher critical current densities.

Deposition of magnetic materials and their subsequent patterning and treatment to impart to them desired characteristics are quite challenging in an IC microfabrication environment. However, due to the importance of these films in cellular and mobile communication applications, they are finding a larger support base in IC microfabrication communities [1,3,4,10]. By depositing magnetic materials over planar inductors, relatively large inductances with high-Q's can be fabricated [1,4]. Most magnetic actuators that are reported in the literature take advantage of attractive forces that can be generated between a ferromagnetic material and a current-carrying coil (an electromagnet), as shown in figure 3.1 [10,12-18].

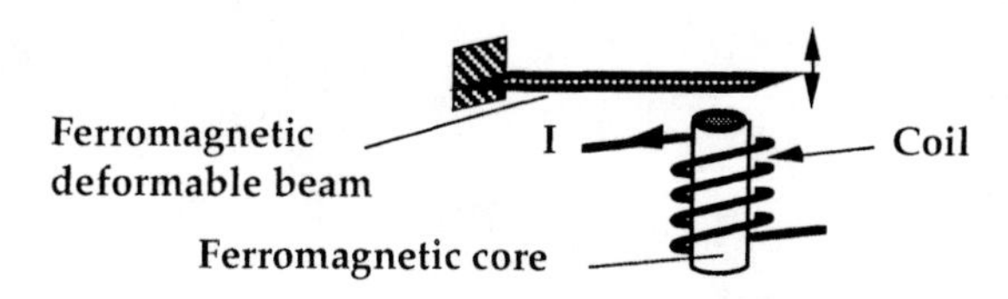

Figure 3.1 A schematic of a magnetization-type actuator. The current through the coil generates a magnetic field that magnetizes a deformable beam and attracts it.

Related to the above microactuator are devices that use a permanent magnet instead of the magnetizable part. In these devices, schematically shown in figure 3.2, the actuator can generate both attractive and repulsive forces [3,21-27].

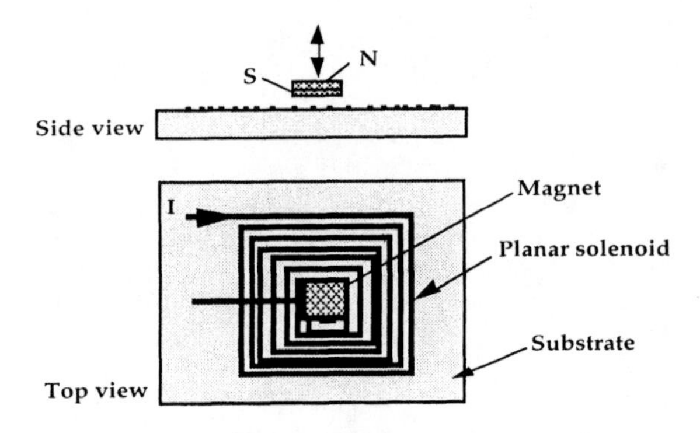

Figure 3.2 A schematic of a planar magnetic microactuator using a spiral inductor to generate a magnetic field and actuate a small magnet situated directly above the inductor [3,10].

Lorentz forces can also be used in microactuators [26,27]. In these devices, the force is generated by a current-carrying conductor and the magnetic field is usually generated by a permanent magnet, as shown in figure 3.3.

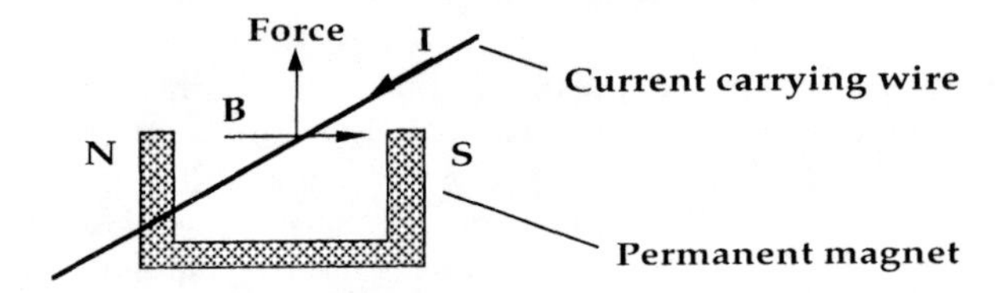

Figure 3.3 A schematic representation showing Lorentz force generated on a current-carrying wire in the presence of a magnetic field.

Magnetostrictive effects, as discussed in section 3.2.6, refer to the dimensional changes that occur in the magnetic material when magnetized [28-30]. Figure 3.4 schematically shows how a magnetostrictive material can be incorporated with a coil to generate force and displacement.

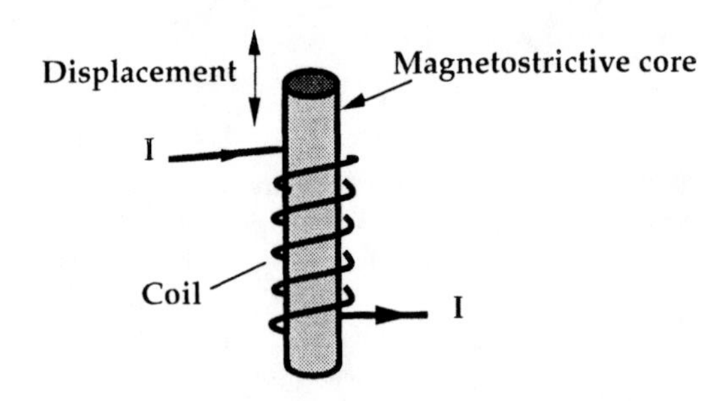

Figure 3.4 A schematic of a magnetostrictive actuator. A coil is used to generate a magnetic field causing the magnetostrictive core to change length.

One of the problems associated with electrostatic motors is friction. There have been efforts to reduce friction by levitating the rotor (or the moving parts) using electric means, as discussed in Chapter 2. Magnetic levitation, however, appears more practical as far as implementation is concerned. Section 3.7 discusses linear motors that take advantage of magnetic levitation to reduce wear and friction [31-34]. So far, only magnetic field repulsion in superconductors (Meissner effect) has been employed to design and fabricate magnetically-levitated linear motors. Repulsion between the similar poles of a magnet can also be used to levitate moving parts at room temperature and is used in magnetic bearings.

The underlying principles and even the device geometry between the magnetic macro-actuators, such as electromotors, electromagnets, relays, etc., and their micro counterparts are the same. However, in microdevices the available volume is quite small and some of the losses associated with the surface effects, which can be ignored in larger-sized devices, should be taken into account. Other differences are that the permeability of the thin-film magnetic materials may not be as large as that of their bulk counterparts and that there may be fabrication issues regarding depositing thick layers (>10 μm) of very high-permeability materials. Thus, the device geometry and topography is more critical in microactuators.

The deposition of magnetic materials and their patterning and subsequent processing should be compatible with the integrated circuit fabrication processes. This requirement can be relaxed in exploratory actuators, but it is an important requirement that should be satisfied to lower cost and enable integration of actuators with sensors and electronics.

In all of the above actuators, the magnetic properties of materials come into play although less so in the case of the Lorentz force-type actuators, where the magnetic materials are only used to concentrate the magnetic field. In section 3.2 we review magnetic properties of pertinent materials and discuss some typical materials used in magnetic microactuators. The subject of magnetism is a very interesting and diverse one that has received a great deal of attention because of its application in electromotors and electromagnets. It is clearly outside the scope of a research monograph to cover the topic of magnetism in any great depth, and section 3.2 only surveys the important magnetic properties of materials of interest in microactuators.

3.2 Magnetic Materials for Microactuators

The magnetic properties of materials are much more diverse and less well understood than their electrical properties. These properties can be categorized as follow [11-20]:

Diamagnetism	Paramagnetism
Ferromagnetism	Ferrimagnetism
Antiferromagnetism	Magnetoresistivity
Magnetostriction	

Magnetic susceptibility (χ_{m}) is defined through the following relationship among the magnetic induction (**B**), the magnetic field (**H**):, and the magnetization (**M**):

$$B = (\mu_0+\chi_m)\, H \qquad (3.1a)$$
$$B = \mu_0(H + M) \qquad (3.1b)$$
$$B = \mu_0(1+\chi_r)\, H = \mu\, H \qquad (3.1c)$$

where μ_0 is the permeability of vacuum. The above relationships also define the permeability of the material given by μ. The relative permeability is given by: $\mu_r=\mu/\mu_0$. The relative magnetic susceptibility χ_r is defined similarly by: $\chi_r=\chi_m/\mu_0$. Here, we will be using MKS units where **B** is in Tesla (Wb/m^2), **H** and **M** are in *A/m* , and μ_0 and χ_m are in *Henry/m*.

The key to understanding the magnetic properties of materials is in understanding how internal magnetic moment (spin) of electrons and their orbital magnetic moments interact with each other and with the externally-applied magnetic fields. These couplings explain the magnetic properties of crystals. In polycrystal materials, one also has to consider the interaction between the magnetic moments of grains and their interfaces. In the following section, we discuss different forms of magnetic behavior in materials.

3.2.1 Diamagnetism

This type of magnetism results in a very small negative relative susceptibility. In most diamagnetic materials, the relative magnetic susceptibility is in the range of -10^{-6} to -10^{-5}, except in superconductors, which are perfect diamagnetic materials, with $\chi_r = -1$. The magnetic field becomes zero inside type *I* superconductors below their critical temperature. This phenomena is called the Meissner effect and can be used to levitate magnets over superconducting planes as discussed in section 3.7.

All materials have diamagnetism, which has its origin in Lenz's law. This law states that when a magnetic flux changes in a circuit, a current which opposes the change of flux is induced. The induced current generates a magnetic field that opposes the externally-applied field. In most magnetic materials there are other contributions to the magnetic moment that overwhelm the diamagnetism.

3.2.2 Paramagnetism

The potential energy of a magnetic dipole with a magnetic moment **m** in a magnetic field **H** is - **m** . **H** and, hence, magnetic moments tend to align themselves with the externally-applied magnetic field. In solids with atoms with permanent magnetic moment, it is experimentally found that these permanent magnetic moments tend to align themselves with the external magnetic induction as shown in figure 3.5. This phenomena gives rise to paramagnetism. Paramagnetic materials have positive relative susceptibility in the 10^{-3}-10^{-6} range. In these materials, the effect of the temperature is to misalign the permanent magnetic moments. Thus, as the temperature (T) is increased, the susceptibility goes down, according to

Curie's law: $\chi_r \alpha\ 1/T$. Paramagnetic materials, due to their relatively low susceptibilities, do not have much application in actuators .

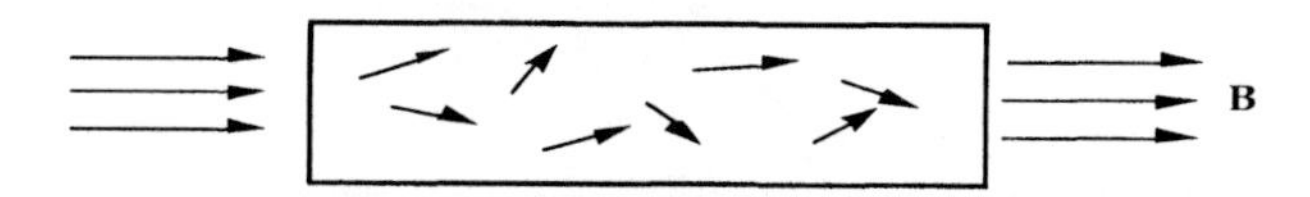

Figure 3.5 In paramagnetic materials, the magnetic dipoles align themselves with the externally applied magnetic field. Dipole-dipole interaction is not very strong in these materials.

3.2.3 Ferromagnetism and Ferromagnets

Ferromagnetic materials have very large positive relative susceptibilities in the range of 1000-500000 [11,16,18]. In these materials, the permanent magnetic dipoles of atoms strongly interact and align themselves with respect to each other (figure 3.6). Above a temperature called Curie temperature, the thermal agitation of atoms reduces this strong interaction between the magnetic dipoles and the material becomes paramagnetic.

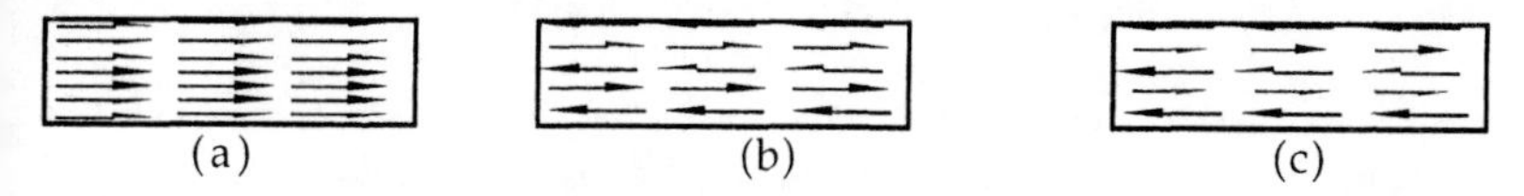

Figure 3.6 a) In ferromagnets, the magnetic dipoles interact with each other strongly, all aligning in the same direction; b) in antiferromagnets, the magnetic dipoles of nearest neighbors are anti-parallel; and c) in ferrimagnets, the magnetic dipoles are anti-parallel, but do not have the same dipole moments, so a net magnetization exists.

The temperature behavior of the susceptibility in ferromagnets is given by the Curie-Weiss law: $\chi_r = C/(T - T_C)$, where C is the Curie constant and T_C is the Curie temperature. According to this law, the ferromagnets behave like paramagnetic materials when the temperature is above their Curie temperature [18].

Figure 3.7 shows a typical magnetization curve. The symmetric major hysteresis loop shown in this figure is traced reproducibly, provided the applied field is sufficient to achieve saturation ($M=M_s$) in each direction. Remnant magnetization (M_r), the width of the hysteresis loop and the coercive field (H_c) are important magnetic material parameters. M_s is an intrinsic property, while H_c and M_r are not. At equilibrium, the magnetization is zero. So-called "soft magnets" have smaller H_c and are easier to magnetize, while "hard magnets" have large H_c and are difficult to magnetize, but, when magnetized, they retain their magnetization better than the soft magnets.

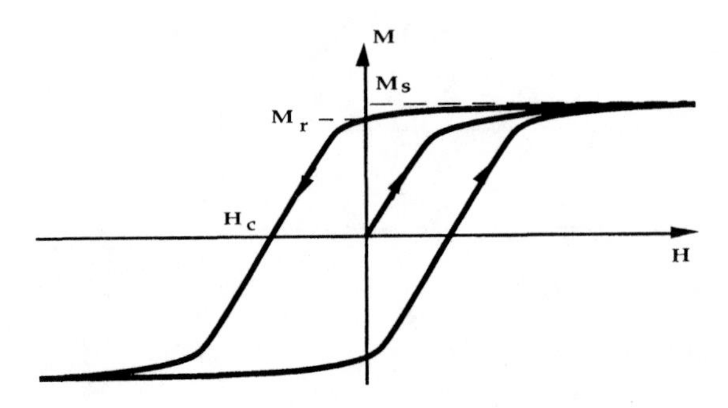

Figure 3.7 A typical magnetization curve *M(H)* of magnetic materials.

The dependence of the magnetic induction on the magnetic field can be calculated from *M(H)*. The remnant magnetic induction (B_r) is just $\mu_0 M_r$, while the coercive magnetic field ${}_BH_c$ is smaller than H_c. Figure 3.8 shows the temperature dependence of the magnetic susceptibility and saturation magnetization in different types of magnetic materials [11]. Of interest here are ferromagnets and ferrimagnets that exhibit paramagnetic properties at $T>T_c$ and large saturation magnetization at $T<T_c$.

Magnetic materials are characterized by six different but interrelated parameters: remnant magnetization, coercive field, saturation magnetization, hysteresis, energy product $(BH)_{max}$, and the Curie temperature. Alnicos (Al-Ni-Co alloys) typically have energy products in the range of 50-100 kJ/m^3. Rare-earth magnets, such as Samarium-cobalt ($SmCo_5$ and Sm_2Co_{17}) have energy products in the range 65 (in $SmCo_5$) -200 (in Sm_2Co_{12}) kJ/m^3. Iron-rich neodymium magnets ($Nd_2Fe_{14}B$) hold the record for energy products of 400 kJ/m^3 [11,16,15]. Table 3.1 shows a list of rare-earth magnets and others in use today.

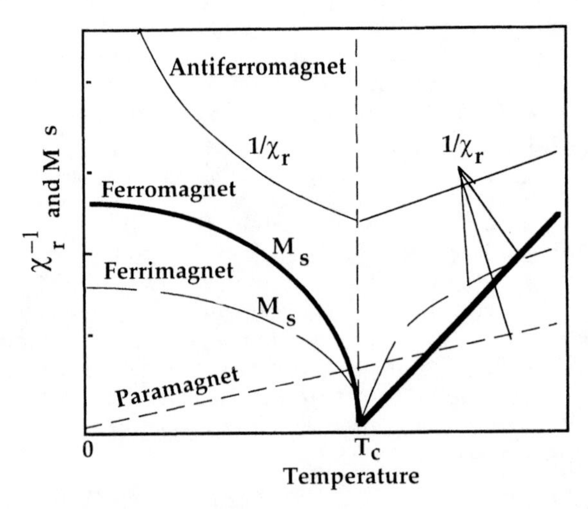

Figure 3.8 The temperature behavior of the relative susceptibility and saturation magnetization of different magnetic materials [16].

Table 3.1 Typical properties of rare-earth and other magnetic materials.

Name	Main phase	B_r (T)	$\mu_0 M_s$ (T)	H_c (kA/m)	$_BH_c$ (kA/m)	$(BH)_{max}$ (kJ/m^3)
Alnico 5	$Fe_{48}Al_{16}Ni_{13}Co_{21}Cu^{+}_{2}$	1.28	1.40	54	52	43
Sm-Co	$SmCo_5$	0.88	0.95	1400	680	150
Sm-Co	Sm_2Co_{17}	1.08	1.15	900	800	220
Nd-Fe-B	$Nd_2Fe_{14}B$	1.25	1.34	1000	920	300
Chrome Steel	96.5 Fe 3.5 Cr	0.95		5.3	920	2.3
Ferrite	$SrFe_{12}O_{19}$	0.39	0.46	275	265	28

Actual ferromagnetic materials are composed of small regions within which the magnetization has its remnant value as shown in figure 3.9. These regions are called magnetic domains, and they are separated from each other by domain or Bloch walls.

The magnetization in different domains points in different directions so that the total magnetization of the sample is zero if the sample has not been in an applied magnetic field. As the externally-applied field is increased, the domains with magnetization that is favorably oriented with respect to the direction of $\boldsymbol{H}$ grow at the expense of the others. At low magnetic fields, this process is reversible, while at higher fields, due to domain wall pinning at imperfection sites, precipitates, and other defects, the process is not reversible. As the magnetic field is increased, the domain walls move inside the sample, and when they are pinned, the magnetization of the sample does not change till the externally-applied magnetic field becomes large enough to unpin the walls. This pinning and unpinning process introduces "step" in the magnetization curve and it is called the Barkhausen effect [16,18]. This effect also causes noise in magnetic circuits.

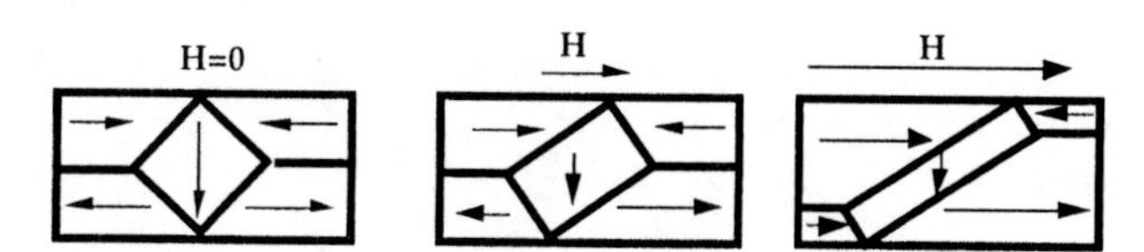

Figure 3.9 A schematic presentation of magnetic domains [18].

In magnetic microactuators, it is necessary to have single-domain magnetic films to eliminate unwanted reduction of magnetization due to multiple domains and also to eliminate the Barkhausen effect. Domain wall movements also give rise to sample-size variations as a function of magnetic field, which is called magnetostriction, as discussed in section 3.2.6.

In ferromagnetic materials, it is found that there is a crystallographic axis, called an easy magnetization axis, along which the magnetic moment tends to align with lower energy than along any other axis. This phenomena stems from the anisotropy of the crystal, which is the largest in Co.

3.2.4 Antiferromagnetism

The coupling between adjacent magnetic dipoles in certain materials, called antiferromagnets, favors anti-parallel alignment of these dipoles (figure 3.6b) [18]. As the temperature of these materials is increased, the thermal agitation of the magnetic dipoles eventually overcome their interaction, resulting in the disappearance of the antiferromagnetic order. The temperature at which such a transition occurs is called the Neel temperature (T_N). The susceptibility in these materials varies according to the modified Curie law: $c_r = C/(T+Q)$, where C is the Curie constant and $Q > T_n$. Antiferromagnets are used in memory devices adjacent to ferromagnetic layers to "pin" the magnetization of these layers.

3.2.5 Ferrimagnetism

Antiferromagnetic materials with unequal magnetic moments of the adjacent dipoles are called ferrimagnets or ferrites. Most permanent magnets, such as Lodestone, before the invention of the rare-earth and Alnico magnets, were of the ferrimagnetic type. Ferrites are usually iron oxides of the form $MOFe_2O_3$, where M stands for different metal ions, or they are of the type MFe_2O_4, in which the iron is trivalent and M stands for a divalent metallic ion. $CoFe_2O_4$ is a notable hard ferrite. Table 3.1 shows the properties of some important ferrites. In ferrites the susceptibility is given by: $\chi_r = \dfrac{(C_A + C_B)T - 2\alpha_E C_A C_B}{T^2 - T_c^2}$, where C_A and C_B are the Curie constants of the two sub-lattices (each containing one of the magnetic dipoles), α_E relates magnetization to an effective magnetic induction inside the material ($\mathbf{B}_E = \alpha_E \mathbf{M}$) [11,18].

3.2.6 Magnetostriction

Magnetostriction is a dimensional variation in the material, caused by magnetization. Figure 3.9 schematically depicts positive and negative magnetostriction effects. Typically, $\Delta l/l$ is around 10^{-6}.

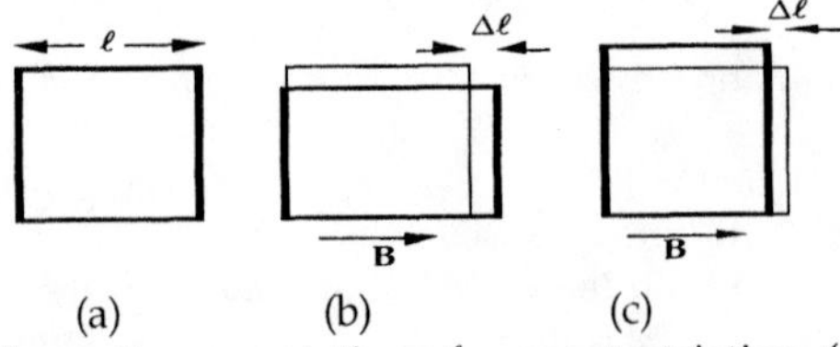

Figure 3.10 A schematic presentation of magnetostriction. (a) The sample with $B=0$, (b) positive magnetostriction, and (c) negative magnetostriction.

Magnetostriction is a complicated tensor property [18,28-30]. Empirically, it is found that under simplified conditions, $\Delta l/l$ is proportional to M^2.

The Weidmann effect is a magnetostrictive effect in materials, caused by excitation of twisting strains in a ferromagnetic rod in the presence of axial and circular magnetization fields [28]. The Weidmann effect is used in torsional and linear displacement actuators and its inverse is used to measure instantaneous torque in the crankshaft of motors [29]. In section 3.6 we discuss a few examples of magnetostrictive actuators.

3.2.7 Magnetoresistivity

Materials with free charge carriers generally exhibit a change in their resistivity in the presence of an externally-applied magnetic field. The origin of this change, which is usually very small except in certain artificially-layered materials, is the interaction between the moving charge carriers and the applied magnetic field through the Lorentz force [19]. Lorentz force causes the charge carriers to bend effectively, increasing the resistivity of the material in most cases (negative changes have also been observed).

Electrons and holes posess spin, which endows them with a built-in magnetic moment that causes them to have different scattering rates when moving through a magnetic material. This phenomena is observed in artificially-layered structures where a non-magnetic layer (such as copper) is sandwiched between two magnetic layers, as shown in figure 3.11. When the film thickness is on the order of the electrons' mean free path, electrons with spin up may experience a lower scattering rate in the upper layer, which is magnetized in the x-direction, and will thus be directed to move in that layer.

On the other hand, electrons with spin down will have a lower scattering rate in the bottom layer, which is magnetized in the x-direction. This behavior results in the non-magnetic (copper) layer acting as a barrier separating electrons moving in their respective layers. When both magnetic layers are magnetized in the same direction, the cross-sectional area doubles (for the geometry shown in the figure) for all the electrons, and electrons with spin up continue their high mobility through these layers, while electrons with spin down experience a higher scattering rate and thus lower mobility. However, the gain in conductance due to spin up electrons is larger than what is lost to these slower electrons. The resistivity of the layered structure thus becomes smaller.

Although magnetoresistive materials are not used in microactuators, they are excellent candidates as sensors. In magnetic microactuators, these devices will find an application as displacement, position, and magnetic field sensors. They already have found applications in magnetic memories and storage devices.

Typical multilayer magnetoresistive materials (3 or 4 NiFeCo or CoFe magnetic layers each 15-40Å thick with 2 or 3 non-magnetic separator regions) have a linear operation range from *0-20* mT, capable of operating in temperatures in excess of *200* °C with a typical temperature coefficient of +0.15%/°C. Properly-designed magnetoresistive devices with a dimension of 80x100 μm can have a frequency response in excess of 100 MHz. Although 12% change in the resistance is typically observed, up to 70% change in resistance is achievable by using insulating separators where charge carriers tunnel between magnetic layers [19].

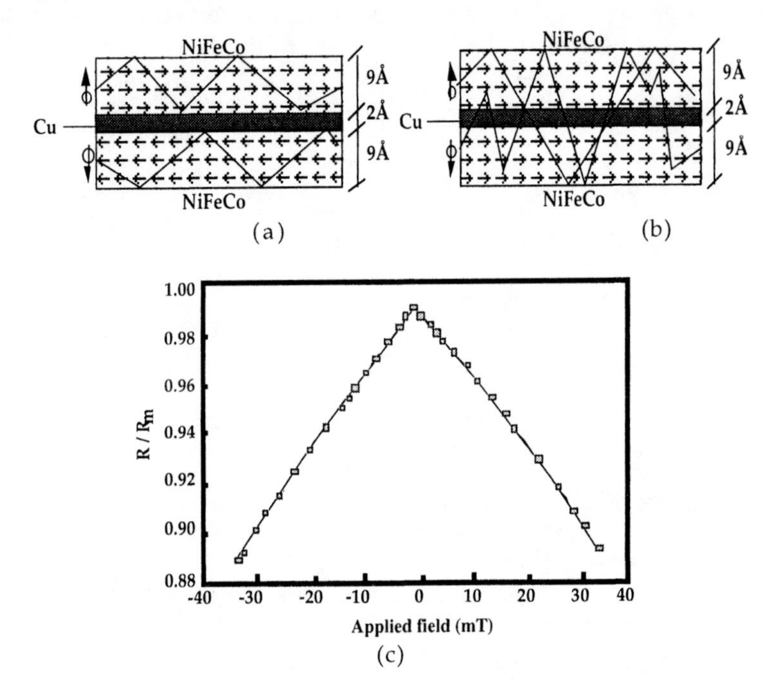

Figure 3. 11 A multilayer structure used to achieve large magnetoresistance. In a) top and bottom layers are magnetized in opposite directions, while in b) they are magnetized in the same direction. c) A typical resistance versus magnetic field in multilayer magnetoresistors.

3.2.8 Molecular and Polymeric Magnets

Polymeric magnets, because of their low weight and mechanical flexibility, are becoming more important. Due to their relative ease of deposition, these materials may gain popularity in microactuators. Polymeric magnets can be processed at low temperatures (<100 °C), they have high strength, and their properties can be varied using organic chemistry. Moreover, they are soluble and do not cause any environmental problems. They can be made transparent and they are usually biocompatible. The coercive fields in some of these materials can be as high as 0.3 T. A review of polymeric magnetic materials is given in [20].

3.3 Force Laws

In magnetic actuators, the source of the magnetic field is either a current-carrying coil (in some case just a wire) or a permanent magnet. The force is generated whenever the magnetic field lines, generated by the above sources, intersect a region of space where a magnetizable material or another magnetic field exist.

We discuss the forces that can be generated by specific magnetic microactautors in the following sections. However, a few considerations common to all these actuators are stated here.

It should be pointed out that the combination of moving magnetic parts that are also possibly conducting and of magnetic fields excited by electromagnets in response to a step in current causes some difficulty in solving for forces and their time dependencies. In most general cases, Maxwell's tensor components should be calculated, from which different components of forces are in turn calculated. Interested readers should consult Chapter 7 of reference [11] for more details.

Equivalent of Coulomb's law in electrostatic is Ampere's law of force in magnetostatic, given by [14,15]:

$$dF = \frac{\mu_0}{4\pi} \frac{I_2 d\ell_2 \times (I_1 d\ell_1 \times R)}{R^2} \tag{3.2}$$

where $\mathbf{d\ell_1}$ and $\mathbf{d\ell_2}$ are, respectively, the length elements at the source and at the test location where the force (dF) is measured, I's are the corresponding currents, and R is the vector connecting $\mathbf{d\ell_1}$ and $\mathbf{d\ell_2}$. Based on equation () a magnetic field (B) is defined that is given by:

$$dB = \frac{\mu_0}{4\pi} \frac{I_1 d\ell_1 \times R}{R^2} \tag{3.3a}$$

or in terms of current density ($\mathbf{J}$):

$$B = -\frac{\mu_0}{4\pi} \int_V J \times \nabla\left(\frac{1}{R}\right) dV' \tag{3.3b}$$

The above equations clearly relate the magnetic field to the current densities.

The basic equations needed to calculate fields in magnetic microactuators are [13]:

<u>Integral form</u>	<u>Differential form</u>	
$\oint_S B \,.\, n \, da = 0$	$\nabla.\, B = 0$	(3.4a)

<u>Integral form</u>	<u>Differential form</u>

$$\oint_C H \cdot d\ell = \int_S J \cdot n \, da \qquad \nabla \times H = J \tag{3.4b}$$

$$\oint_S J \cdot n \, da = 0 \qquad \nabla \cdot J = 0 \tag{3.4c}$$

where da is an element of area, $d\ell$ is an element of the length, n is the unit vector normal to da and having a right-handed relationship with $d\ell$, and the surface S is enclosed by the contour C.

The above equations along with the constitutive relationships expressed by equations 3.1-3.3 are sufficient to determining the magnetic fields in both stationary and dynamic situations where the boundaries and volumes are deforming. In dynamic cases, however, care should be taken regarding the constituent equations since they are only valid in stationary systems. When there is a motion, these equations are only valid for an observer moving with the medium [13].

The boundary conditions for the magnetic field are [17]:

1) The normal component of the flux density is continuos:

$$n \cdot (B_1 - B_2) = 0 \tag{3.5a}$$

2) The normal component of the magnetic field is discontinuous:

$$n \cdot (\mu_1 H_1 - \mu_2 H_2) = n \cdot M_2 \tag{3.5b}$$

3) The tangential component of the magnetic field is discontinuous:

$$n \times (H_1 - H_2) = J_s \tag{3.5c}$$

where subscripts 1 and 2 refer to the two adjacent magnetic media having permittivity μ_1 and μ_2, respectively. n is the unit vector normal to the boundary of region 2, M_2 is the magnetization of region 2 and J_s is the surface current density.

Inside the permanent magnets it can be shown that B and H are antiparallel to each other. According to the equation (3.5a), the normal component of B is continuos while the normal component of H is discontinuous. Outside the magnet, B and H are related through the equation (3.1c).

Permanent magnets are quite unique materials since they are capable of producing magnetic flux with no expenditure of energy. Electronic currents circulating on the atomic scale through resistance's paths are responsible for the permanent production of the magnetic flux. A long cylindrical rare-earth 1 T permanent magnet, which is commercially available, produces a flux that is equivalent to the magnetic flux of a solenoid of several centimeter diameter carrying a current of 796 kA/m [11].

Here we consider some simple cases that are more likely to be encountered in magnetic microactuators at this stage of their development. Future, more sophisticated magnetic microactuators may require the complete treatment mentioned above. We divide the magnetic actuators into two categories: i) actuators with large portions of their magnetic fields existing over regions of free space or low permeability materials, and ii) actuators with only very small portions of their magnetic fields existing in free space or low-permeability materials. In the first category, it is much more convenient to approximate the magnetic field near the deformable part and calculate the force using the potential energy approach described below. In the second case it is much simpler to use the magnetic circuit theory also described below [3,11.b, 21, 22].

The change in the potential energy (W_m) of a magnetic material, with total magnetization $\boldsymbol{M_T}$ (=$\boldsymbol{M}V$, where V is the volume), in an external field $\boldsymbol{B}$, is given by:

$$W_m = -\boldsymbol{M_T} \cdot \boldsymbol{B} \tag{3.6}$$

The net force ($\boldsymbol{F_\alpha}$) in the direction of α, experienced by the magnetic material is:

$$F_\alpha = -\frac{\partial W_m}{\partial \alpha} \tag{3.7}$$

We will use the above equations to calculate the force that can be generated in the configurations shown in figures 3.1, 3.2, and 3.12.

In the second category of microactuators, where a large portion of the magnetic fields exist in regions with very high permeabilities, and in most other cases, the magnetic circuit approach is more convenient. In this approach, magnetomotive force sources along with reluctances of different parts of the magnetic device are used to find the force and the coil voltage-current relationships as described below.

In magnetic circuits, the magnetomotive potential (F_{ab}) between points "a" and "b" is given by [11.b,12]:

$$F_{ab} = \int_a^b \boldsymbol{H} \cdot d\ell \tag{3.8}$$

In these circuits, F_{ab} is the equivalent of voltage, magnetic flux (f) is the equivalent of current, reluctance (R_{ab}), defined as F_{ab}/ϕ, is equivalent to resistance, and permeance (P=$1/R$) is equivalent to conductance. Reluctance in ampere/Weber is defined through the following relationship:

$$R_{ab} = \frac{\ell_{ab}}{\mu A} \tag{3.9}$$

where ℓ_{ab} is the length between "a" and "b" and A is the area of the part where the magnetic flux exists. Figure 3.12 shows the μR of some standard parts. The magnetic circuit analogy and solution is valid only if the permeability of the media being considered is large compared with that of the free space, or if the regions of low-permeability space accessible to the magnetic field are small compared with the rest of the circuit with regions with very large permeabilities. Moreover, the magnetic circuit concepts are only valid in a steady state.

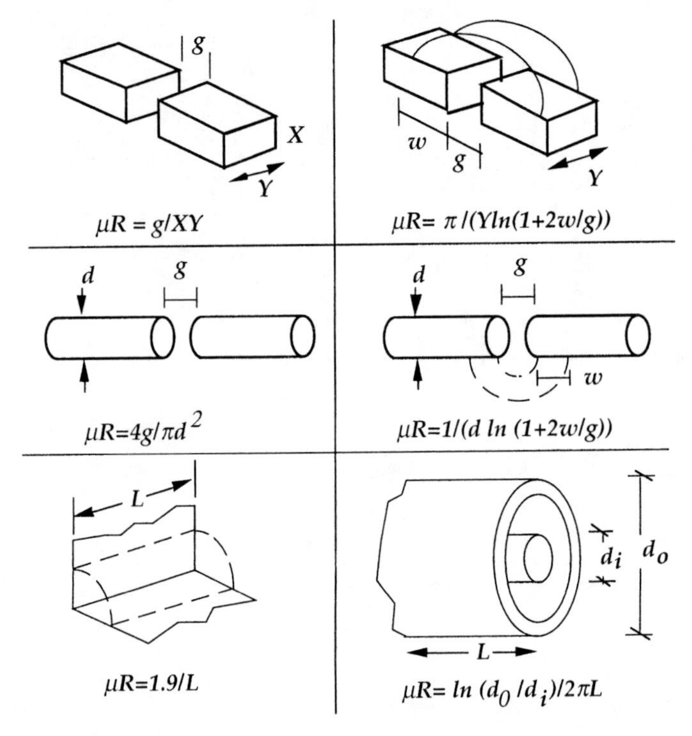

Figure 3.12 Common magnetic elements used in magnetic devices and their reluctances [11.c].

To calculate the force, we need to find the energy stored in the magnetic circuit. In general, the stored energy can be calculated from the Poynting vector, but it is more convenient to calculate it directly from the magnetomotive force and flux. It can be shown that under the simplified conditions of magnetic circuit theory, the stored energy is given by:

$$W_m = \frac{1}{2} F \phi = \frac{1}{2} \frac{F^2}{R} \tag{3.10}$$

The above relationship can also be derived from the inductance of the circuit defined as the ratio of the magnetic flux to the current (I):

$$L = \frac{1}{I}\int_S B \, . \, ds \tag{3.11}$$

Using the definition of reluctance (=F/ϕ), we find an expression for the flux in terms of the reluctance (ϕ=F/R), and we relate the inductance to the magnetomotive force and the reluctance:

$$L = \frac{F}{I\,R} \tag{3.12}$$

The stored energy, thus, is also given by:

$$W_m = \frac{1}{2}\,L\,I^2 \tag{3.13}$$

The force can be calculated from the derivative of the energy with respect to a position variable (a) that defines the component of the force (F_a) as before:

$$F_\alpha = -\frac{\partial W_m}{\partial \alpha} \tag{3.14}$$

where a minus sign is included to indicate that if the stored energy increases as a is increased, then a force should be applied to the system. Otherwise, the system applies the force and performs the work.

In the following sections, we apply the above equations to obtain the forces generated by various magnetic microactuators.

3.4 Magnetization-Type Microactuators

These devices are divided into three different types of actuators: i) thermo-magnetic, ii) magnetostatic, and iii) electromagnetic microactuators. In thermo-magnetic actuators, the magnetization of a ferromagnetic film is "wiped out" or reduced by heating it locally and causing a permanent magnet to attract adjacent parts of the film, which is cold and has larger susceptibility. The Curie wheel discussed in section 3.3.1 is an example of a thermo-magnetic actuator. Magnetostatic microactuators (for lack of a better term) are types of actuators that rely on the attraction between a permanent magnet, or an electromagnet, and a ferromagnetic (or ferrimagnetic) part. These actuators are discussed in section 3.3.2. Electromagnetic microactuators rely on the attraction or repulsion between an electromagnet and a permanent magnet. Loudspeakers are an example of electromagnetic actuators and are discussed in section 3.4.4.

3.4.1 Thermo-Magnetic Microactuators

Figure 3.13 schematically shows a simple example of a thermo-magnetic actuator, where illumination is used to heat a section of a ferromagnetic film adjacent to a permanent magnet. As the film heats up above its Curie temperature, it becomes paramagnetic, and the magnetic field lines move closer to the ferromagnetic part of the film, exerting a net tangent force that results in the rotation of the wheel. This actuator is called a "Curie wheel" for obvious reasons. The rotation speed and the amount of tangent force that can be generated both depend on how fast the spot can be heated, how the magnetic field lines, what the strength of the magnetic field is, and what the magnetic susceptibility of the film and its Curie temperature are.

Magnetic circuit theory can be used to find the torque generated by this motor. We assume that the thin magnetic film has a very large permeability ($\mu \approx \infty$) and that upon heating above its Curie temperature, its permeability reduces to that of the free space. The magnetomotive force generated by a magnet of length ℓ_m having a magnetic field of H_m is given by equation (3.8):

$$F_m = H_m \ell_m \tag{3.15}$$

The reluctance of the airgap (R_{gap}) is $R_{gap} = \ell_{gap}/\mu_0 A$, where A is the area of the pole of the magnet (= wd). The energy stored in the magnetic field magnetic is:

$$W_m = \frac{1}{2}\frac{F_m^2}{2R_{gap}} \tag{3.16}$$

The reluctance of the gap when the thin-film magnetic material is heated above its Curie temperature (R'_{gap}) (see figure 3.13b) is $R'_{gap} = (\ell_{gap} + d\ell)/\mu_0 A$, with $d\ell$ is approximately given by $1.9d$, according to figure 3.12 (bottom corner), where d is the length of the magnet pole. In this case the stored energy becomes:

$$W'_m = \frac{1}{2}\frac{F_m^2}{R_{gap} + R'_{gap}} \tag{3.17}$$

To calculate the force in the x-direction (figure 3.13b), we note that the displacement of the rotor in the negative x-direction in figure 3.13b, changes the stored energy from W'_m to W_m. To accomplish this, according to the geometry of the device shown in figure 3.13b, the thin film has to move by w. Thus, the force is given by:

$$F_x = \frac{\Delta W_m}{\Delta x} = \frac{W_m - W'_m}{-w} \approx -\frac{\mu_0 F_m^2 d}{4\ell_{gap}} \tag{3.18}$$

where we have assumed that w *and* t are much larger than ℓ_{gap}. Taking $H_m \approx 0.6$ Tesla (rare-earth magnets can easily produce such a field), $\ell_m \approx 5$ mm, $\ell_{gap} \approx 25$ µm, and $d \approx 5$ mm, we obtain $F_x \approx 0.57$ nano-N, which produces a torque of 5.7 pN.m for a rotor of 1 cm radius.

A variety of methods can be used to heat the magnetic thin film in the above device, including illumination, resistive- and eddy-current heating. The generated force and stepping action can also be improved by corrugating the magnetic thin film as shown in the bottom of the above figure.

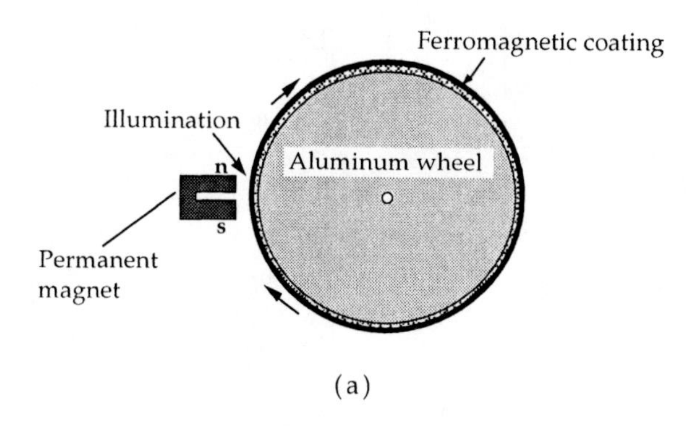

(a)

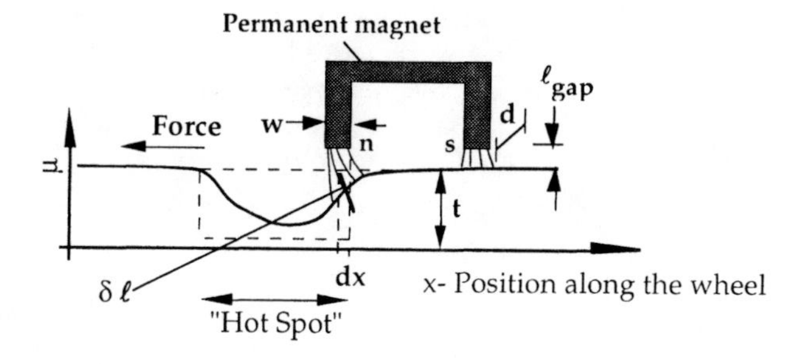

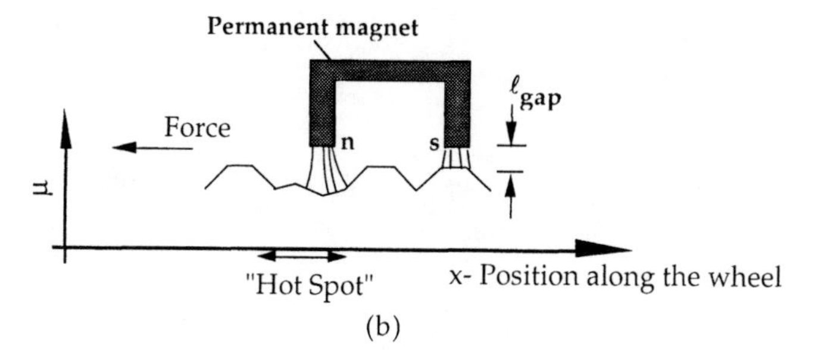

(b)

Figure 3.13 a) A Curie wheel. b) A simple linear model to calculate the force generated by the thermo-magnetic method.

3.4.2 Magnetostatic Microactuators

These actuators are schematically shown in figure 3.14. Two configurations are possible, which are denoted by type A and type B structures. In type A structures, when the magnetic slider moves towards the electromagnet, it reduces the air gap. In type B structures, the magnetic slider moves into the air gap. Magnetic circuit theory can be used to find the force that can be generated by these actuators, as discussed next [12, 21,22].

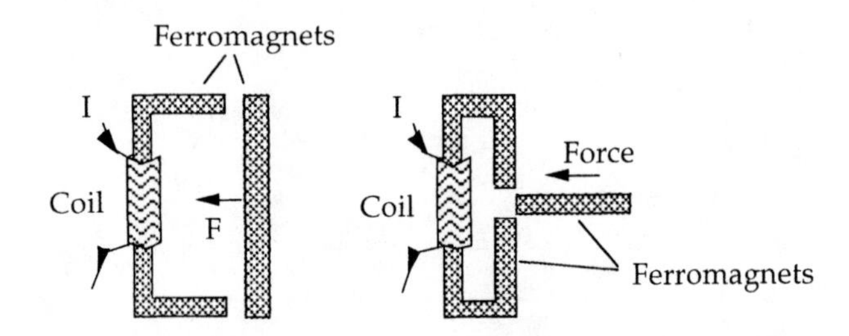

Figure 3.14 A schematic diagram of two types of magnetic actuators. a) The overlap area is fixed and the distance between the movable part and the electromagnet is affected by actuation (type A), and b) the overlap area is affected by the motion of the movable part (type B).

Figure 3.15 shows the schematic structure of a microfabricated magnetic microactuator. In this device, the overlap area between the deformable part and the magnet is fixed and only the distance varies (type A). The force that can be generated by this actuator can be calculated by using equations (3.8)-(3.14). The gap and core reluctances, respectively R_{gap} and R_{core}, are given by:

$$R_{gap} = (g-x)/\mu_0 A_g \tag{3.19}$$

and,

$$R_{core} = \ell_c/\mu A_c \tag{3.20}$$

where A_g is the gap area ($\ell_{ol}.w$), A_c is the core area ($w.t$) and ℓ_c is the total core length, as shown in figure 3.15.

The corresponding stored energies (W_{gap} and W_{core}) and the total energy ($W_m = W_{gap} + W_{core}$) are:

$$W_{core} = \frac{1}{2}\frac{N^2 i^2 R_{core}}{(R_{core} + R_{gap})^2} \tag{3.21}$$

and,

$$W_{gap} = \frac{1}{2}\frac{N^2 i^2 R_{gap}}{(R_{core} + R_{gap})^2} \tag{3.22}$$

and W_m is given by:

$$W_m = \frac{1}{2}\,\mathrm{L}i^2 = \frac{N^2 i^2 R_{gap}}{(R_{core} + R_{gap})^2} + \frac{N^2 i^2 R_{core}}{(R_{core} + R_{gap})^2} \tag{3.23}$$

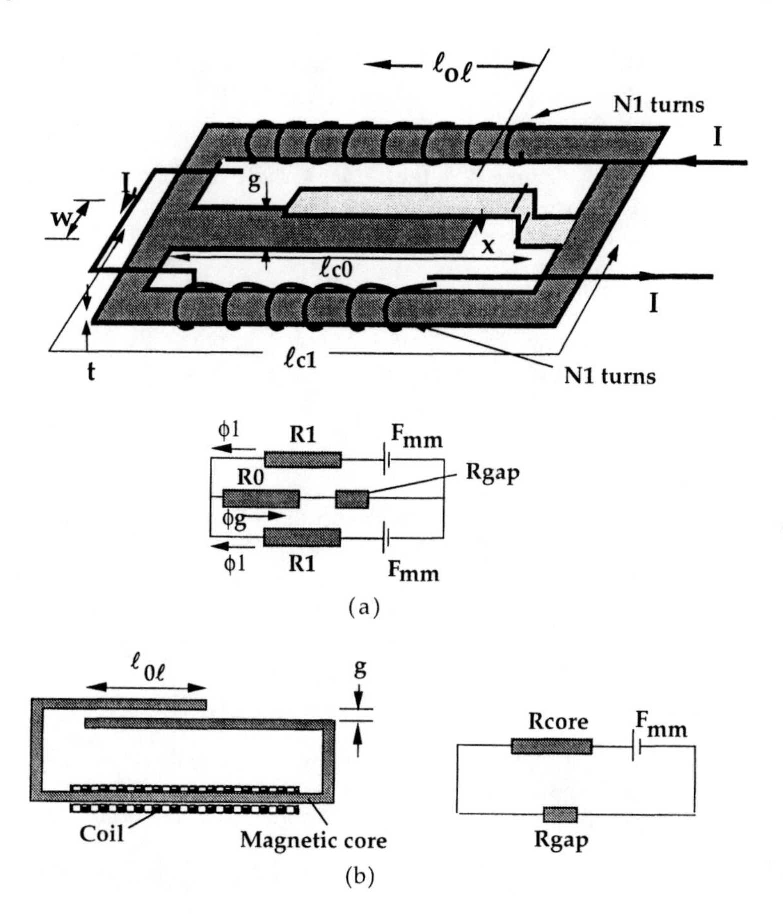

Figure 3.15 (a) A schematic of a type A magnetic microactuator and its magnetic circuit model. (b) A simplified magnetic actuator and its circuit model [22].

The force in the x-direction (figure 3.15) is calculated by differentiating W_m with respect to x:

$$F_x = \frac{\partial W_m}{\partial x} = \frac{1}{2}\frac{N^2 i^2}{\mu_0 A_g (R_{gap} + R_{core})^2} \tag{3.24}$$

Using equations (3.15) and (3.16) in 3.20 we obtain:

$$F_x = \frac{1}{2}\frac{N^2 i^2}{\mu_0 A_g \left(\frac{\ell_c}{\mu A_c} + \frac{g - x}{\mu_0 A_g}\right)^2} \tag{3.25}$$

To optimize force we can examine the variations of F_x with respect to parameters that can be varied in a given actuator. In the above example, A_g can be easily adjusted. Thus, differentiating equation (3.21) with respect to

A_g, we find that F_x is maximized when $R_{core}=R_{gap}$. This is equivalent to the requirement of $R_s=R_{load}$, where R_s is the source resistance and R_{load} is the load resistance, in electric circuits.

To apply the above analysis to the actuator shown in figure 3.15, we note that R_{core} is given by:

$$R_{core} = \frac{R_1}{2} + R_0 \qquad (3.26)$$

where R_1 is the reluctance of two outside arms that contain the coils and R_0 is the reluctance of the central part:

$$R_{core} = \frac{(0.5\ell_{c1} + \ell_{co})}{\mu A_c} \qquad (3.27)$$

The air gap reluctance, as before, is given by equation (3.15). Using R_{core} and R_g in equation (3.19), and differentiating W_m with respect to A_g, an expression for the force acting on the deformable part can be found:

$$F_x = \frac{A_g i^2 N_1^2}{2\mu_0}\left(\frac{\mu_0 \mu A_c}{0.5\,\mu_0 \ell_{c1} A_g + \mu_0 \ell_{co} A_g + \mu A_c (g - x)}\right)^2 \qquad (3.28)$$

Figure 3.16 shows a plot of F_x versus position x superimposed on F_x-x of the cantilever that forms the deformable part of the structure. In that plot we have assumed that the core is composed of NiFe material having a permeability of $\mu=5000\mu_0$, $N_1=100$ turns, $i=1$ mA, $t=10$ μm, $w=1$ mm, $\ell_{c1}=\ell_{c0}=2000$ μm, $\ell_{ol}=200$ μm and $g=20$ μm. We have also taken the effective Hooke's constant of the cantilever beam to be 10 N/m.

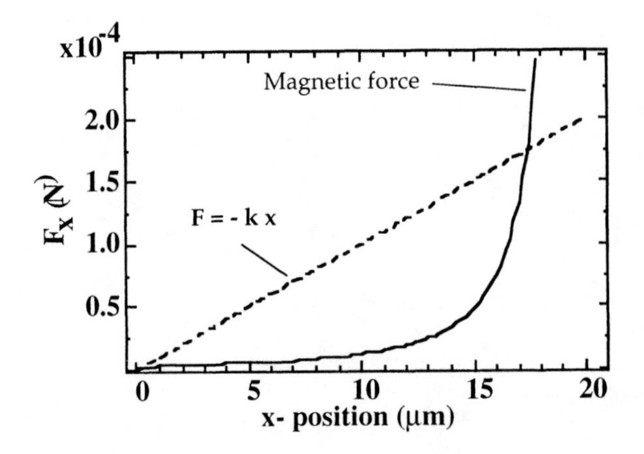

Figure 3.16 The magnetic force and the cantilever beam reaction force as a function of the position of the cantilever beam tip.

3.4.3 Design of Magnetic Microactuators with Planar Coils

Due to the importance of inductors with high inductance values in communication circuits [1] and in telemetry [4], planar inductors with magnetic core materials, such as NiFe, have been designed and fabricated. Electroplating is usually used to deposit up to 10 μm of NiFe either underneath the coil or at its center. It is found that a magnetic material layer covering the area underneath the planar coil results in the best performance. A 2 mm x 10 mm x 10 μm (thick), 10 turns Cu coil deposited over an insulating 6 μm thick polyimide layer over 10 μm of NiFe magnetic layer has been fabricated and tested. This coil had 2.70 μH inductance at 4 MHz and had a series resistance of 10 Ω [4].

Using planar magnetic coils and films, a variety of interesting and useful devices can be designed and fabricated [25.b]. One such example is a microrelay, schematically shown in figure 3.17. It is composed of a magnetic planar spring and a coil. When a magnetic field is generated by the coil, the spring is pulled down. The physical contact between the spring and the top of the coil provides an electrical contact: the "on" position of the relay. The interesting part of this design was the inclusion of a small magnet at the center of the planar spring that enabled the spring to be latched to the "on" position when the current through the coil was turned off. To un-latch the spring, the coil was excited with a current polarity that generated a repelling magnetic field.

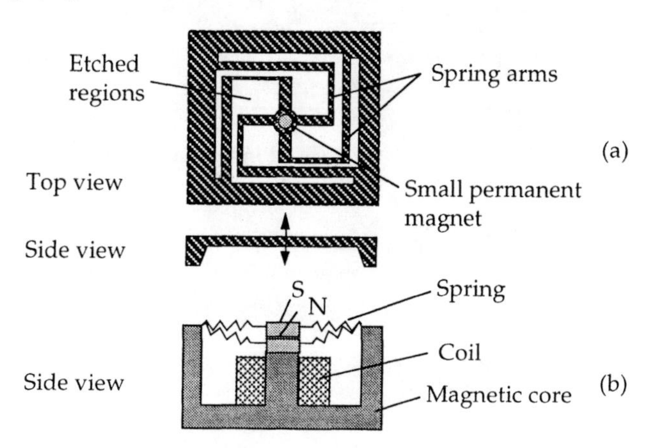

Figure 3.17 (a) A schematic of a flat ferromagnetic spring used in microrelays and switches. (b) The latching mechanism [25.b].

Microrelays have lower on-resistance, higher off-resistance, higher dielectric breakdown strength, lower power consumption and lower cost compared with semiconductor switches. Furthermore, by using MEMS technology, the size, cost, density (device per unit area) and speed of these relay can be greatly improved. Moreover, by combining with electronics, multiplexing can be performed to reduce the number of electrical lines in relay matrix arrays; an 8x8 array will require 16 control lines, as opposed to

64 control lines, when the relays are packaged separately without the multiplexing electronics. The reliability of microrelays, though, should be examined very carefully because it is not clear whether these relays can be designed to handle very large power loads.

The microrelay structure shown in figure 3.17 had an on resistance of 2 W using either gold, palladium or silver contacts. Its breakdown field was on the order of 650 Volts over a 6 μm contact gap. Switching speeds as high as 0.05 were also obtained [25.b].

3.4.4 Electromagnetic Microactuators

One of the important examples of electromagnetic actuators is the loudspeaker. The magnet and coil arrangement in loudspeakers is shown in figure 3.18 where a coil is located above a magnet. A magnetic core is used to concentrate the magnetic flux in the airgap surrounding the coil. The coil can move up and down, depending on the polarity of the excitation current.

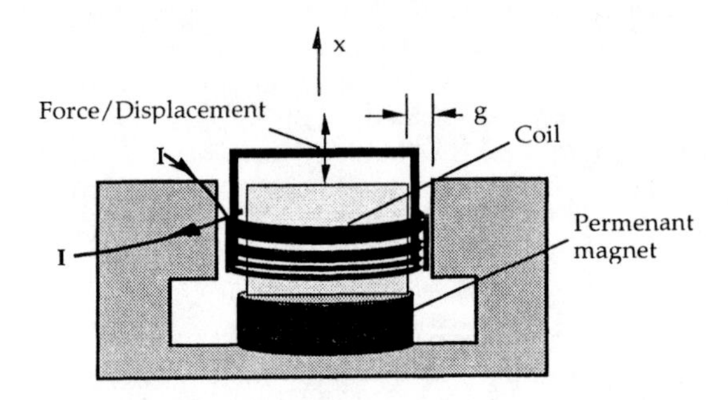

Figure 3.18 The electromagnet structure used in loudspeakers.

Since in this design the magnetic flux is confined to high permeability regions (presumably with no magnetic leakage), we can use magnetic circuit theory to solve for the force and displacement that can be generated with this electromagnet.

It can be shown that the electromagnetic force exerted on the coil (N-turns carrying current i) is given by [13; page 529]:

$$F_x = 2\pi RN(\mu_0 H_0 i + \frac{\mu_0 N i^2}{2g}) \tag{3.29}$$

where g is shown in figure 3.18, H_0 is the radial magnetic field in the gap, and R is the radius of the coil. The first term in the above equation is due to the interaction of current-carrying coil with H_0 while the second term is due to the self-interaction of the coil (i.e.; interaction between current of one wire with the current of the rest of the wires). To achieve linear operation, usually $|H_0|$ is made much larger than $|Ni/2g|$ to reduce the importance of the second term. Thus, F_x becomes simply $2\pi RNB_0 i$.

Conventional devices suffer from sound distortion and electro-acoustic and mechanical feedback, and they cause discomfort and have undesirable cosmetic appearance. To address these issues and also to assist those for whom conventional hearing aids cannot work, implantable hearing aids that can directly stimulate the inner ear are considered [10]. Various device structures using both electromagnetic and piezoelectric actuation mechanisms are proposed for implantable hearing aids. Piezoelectric implantable hearing aids suffer from poor frequency response and sound quality [10].

Implantable hearing aids with high-fidelity and with a wider frequency response have been demonstrated using magnetic actuation mechanisms. Figure 3.19 schematically shows a diagram of a middle-ear implant hearing aid that uses a coil to drive a permanent magnet. The required driving force in these devices is 0.16-16 μN with a displacement range of 1 to 100 Å with a frequency range of 100 Hz- 7 kHz. Design criteria as well as preliminary design and fabrication of the device are discussed in ref. [10].

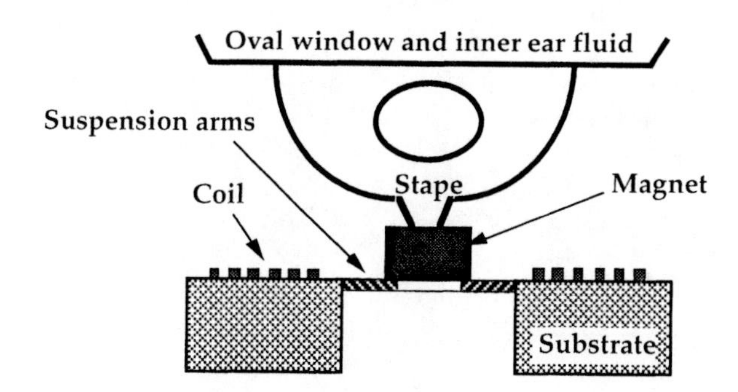

Figure 3.19 A schematic diagram of a middle-ear implant hearing device [10].

It can be shown that the z-component of the magnetic flux of a planar coil of inner and outer radii r_{in} and r_{out}, respectively, and effective thickness b is given by [3]:

$$B_z = \frac{\mu_0 N i}{4\ell(r_{out} - r_{in})} \left\{ (b+Z) \ln \left[\frac{r_{out} + (r_{out}^2 + (b+Z)^2)^{1/2}}{r_{in} + (r_{in}^2 + (b+Z)^2)^{1/2}} \right] + (b-Z) \ln \left[\frac{r_{out} + (r_{out}^2 + (b-Z)^2)^{1/2}}{r_{in} + (r_{in}^2 + (b-Z)^2)^{1/2}} \right] \right\} \quad (3.30)$$

where i is the current, and ℓ is the effective length of the coil wire given by $2\pi N r_{out}$. The dipole moment of the rare-earth permanent magnet of volume V and remnant magnetization B_r ($\approx$ 1-2 Tesla) is given by:

$$m_z = VB_r / \mu_0 \tag{3.31}$$

Then the force on the magnet due to the presence of B_z is simply given by:

$$F_z = m_z \frac{\partial B_z}{\partial Z} \tag{3.32}$$

The magnetic actuator shown in figure 3.19 using a coil with r_{in}=100 μm, r_{out}= 1000 μm , and b=0.25 μm, and a magnet with B_r=1.15 T, V=1500 μm^3, and carrying a current i= 3 mA, produces a peak force of $3x10^{-5}$ N at Z=400 μm. The force is very small at Z=0 and at Z>800 μm. Note that the maximum current that can be carried by the coil is limited to around $5\text{-}6x10^6$ A/cm^2 due to the heating/melt-down of the wire caused by the limitations in the thermal conductivity of the Si_3N_4 or SiO_2 structure used in these actuators. Electromigration is another consideration in long-term operation of thin-film aluminum lines.

3.5 Lorentz Force Actuator

An example of a Lorentz force actuator is schematically shown in figure 3.20, where a force is generated between a current-carrying wire and the magnetic field of a permanent magnet [26]. The Lorentz force (F_L) when the magnetic field and current are perpendicular is given by $\ell_c IB$, where I is the current, B is the magnetic field, and ℓ_c is the effective length of the conductor.

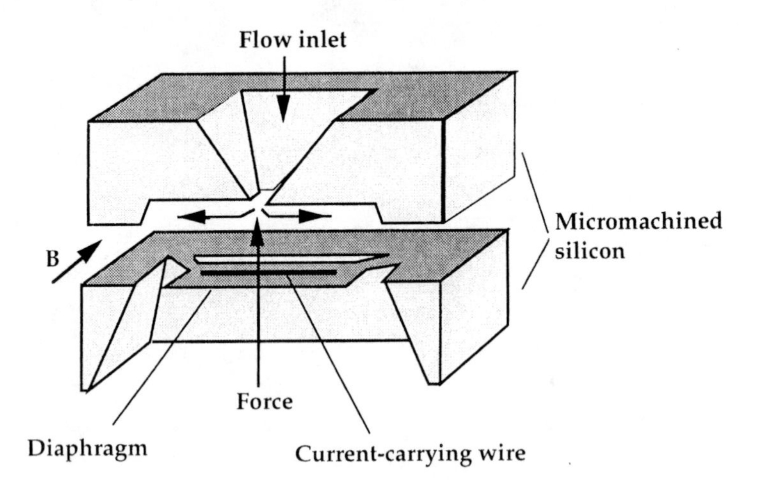

Figure 3.20 A schematic cross-section of a microvalve showing the generation of the Lorentz force F_L acting on the deflectable membrane [26].

The Lorentz force in the above structure does not depend on the deflection of the diaphragm in contrast to other forces, such as those which are electrostatic. This is because it is assumed that the external magnetic field is uniform over the span of the diaphragm. The mechanical reaction force, against which F_L acts, is composed of two parts: the diaphragm-bending force and a tensile force, both of which depend on the displacement of the diaphragm as well as a gas or fluid back-pressure. These are all discussed in [26]. Lorentz forces on the order of 2×10^{-4} N were generated in their device with $B=1.4$ kG, $\ell_c = 3$ mm, and $I=0.2$ A. They were also able to control flows of up to 3 ml/min at flow pressures of up to 160 mbar.

Figure 3.21 schematically shows the structure of a Si_3N_4 diaphragm with a meander-line transducer (MLT). By passing an a.c. signal through the MLT, Lorentz forces, which were perpendicular to the plane of the Si_3N_4 diaphragm, were generated in the presence of an in-plane d.c. magnetic field. The magnetic field was externally applied, as shown in the figure. The perpendicular Lorentz forces excited the flexural plate waves in the Si_3N_4 diaphragm [27].

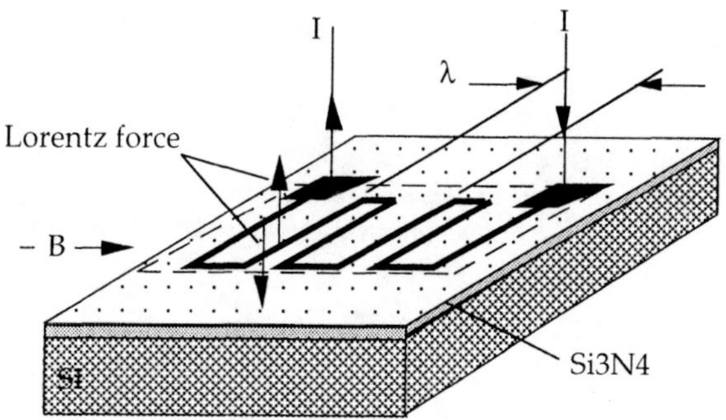

Figure 3.21 A schematic of a flexurally actuated Si_3N_4 diaphragm [27].

The impedance of the MLT in the above device was shown to be given by:

$$Z(\omega) = R_0 + \frac{c_2 w B^2 N}{\lambda \left(Z_a + j\omega\rho_s + \frac{T h k^2}{j\omega} + \frac{Dk^4}{j\omega} \right)} \tag{3.33}$$

where D, T, h, w and r_s are the membrane's bending moment, in-plane stress, thickness, width, and aerial mass density, respectively. Also, R_0 is the resistance of MLT, B is the magnetic applied field, N is the number of MLT periods, Z_a is the mechanical impedance arising from the conducting fluid (gas or fluid), and λ is the MLT wavelength. The resonances occur at:

$$\omega = k \left(\frac{Dk^2 + T h}{\rho_s} \right)^{1/2} \tag{3.34}$$

For a 2.2 x 3.4 mm low-stress (6.1×10^8 dynes/cm^2) Si_3N_4 (2μm thick) with a 0.5 μm thick gold (over Cr) MLT of 320 μm wavelength, the resonance frequency was 406.2 kHz. The resonance frequency (f_0) increased as a function of the a.c. power due to the membrane stiffening at higher flexing amplitudes. f_0 also depends on the Si_3N_4 residual stress and, by heating the membrane (i.e., by passing a d.c. current through the MLT), f_0 shifted to lower frequencies.

When gases are introduced to MLT, its resonance frequency changes due to mass loading. Using the above MLT, Argon, Nitrogen and Helium gases at pressures ranging from 50 Torr to 700 Torr were detected. These gases shifted the f_0 differently and the device had good selectivity [27].

3.6 Magnetostrictive Microactuators

Magnetostrictive actuators can have a very high frequency response, usually have a very simple device structure and design, can be actuated remotely and are cost-effective from a manufacturing point of view.

There are many different methods for utilizing magnetostrictive materials in microactuators. We discuss two of these methods [28-30].

Figure 3.22 shows the structure of a magnetostrictive actuator that operates based on the Wiedmann effect. The Wiedmann effect refers to the excitation of twisting strains in a ferromagnetic rod by the axial (B_l) and circular (B_c) magnetic fields applied to the rod.

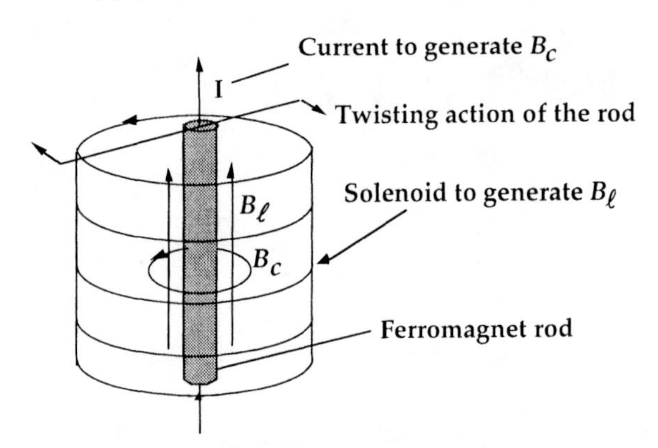

Figure 3.22 A schematic of a Wiedmann type magnetostrictive actuator [28].

B_l and B_c are both generated using inductor coils. The twisting of the magnetic rod is also accompanied by a change in its length. Precision displacements of 10-100 μm with accuracy of positioning up to 0.1 μm or better at 1 ms time scale are easily obtained with a permendur magnetic rod of 10 cm length and 1 cm diameter. Permendur is an alloy of 49% Fe, 49% Co and 2% V [28]. It is also interesting to note that this device can be used as a torque sensor when the signal in one of the coils is monitored as a function of the twisting of the magnetic rod.

Figure 3.23a shows another magnetostrictive device structure that is more suitable for microactuator applications. It consists of a magnetostrictive thin film deposited over a silicon micromachined cantilever beam. When a magnetic field is applied, as shown in the figure, the magnetostrictive film expands, bending the composite beam downward. For a 10 μm thick TbDyFe film over 50 μm thick Si cantilever beam of 20 mm, the beam deflection was 230 μm at 0.05 T, as shown in figure 3.20b. The Young's modulus of the TbDyFe film was 50 GPa and its magnetostriction was 200×10^{-6} at 0.05 T. This actuator could be driven at frequencies as high as 500 Hz, and no degradation in its performance was observed after more than 10^7 deflection cycles [30].

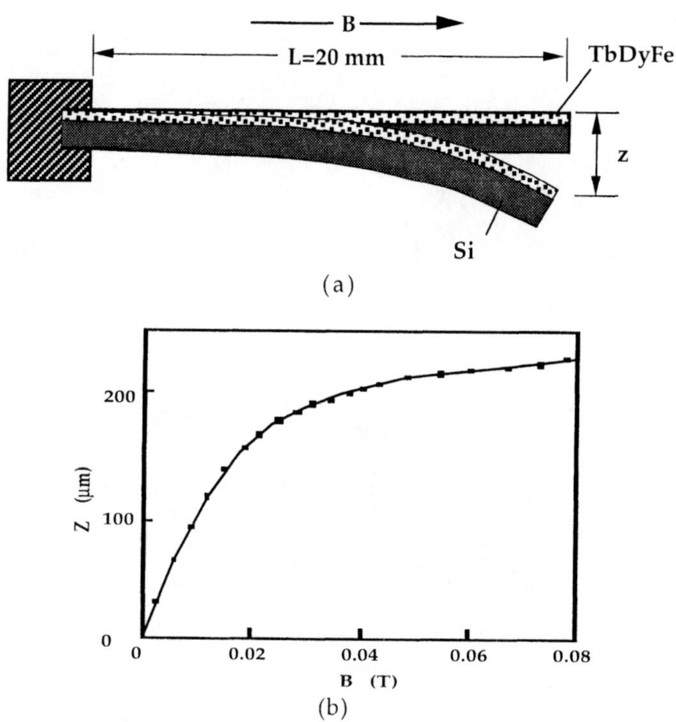

Figure 3.23 (a) A schematic of a magnetostrictive silicon-based microactuator. (b) Displacement of the actuator as a function of the applied magnetic field [30].

One of the requirements for fabricating magnetostrictive actuators using micromachining methods is the requirement of the processibility of the magnetic film using methods that are compatible with silicon device and IC fabrication. Sputtering of $SmFe_2$ (a negative magnetostrictive material) and $TbFe_2$ or $Tb_{0.3}Dy_{0.7}Fe_2$ (positive magnetostrictive materials) has been shown to result in good quality magnetostrictive films. $Tb_{0.3}Dy_{0.7}Fe_2$ is known by its trade name of Terfenol-D. It is also shown that the anisotropy and the easy-magnetization axis in these films can be adjusted by changing the sputtering parameters (i.e., power, d.c. bias and ambient Argon pressure).

3.7 Magnetic Levitation

Repulsion between magnetic poles of the same polarity is the underlying principle of levitation of objects using magnetic methods. Magnetic bearings use this principle to reduce friction. In superconductors, complete repulsion of the magnetic field lines from the interior results in generation of repulsive forces between the superconductor and the magnetic field, resulting in magnetic levitation. Figure 3.21 shows a device structure that uses a permanent magnet slider situated above a YBaCuO high T_c substrate. Series of conducting lines are also deposited over an insulating layer over the YBaCuO substrate. These conducting lines are used to generate Lorentz force to move the slider left or right, parallel to the plane of the device [31,32].

In these linear micromotors, a Nd-Fe-B 8 mg permanent magnet was used for the slider, which levitated 1 mm above the stator and was moved by a velocity of 7 mm/s and a driving force of 30 μN. The stator current was manually switched to generate a Lorentz force on the slider. The slider width was 1 mm and the stator wavelength was 0.4 mm. The drag force, including that produced by eddy currents in the stator, was estimated to be approximately 10 μN [32].

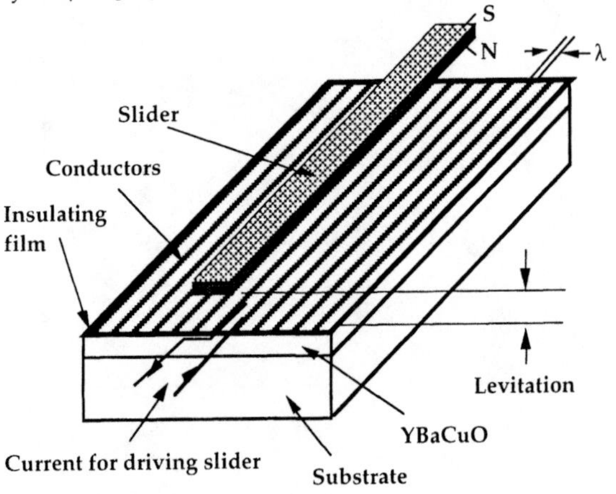

Figure 3.24 A schematic of a linear synchronous microactuator using the Meissner effect [31,32].

In a slightly different configuration, the conductor strips in the above micromotor were replaced with superconducting lines that turned normal when sufficiently large currents, in excess of their critical current, were passed through them. Then, the slider moves over the strip that became normal. This mechanism provided a driving force to move the slider in a desired direction. The design parameters were the wavelength of the strips and the width of the slider. For wavelength and levitation height of 100 μm, driving force of 0.26 N m^{-2} A^{-2} and levitating force of 1.7 N m^{-2} A^{-2} were obtained [32].

3.8 Fabrication of Magnetic Actuators

As is evident from the above sections, magnetic actuators usually have three different parts: i) a high permeability core region (stator), ii) a plunger (rotor), and iii) an excitation coil that is usually wound around the core. The first two parts can be readily fabricated by a variety of methods, discussed at the end of Chapter 1. The most problematic part of microfabricated magnetic actuators is the design and fabrication of the excitation coil.

There are three basic approaches to designing and microfabricating inductors [4,-9,32-36].

The first approach is based on using the standard IC fabrication process. In this method, metallic thin-films are evaporated or sputtered over the substrate and patterned to form the desired coils. Two levels metallizations are used to contact the inner port of the inductor. An interesting variation on this approach where a magnetic film was incorporated below or above the inductor plane to increase its inductance [4], has been reported recently .

The second approach uses a permalloy mandrel fabricated using the LIGA method. The mandrel is used as a magnetic core, and magnet wires are either manually or automatically wound around it. Then the coil is inserted back into the magnetic actuator co-fabricated using LIGA [25].

The third method uses the magnetic core wrapped around a meander line coil, as shown in figure 3.26. This is an interesting method that takes advantage of the fact that the reluctance of the magnetic core is probably less affected by the fact that different parts of the core are located at two planes above and below the meander coil. In contrast, the electrical resistance between the top and bottom-plane metal lines of two-level metallization can be quite large. Thus, instead of wrapping the coil around the core, the core is wrapped around the coil [36].

Polyimides are used in various aspects of IC fabrication as dielectric layers in multilevel interconnect technology or in multi-chip packaging. In micromachining, it has gained some popularity due to its ability to form similar structures to those fabricated using LIGA [36].

To fabricate magnetic microstructures, a plasma-based polyimide process is used. After deposition of a seed layer over an appropriate substrate, a polyimide adhesion promoter layer and a polyimide layer is deposited. For thinner polyimide layers (≤20 μm), multiple spin-coating is used. For thicker layers (>20 μm), screen printing or spray are more suitable.

The seed layer usually has three metal layers. Its first layer is to increase the adhesion between the seed layer and the substrate. The second layer is the actual seed layer, which is topped by a third metal layer that acts as a protective layer to prevent damage to the seed layer during polyimide deposition and etching.

After deposition of the polyimide, a suitable plasma-etch masking layer, usually Al, is deposited and patterned. Plasma etching of polyimide can be achieved using different gases such as O_2, CF_4, CHF_3 and SF_6. The slope of the etched side walls can be controlled by varying the plasma parameters, such as the relative concentration of gases, the power and the chamber pressure. After etching, the mask layer is removed and the etched regions of the polyimide are used as a mold and, using electroplating, the microstructures are formed inside this mold. Nickel, nickel-iron alloys, copper and gold microstructures have been fabricated using this method.

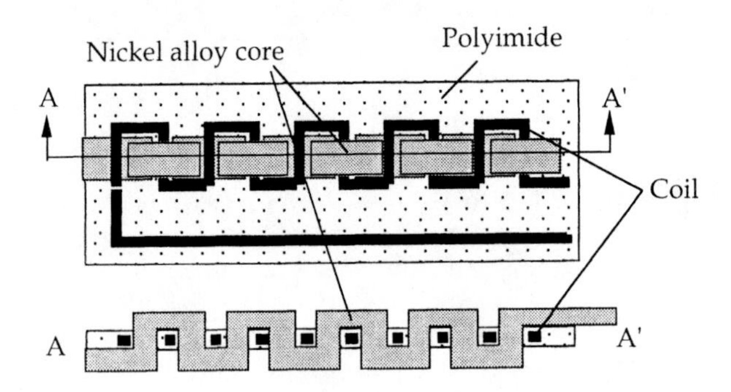

Figure 3.25 A schematic of an electromagnet with a nickel-iron core wrapped around a meander line coil [36].

3.9 References

1. D. C. Webb, "Status of Ferrite Technology for High Volume Microwave Applications." In: Materials and Processes for Wireless Communications. Edited by T. Negas and H. Ling, Ceramic Transactions, Volume 53, The American Ceramic Society, pp. 11-26 (1995).
2. M. Tabib-Azar and A. Garcia, "Sensing Means and Sensor Shells: A New Method of Comparative Study of Piezoelectric, Piezoresistive, Electrostatic, Magnetic, and Optical Sensors." Sensors and Actuators A, Vol. 48, pp. 87-100 (1995).
3. I. J. Busch-Vishniac, "The Case for Magnetically Driven Micro-actuators." Sensors and Actuators A Vol. 33 (3), pp. 207-220 (1992).
4. J. A. Von Arx and K. Najafi, "On-Chip Coils with Integrated Cores for Remote Inductive Powering of Integrated Microsystems." Transducers '97, pp. 999-1002 (1997).
5. P. A. Neukomm and H. Kundig, "Passive Wireless Actuator Control and Sensor Signal Transmission." Sensors and Actuators, A21-A23, pp. 258-262 (1990).
6. J. R. Long and M. A. Copeland, "The Modeling, Characterization, and Design of Monolithic Inductors for Si RF IC's." IEEE Journal of Solid-State Circuits, Vol. 32 (3), pp. 357-369 (1997).
7. (a) J. N. Burghartz, K. A. Jenkins and M. Soyuer, "Multilevel-Spiral Inductors Using VLSI Interconnect Technology." IEEE Electron Devices Letters, Vol. 17 (9), pp. 428-430 (1996).
   (b) C. P. Yue, C. Ryu, J. Lau, T. H. Lee and S. S. Wong, "A Physical Model for Planar Spiral Inductors on Silicon." Proceedings of International Electronic Devices Meeting '96, pp. 155-158 (1996).

8. H. M. Greenhouse, "Design of Planar Rectangular Microelectronic Inductors." IEEE Transaction on Parts, Hybrids, and Packaging, Vol. PHP-10 (2), pp. 101-109 (1974).
9. A. A. Abidi, "Low-Power Radio-Frequency IC's for Portable Communications." Proceedings of the IEEE, Vol. 83 (4), pp. 544-569 (1995).
10. W. Affane and T. S. Birch, "A Microminiature Electromagnetic Middle-Ear Implant Hearing Device." Sensors and Actuators A 46-47, pp. 584-587 (1995).
11. J. M. D. Coey, "Introduction." In: Rare-Earth Iron Permanent Magnets. Edited by J. M. D. Coey, Oxford University Press, Oxford GB, pp. 6-10 (1996).
12. H. H. Woodson, J. R. Melcher, Electromechanical Dynamics, Part I: Discrete Systems. John Wiley & Sons, Inc., New York (1968).
13. H. H. Woodson, J. R. Melcher, Electromechanical Dynamics, Part II: Field, Forces and Motion. John Wiley & Sons, Inc., New York (1968).
14. W. K. H. Panofsky and M. Phillips, Classical Electricity and Magnetism. Addison-Wesley Publishing Company Inc., Reading MA, (1955).
15. J. C. Slater and N. H. Frank, Electromagnetism. McGraw-Hill, New York (1947).
16. P. Robert, Electrical and Magnetic Properties of Materials. Artech House, Boston, MA (1988).
17. S. Ramo, J. R. Whinnery and T. Van Duzer, "Fields and Waves in Communication Electronics." 3rd edition, John Wiley & Sons, Inc., New York (1994).
18. G. Burns, "Solid-State Physics." Academic Press, San Diego (1985).
19. R. Neale, "Taming the Giant Magneto Resistance (GMR) Effect." Electronic Engineering , April issue, pp. 36-40, (1996).
20. Joel S. Miller and Arthur J. Epstein, "Molecular and Polymeric Magnets."
21. R.L. Smith, R.W. Bower, and S.D. Collins, "The Design and Fabrication of a Magnetically Actuated Micromachined Flow Valve." Sensors and Actuators, A24, pp. 47-53 (1990).
22. Z. Nami, C.H. Ahn and M.G. Allen, "An Energy-Based Design Criterion for Magnetic Microactuators." J. Micromech. Microeng. Vol. 6, pp. 337-344 (1996).
23. H. Hauser, "Automatic Quality Control of Small Relays and Their Magnetic Parts." Sensors and Actuators A 46-47, pp. 588-592 (1995).
24. J. W. Judy and R. S. Muller, "Magnetic Microactuation of Torsional Polysilicon Structures." Sensors and Actuators A 53, pp. 392-397 (1996).
25. H. Hosaka, H. Kuwano and K. Yanagisawa, "Electromagnetic Microrelays: Concepts and Fundamental Characteristics." Sensors and Actuators A, 40, pp. 41-47 (1994).
26. D. Bosch, B. Heimhofer, G. Muck, H. Seidel, U. Thumser and W. Welser, "A Silicon Microvalve with Combined Electromagnetic/Electrostatic Actuation." Sensors and Actuators A Vol. 37-38, pp. 684-692 (1993).
27. M. A. Butler, S. J. Martin, J. J. Spates and M-A Mitchell, "Magnetically-Excited Flexural Plate Wave Devices." Transducers '97, pp. 1031-1034 (1997).
28. V. I. Aksinin, V.V. Apollonov, V.I. Borodin, A.S. Brynskikh, S.A. Chetkin, S.V. Murav'ev, V.V. Ostanin, and G.V. Vdovin, "Spring-type

Magnetostriction Actuator Based on the Wiedmann Effect." Sensors and Actuators, A21-A23, pp. 236-242 (1990).
29. J. A. Granath, "Instrumentation Application of Inverse-Wiedemann Effect." Journal of Applied Physics Vol. 31 (5), pp. 178S-180S (1966).
30. E. Quandt and K. Seemann, "Fabrication and Simulation of Magnetostrictive Thin-Film Actuators." Sensors and Actuators A 50, pp. 105-109 (1995).
31. Yong-Kweon Kim, M. Katsurari, and H. Fujita, "A Superconducting Actuator Using the Meissner Effect." Sensors and Actuators, Vol. 20, pp. 33-40 (1989).
32. Yong-Kweon Kim, and M. Katsurari, "A Levitation-type Synchronous Micro-Actuator Using the Meissner Effect of High-T_c Superconductors." Sensors and Actuators, A29, pp. 143-150 (1991).
33. A. B. Frazier, C. H. Ahn and M. G. Allen, "Development of Micromachined Devices using Polyimide-Based Processes." Sensors and Actuators A 45, pp. 47-55 (1994).
34. F. J. Cadieu, H. Hegde, A. Navarathna, R. Rani and K. Chen, "High-Energy Product $ThMn_{12}$ Sm-Fe-T and Sm-Fe Permanent Magnets Synthesized as Oriented Sputtered Films." Appl. Phys. Lett. Vol. 59, pp. 875-877 (1991).
35. H. Guckel, T. Earles, J. Klein, J. D. Zook and T. Ohnstein, "Electromagnetic Linear Actuators with Inductive Position Sensing." Sensors and Actuators A 53, pp. 386-391 (1996).
36. A. Bruno Frazier, C. H. Ahn and M. G. Allen, "Development of Micromachined Devices Using Polyimide-Based Processes." Sensors and Actuators A 45, pp. 47-55 (1994).

Thermal and Phase-Transformation Microactuators

4.1 Introduction

In this chapter we explore devices that utilize direct thermal schemes to achieve actuation. Heat can be generated by different means that include passing current through a resistor, using light, and employing junctions between dissimilar regions. The resistive heating is by far the most popular method that is used in actuators. Thermal actuators are usually simpler and easier to construct than their other counterparts such as electrostatic actuators. The main disadvantage of thermal actuators is their slow speed. Therefore, these types of actuators stand to gain substantially from miniaturization and they have received renewed interest in the microdevice community. Thermal actuators employing gas, liquid, and solid expansions are reported in the literature as discussed in the following sections.

The operation of thermal actuators is along the following sequence of events. First a control signal in the form of electric or optical power is applied to a heating element. Upon heating, the heater transfers thermal energy to a gas or a liquid or a solid that subsequently expands and performs the desired work. Or the heating element may expand and perform the work itself in a bimorph configuration. The following block diagram schematically shows these steps.

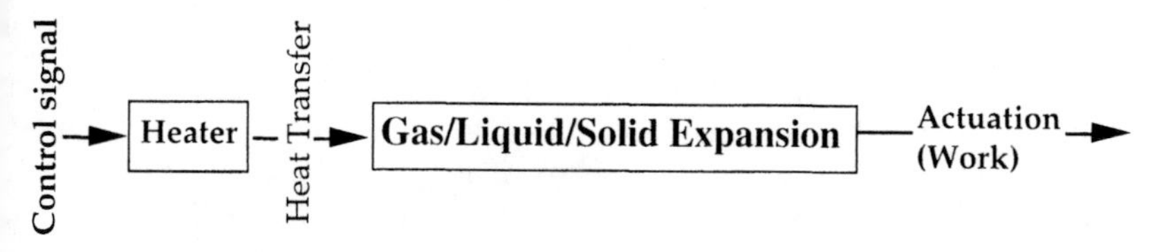

Figure 4.1 Schematic representation of the thermal actuators.

In the case of thermal microactuators there is a critical issue that should be addressed: i) how fast can it respond to a control signal and how long it takes to recover? Although to answer this question the details of the actuator structure should be considered, an overall analysis can be performed that apply to different structures.

To obtain an insight into the operation of thermal microactuators, let us consider the simple structure shown in figure 4.2 where a bimorph over the cantilever beam produces a bending of the beam upon heating. The metallic strip may be heated by an electrical or an optical signal. The cantilever beam can be fabricated using micromachining methods with typical dimensions of 600 μm x 50 μm x 5 μm [see Chapter 1]. We denote the total input control power as P_{in}. Three things can happen to the heat generated in the metallic strip [1]: i) it can be transferred to the surrounding gas/liquid or solid environment through conduction, ii) through convection, and or iii) through radiation into the surrounding mater or vacuum. Thus, the following power balance equation can be written:

$$P_{in} = P_{ht} + P_t + P_w \tag{4.1}$$

where P_{ht} is the thermal power transferred to the surrounding, P_w is the power converted to the mechanical work that goes into bending the cantilever beam, and P_t is the thermal power that remains in the heater and goes into increasing its temperature as a function of time.

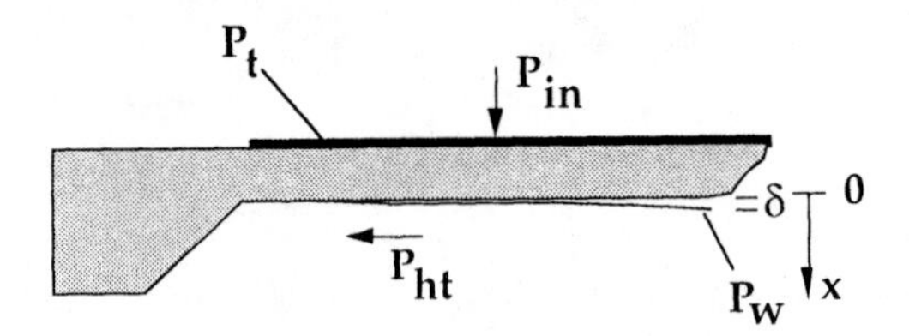

Figure 4.2 A simple actuator geometry with a bimorph thermal actuation mechanism [2].

P_{ht} can be written as a sum of three terms associated with different heat transfer mechanisms.

$$P_{ht} = P_{cd} + P_{cv} + P_r \tag{4.2}$$

where P_{cd} is the power transferred due to thermal conduction (thermal diffusion), P_{cv} is the power transfer due to convection, and P_r is the radiative power transfer. We discuss each of these components in detail.

Transfer of heat through conduction is a diffusion process. Transfer of heat through one-dimensional conduction in the solid, is given by the Fourier rate equation:

$$P_{cd} = -k\, A\, dT \,/\, dx \tag{4.3}$$

where P_{cd} is the rate of conductive heat transfer (W), T is the temperature distribution (K), k is the coefficient of heat conductivity (W/cm °C), and A is the area (m^2). To obtain $T(x)$, the equation of heat conduction in homogeneous materials is solved [1]:

$$\frac{\partial T(x,t)}{\partial t} = \kappa \frac{\partial^2 T(x,t)}{\partial x^2} \tag{4.4}$$

where $\kappa = k/\rho c_p$ is the coefficient of thermal diffusivity (cm^2/s); ρ is the material density and c_p is the heat capacity at constant pressure.

Microactuators are three dimensional structures with multiple layers of different thermal properties. The heat conduction equation in these structures can be simplified and written in one dimensional form provided that a correction factor (ε) is introduced:

$$\frac{\partial T(x,t)}{\partial t} = \kappa\left[\frac{\partial^2 T(x,t)}{\partial x^2} - \varepsilon(T - T_r)\right] \tag{4.5}$$

where ε and T_r depend on the structure , dimension, and thermal properties of the actuator. In general, ε is given by:

$$\varepsilon = \frac{k F}{k_p h(t_n + t_0 + \ell)} - \frac{J^2 \rho_0 \xi_p}{k_p} \tag{4.6}$$

$$T_r = T_\infty + \frac{J^2 \rho_0}{k_p \varepsilon} \tag{4.7}$$

where k is the combined thermal conductivity and F is the excessive heat-flux shape factor. Table 4.1 shows k, c_p, κ, and the coefficient of linear expansion ($\Delta\ell/\ell$) for selected materials.

Table 4.1 Thermal properties of selected materials.

Material	c_p (cal/g °C)	κ (cm²/s)	k (Watt/cm°C)	$\Delta\ell/\ell$ (°C⁻¹)
Al	0.215		2.37	25×10^{-6}
Ni	0.1061		0.899	13×10^{-6}
Mo	0.0597		1.4	5×10^{-6}
Cu	0.0924		3.98	16.6×10^{-6}
Au	0.0308		3.15	14.2×10^{-6}
Pb	0.0305		0.346	29×10^{-6}
Pt	0.0317		0.73	9×10^{-6}
Ag	0.0566		4.27	19×10^{-6}
Zinc	0.0928		1.15	35×10^{-6}
Si	0.168	0.9	1.5	3×10^{-6}
Ge	0.074	0.36	0.6	5.8×10^{-6}
SiC	-	-	5	-
C (diam.)	-	-	20	-
GaN	-	-	1.5	-
GaAs	0.084	0.44	0.46	6.86×10^{-6}
SiO_2	0.239	0.006	0.014	5×10^{-7}
Si_3N_4	-	-	-	4×10^{-6}

Transfer of heat through convection is a drift process where the surrounding air or liquid physically move and carry the thermal energy with it. Two cases of natural convection and forced convection are distinguished from one another. In the forced convection the surrounding fluid is set into a motion externally while in the natural convection the temperature gradient at the surface of the solid gives rise to the fluid flow. The rate equation for one-dimensional convective heat transfer, referred as the Newton rate equation, is given by:

$$P_{cv} = h_{cv} A \, \Delta T \tag{4.8}$$

where P_{cv} is the rate of convective heat transfer (W), A is the area normal to direction of heat flow (m^2), and ΔT is the temperature difference between surface and fluid (K). The coefficient h_{cv} depends on the properties of the solid surface and its geometry, the properties of the adjacent fluid, and ΔT. Derivation of h_{cv} from first principles is quite involved because of the contribution of different factors such as fluid viscosity, thermal conductivities and expansion coefficients, flow rates and other parameters. In air, it is between 5-50 W/m^2 K for natural convection and 25-250 W/m^2 K for forced convection.

Transfer of heat by radiation is through generation of infrared electromagnetic waves that radiate away from the heated body into the surroundings. According to Stefan-Boltzmann law the total energy emitted by a black-body (E_b) is given by:

$$E_b = \sigma T^4 \tag{4.9}$$

where T is the temperature of the black-body in Kelvin and σ is called Stefan-Boltzmann constant and has the value of $5.676\text{x}10^{-8}$ W/m^2 K^4. Solid surfaces, are characterized by their emissivity (ε) that is defined through the following relationship:

$$E_s = \varepsilon E_b \tag{4.10}$$

Clearly the emissivity of a black-body is unity and that of the solid surfaces is usually less than one. The above relationship defines the total emissivity which is related to monochromatic emissivity (ε_λ) and directional emissivity (ε_θ) through the following integrals:

$$\varepsilon = \int_0^\infty \varepsilon_\lambda d\lambda \text{ and } \varepsilon = \int_0^{\pi/2} \varepsilon_\theta \sin 2\theta \, d\theta \tag{4.11}$$

where the angle θ is shown in figure 4.3. As a rule of thumb, the total emissivity is 1.2 time the normal emissivity (ε_n) in polished metals, and it is slightly less than the ε_n in nonmetallic and other surfaces.

In general the emissivity depends on the surface conditions and the emissivity of highly polished surfaces are very low. The emissivity of all metallic surfaces increase with temperature while that of the nonmetallic surfaces decrease as temperature increases. Formation of a thick oxide layer on metals and roughening of their surfaces increase their emissivity appreciably and the emissivity of nonmetallic surfaces are much higher than metallic surfaces. The net amount of heat that is radiatively transferred between a physical body and its surrounding also depends on its absorptivity (α) that we will take it to be equal to its emissivity

(Kirchhoff's law which holds under equilibrium condition). (In cases where the incident radiation is from sources at very high temperatures the absorptivity and emissivity of surfaces can differ greatly.) Table 4.2 shows emissivity of selected materials.

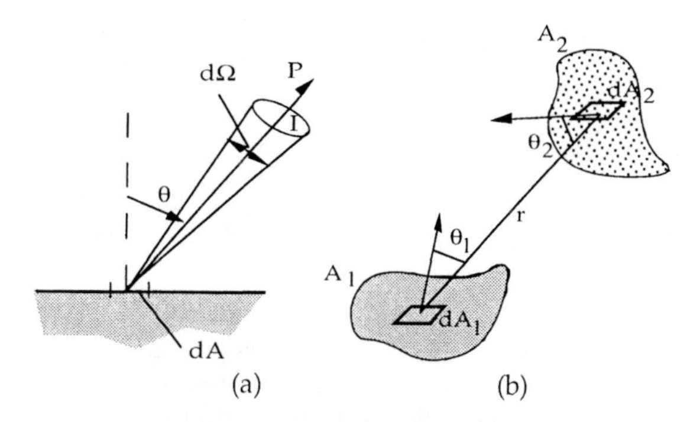

Figure 4.3 (a) The intensity of radiation from an element of area dA. (b) Radiant heat transfer between two surfaces A_1 and A_2 [1].

Table 4.2 Emissivity of selected materials [1].

Surface	T, °C	Emissivity
Polished Cr	37-1090	0.36-0.08
Polished Cu	100	0.052
Polished Au	227-627	0.018-0.035
Chromnickel	52-1030	0.64-0.76
Polished Pt	227-627	0.054-0.104
Polished Ag	37-371	0.022-0.031
W (Filament)	3320	0.39
Polished W	100	0.032-0.35
Lampblack C	98-227	0.95-0.96
SiO_2	22-500	0.85-0.95

The net heat transfer ($dP_{r1\text{-}2}$- $dP_{r2\text{-}1}$) between surface dA_1 to surface dA_2, shown in figure 4.3, is:

$$P_{r1\leftrightarrow 2} = A_1(\varepsilon_1 E_{b1} - \varepsilon_2 E_{b2})\, F_{12} \tag{4.12}$$

$$F_{12} = [\int_{A_1}\int_{A_2} \cos(\theta_1)\cos(\theta_2) dA_1 dA_2 / \pi r^2] \,/\, A_1 \tag{4.13}$$

It can be shown that a reciprocity relationship exist between F_{12} and F_{21}: $A_1F_{12}=A_2F_{21}$. The view factor takes into account the shape of the emitting surfaces and for selected geometries they are tabulated in the literature.

In engineering applications, the radiative heat transfer is usually accompanied by the convective component and the heat transfer coefficient

(h_t) is written as:

$$h_t = h_{cv} + h_r \tag{4.14}$$

where h_r is given by:

$$h_r = 4\sigma T_1^3 F_{12} \tag{4.15}$$

It should be noted that in this approximation the total heat transfer rate (P_t) is $\approx h_t \Delta T$. The power converted to useful work is the desirable output of the actuator.

$$P_w = dW/dt = P_{in} - P_{ht} - P_t \tag{4.16}$$

To estimate the speed of the actuation two things have to be considered: how fast the resistor can be heated to a given temperature, and how fast the heat can be removed from the device upon cooling. In bimorph configuration, the heater is usually a polysilicon or metallic strip and the heating is accomplished by passing a current through it. In these structures, the heating can be assumed to occur very fast (at the rate of energy transfer between the electron gas and the lattice vibration). If other power sources, such as optical signals, are used for heating, the heating is accomplished through the interaction of the photons with the lattice in the long-wavelength part of the spectrum and free carrier absorption and subsequent transfer of energy between carriers and the lattice vibration. Thus, electric and optical heating can have different speed and we consider these two cases separately.

In other cases, the heater heats gases or liquids that upon expansion actuate a diaphragm and drive a pump. These cases are discussed in the following sections.

i) Electric Control Signals: In actuators where the control signal is electrical, a heater is used to convert the electrical power to heat. P_{in} is the electric power input ($=V^2/R$), P_{out} is the thermal power flowing out of the heater into the surrounding, and P_{heater} is the power that is left behind in the heater and it increases its temperature. The thermal energy (ΔQ) stored in a heating element having mass "m", specific heat at constant pressure "c_p", and temperature T is $\Delta Q = mc_p\Delta T$, where ΔQ is in calories (1 calorie = 4.186 Joules), and ΔT is in Kelvin. According to equation (4.1), we have:

$$P_{heater} = P_{in} - P_{out} = d(4.186\ \Delta Q)/dt = 4.186\ m\ c_p\ d\Delta T/dt \tag{4.17}$$

where "t" is time (s). Thus, when P_{out} is zero the temperature of the heater increases as a function of time with the slope of $P_{in}/(4.186\ \mathrm{m}\ \mathrm{c_p})$ for a constant P_{in}. Three processes contribute to P_{out}: convective, conductive, and radiative heat transfers. In vacuum, the radiative heat transfer may dominate, while in solids usually conduction is much larger. In gases and liquids both convective and conductive processes are important.

ii) Optical Control Signals: In these actuators, the optical power is absorbed by an absorber that subsequently heats up an adjacent region as shown in figure 4.4.

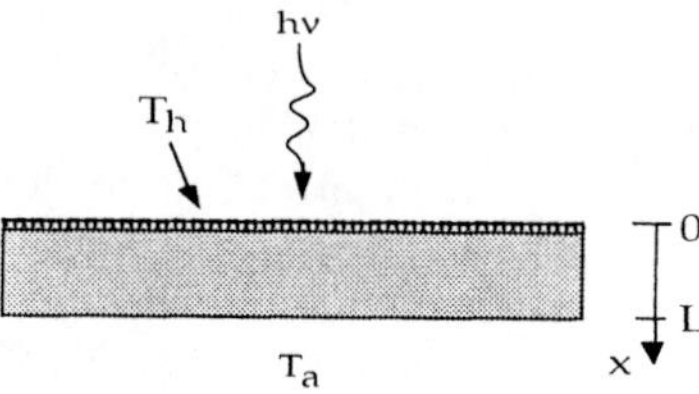

Figure 4.4 Optical absorber.

Inside the optical absorber, the light intensity falls down exponentially as $I_0e^{-\alpha x}$ and the generated heat is proportional to $I(x)$. To calculate the speed of response we consider a simple geometry shown in the following figure.

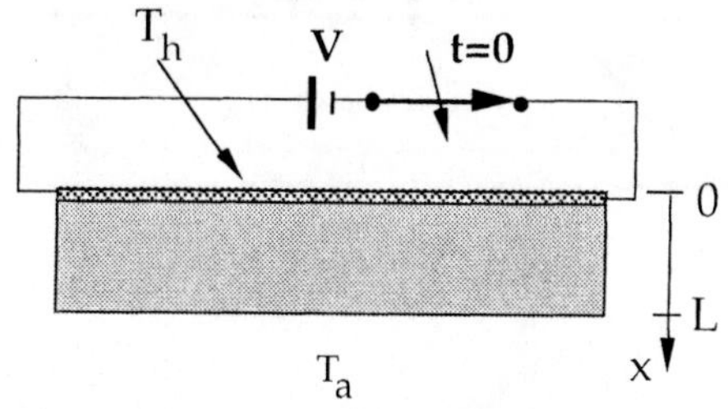

Figure 4.5 A simple heater geometry to calculate the time response of the thermal actuator.

At time t=0, the power is turned on and T(x,t) is determined in the substrate. At time $t=t_0$ the power supply is turned off and T(x,t) in the substrate is again determined. The initial condition is:

$$T(x,0)= T_0 \quad \text{for} \quad t<0$$

$$T(x,t) = T_0 + \sum_{m=1}^{\infty} H_m \sin \frac{m\pi x}{L} \exp\left(-\frac{m^2\pi^2}{L^2}\kappa t\right) \qquad \text{for } t>0 \qquad (4.18)$$

The boundary condition at x=L is approximately insulating and we use $\partial T/\partial x=0$ at x=L. If it was thermally conducting, we use $k_1\partial T/\partial x=k_2\partial T/\partial x$ at x=L. The above equations are sufficient to solve most of the heat conduction problems in thermal microactuators.

4.2 Gas Expansion

When simple gases are heated they expand according to the well known ideal gas law: $PV = nRT$, where P is the gas pressure, V its volume, T its absolute temperature, R is the gas constant (0.0821 liter atm/mole K) and n is the number of moles. Assuming that the silicon (p^+ boron-doped) square diaphragm is 5 μm thick, has 2 mm x 2 mm area, and the height of the cavity is 100 μm, the cavity volume is approximately 0.4 mm^3. At 1 atm and at room temperature, the number of moles of air inside this cavity is simply $n \approx 1.6 \times 10^{-8}$. Up on heating to 400 K, the air in this cavity would apply a pressure of 1.3 atm to the diaphragm. This pressure inside the cavity will be counteracted by the diaphragm which experiences a difference of 0.3 atm of pressure. To find the displacement of the diaphragm the equation (1.30) in Chapter 1, can be used. A more accurate finite-element analysis using ABAQUS yields 20 μm displacement at the center of the p^+-silicon diaphragm. The Young's modulus of 1.3×10^5 atm and Poisson's ratio of 0.3 were used as the properties of heavily boron doped (p^+) silicon diaphragms in the simulation. There is always built-in stress in p^+ silicon diaphragm. The distribution of the stress is non-uniform due to process and the stress level is also process dependent. The maximum tensile stress on the surface of the p^+ silicon is around 6.8×10^3 atm.

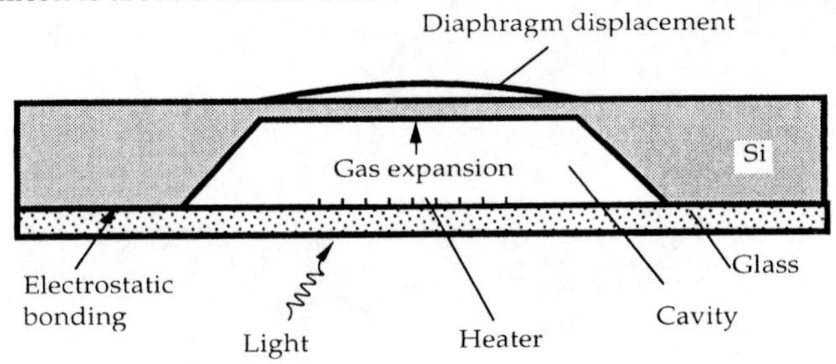

Figure 4.6 Schematic cross-section of an actuator using gas expansion to deflect a diaphragm. Various methods can be used to heat the gas including: an electric heater (resistor) and light absorption.

4.2.1 Thermopneumatic Micropump

There is a great interest in the pharmaceutical applications to develop a liquid pump that is capable of delivering microlitter per minute of fluids. Pumps with such characteristics are not easy to construct. Operation of most micropumps are hampered by the viscosity of the liquid (which is temperature dependent) and the pressure on their input-output ports that result in unacceptable variations in their flow rates.

An example of such a micropump that is fabricated using micromachining technology is shown in figure 4.10. A gas filled chamber separates the heater from the flow chamber. When the heater is energized, its surrounding gas is heated. The expanding gas displaces a diaphragm that exerts a pressure to the fluid in the flow chamber. The pressure on the fluid causes the valve #1 to close and valve #2 to open. Upon cooling, the chamber retracts closing valve #2 and opening valve #1 drawing the liquid into the pump. This action pumps the fluid through the input/output ports of the device [3].

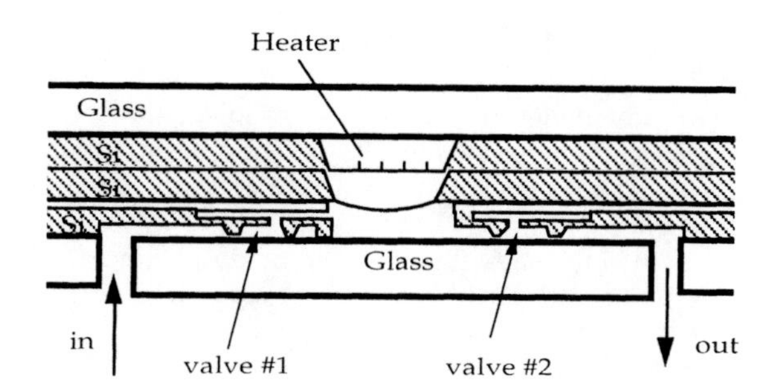

Figure 4.7 Schematic cross-section of a thermopneumatic pump. The heater is used to heat the air trapped in the chamber located above the pump. Upon heating, the chamber expands opening the valve #2 and closing the valve #1. Hence the fluid that was trapped in the pump is driven out. Upon cooling, the chamber retracts closing valve #2 and opening valve #1 drawing the liquid into the pump [3].

Figure 4.8 shows the flow rate as a function of the excitation frequency of the heater. The excitation voltage was 6 volts at 100 mA. Peak flow rate of 34 μℓ/min was obtained at 1 Hz. The built-up pressure inside the fluid camber was 0.05 atm.

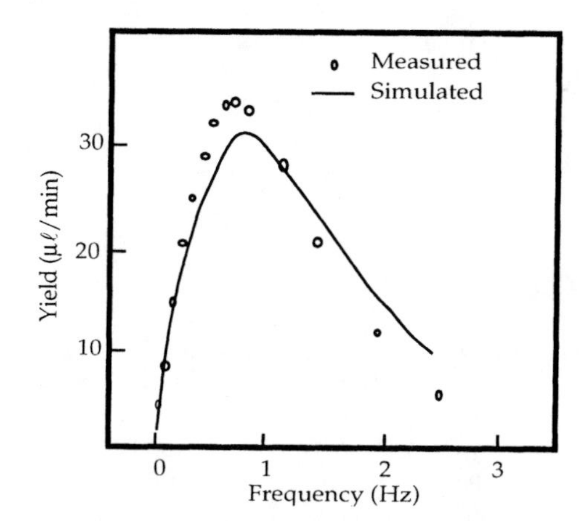

Figure 4.8 Flow rate (yield) as a function of excitation frequency of the pump shown in figure 4.7. The back pressure was zero and the excitation voltage was 6 V [3].

The above pump is characterized by the thermal response time of its gas filled chamber ($\tau_t \approx 0.1$ s), the pneumatic relaxation time constant of gas pressure inside that chamber ($\tau_p \approx 5$ s), and hydraulic relaxation time constant which is determined by the time it takes for the fluid pressure to increase or decrease inside the flow channel due to the expansion of the diaphragm.

4.2.2 Thermal Inkjet Printhead

Figure 4.9 shows schematic cross-section of an inkjet that was fabricated using bulk micromachining technology. It uses short voltage pulses (<6μm, 6 mW) to produce a vapor bubble that creates a pressure wave generating a single droplet of ink. This thermal inkjet is called "backshooter" since it is based on the growth of the vapor bubble and ink droplet injection in opposite directions. Edgeshooter printhead is also designed and fabricated [4].

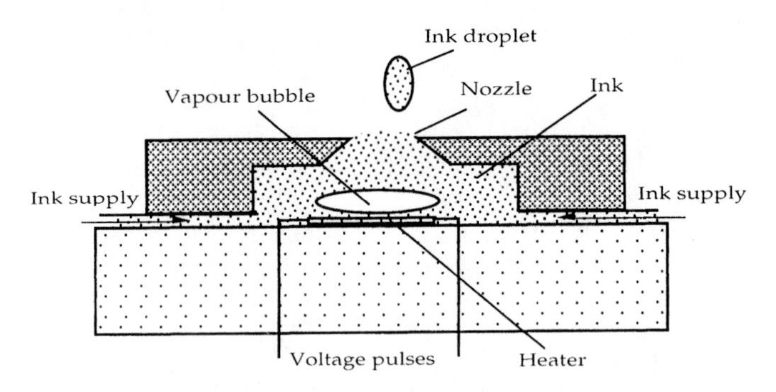

Figure 4.9 Schematic of a "backshooter" inkjet printhead [4].

Depending on the design parameters, reproducible droplet velocities between 10 and 15 m/s and a droplet mass of 110 ng with 300 dpi and of 60 ng with 600 dpi were obtained. The operation frequency of 5 kHz (corresponding to a refill time of 200 μs) was achieved that can be improved further by optimization of the ink supply throttle [4]. These printheads were integrated with power electronics and control circuitry to reduce cost and input/output wires.

4.2.3 Stirling Microengines

Figure 4.10 schematically shows a diagram of a Stirling engine that has efficiencies close to the Carnot engine [5]. It consists of the following essential parts: a displacer, a piston, cold/hot sources, crank shaft and a flywheel. It operates by using the displacer to allow the gas in the piston cylinder to be either heated or cooled. The motions of displacer and the piston are 180 out of phase with respect to each other. In the cycle shown in figure 4.10, the displacer is up and it allows the hot air to enter the piston cylinder. As the gases expand the push the piston up. The crank shaft translates the piston's up motion and pushes down the displacer. Once the displacer is pushed down, the gases in the piston cylinder are exposed to the cooler. As the volume of the gas decreases up on cooling, the piston is pulled down and the cycle repeats itself. For the engine to work, the displacer should have only two resting positions at the top and the bottom of the cylinder. This snap action can be produced by a variety of methods using electrostatic, magnetic and bistable solid links.

The Stirling engine does not require combustion for its operation and can be powered by a variety of methods. A miniature Stirling engine was designed and built that with only a small temperature difference of 100 K (cold wall at 273 K and heater at 373 K) produced 20 mW of rotational power at 10 Hz. The overall size of the engine was 0.11 cm^3.

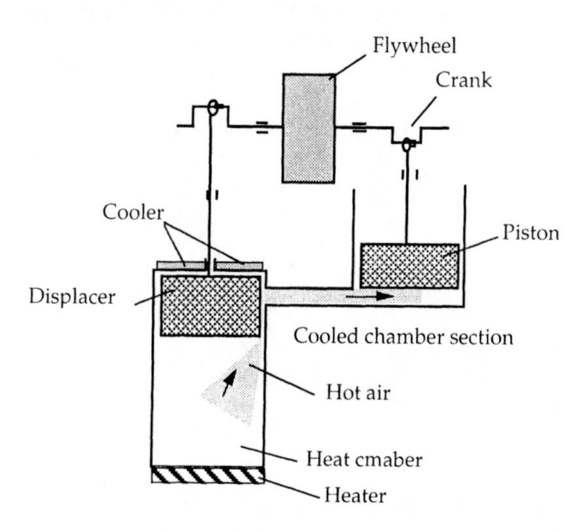

Figure 4.10 Schematic of the Stirling engine [5].

4.2.4 Power MEMS and Microengines

Spontaneous oxidation of fossil fuel and highly reactive gases such as H_2 generates heat and produces other expanding gases that can be used to generate mechanical and electrical power. It is well know that oxidation of one gram of H_2 at 25°C and at 1 atm produces 143 kJ of energy. Hydrocarbons, such as C_8H_{18}, produce approximately 48 kJ per gram. MEMS Microengines are being developed and designed to take advantage of these high energy fuels. The goal is to produce engines with around 1-10 gram of overall weight that can produce electricity of 10-100 W and 0.05-15 N trust and consume only 10 gr/hr of H_2 or similar fuels [6,7].

Large numbers of these modular MEMS engines can be integrated together on appropriate surfaces to produce large powers to propel 100-1000 kg objects. Since these microengines will be batch fabricated, they will be very-inexpensive.

Due to their small size, the MEMS engines can be placed over the boundaries of the propelled object. Thus, these microengines can directly modify the boundary layer enabling invention and development of entirely new classes of self-propelled crafts.

A macro-scale gas turbine with an air intake opening of one meter, for example, generates power on the order of 10^8 W. Scaling down such an engine to a millimeter scale power production on the order of 10 W can be expected provided that the power per unit volume of intake air is maintained.

Figure 4.11 schematically shows a turbine engine that uses the combustion exhaust to power a turbine that in turn drives the compressor and the electric power generating dynamo. It consists of three main parts: i) a compressor, ii) a combustion chamber, and iii) turbine. A thin film electric induction starter-generator is mounted on a plate over the compressor blades. The supersonic radial flow compressor and the turbine are connected by a hollow shaft. The gaseous fuel is mixed by the air at the compressor exit. The air/fuel mixture flows radially outward through the flame holders into the combustion chamber. The combustor discharges radially inward toward the turbine. The exhaust gases exit through the engine nozzle. The engine is a cylinder with approximately 12 mm diameter and 5 mm height. Its rotor speed (2-3 mm in diameter) 2400,000 rpm.

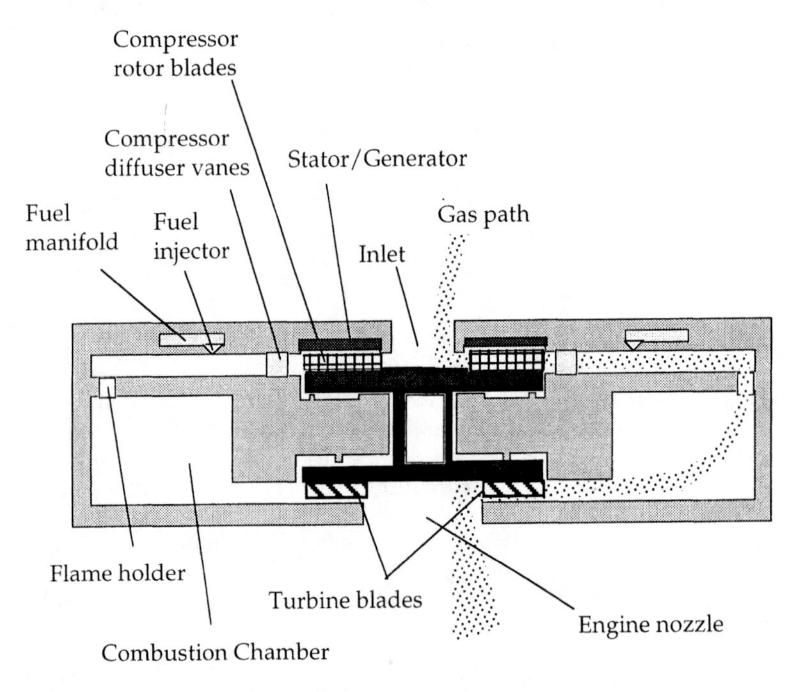

Figure 4.11 Schematic cross-section of a microgas turbine power generator [6,7].

There are many material and design issues that are being addressed. For example, the combustor temperature is expected to reach 1300-1700 K while the compressor may stay at 600 K. Silicon can withstand the compressor temperatures and SiC is needed for the combustor.

4.3 Solid Expansion

In these actuators one takes advantage of un-equal thermal expansion coefficients of two dissimilar material deposited over each other. As the temperature increases, because of the different expansion coefficients of various layers, the structure bends and performs the desired work.

Many different types of bimorph and multilayer cantilever beam thermal actuators have been fabricated and reported [8-10]. The most important advantages of these actuators are: i) They do not need any special fabrication step other than those used in electronic microdevice fabrication. Other actuators such as piezoelectric or magnetic all require special non-standard material deposition and processing. ii) They can generate large forces and if properly designed they can produce large displacements as well. iii) Their structure is quite simple enabling the designer to have many different design options. On the other hand, these actuators can get quite hot and need proper heat sinks. They also consume large amount of power.

The basic structure of these actuators are shown in figure 4.12. Here we assume that different layers can have different uniform widths of b_i, where i refers to the ith layer, and uniform thickness of t_i. z_0 denotes the plane of zero strain, r is the bending radius, L is the beam length and z is the distance to the bottom of the cantilever beam.

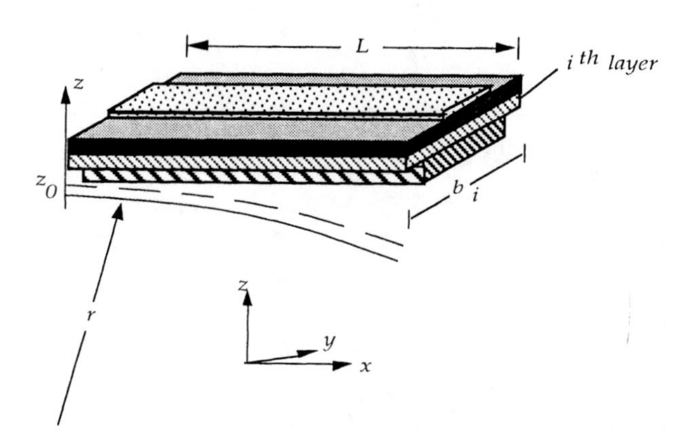

Figure 4.12 Schematic structure of thermal actuator based on multilayer cantilever beams. One of the layers form a resistor that upon excitation is heated and due to the difference in the thermal expansion coefficients of different layers, the beam bends.

To calculate the amount of bending we note that a temperature difference of ΔT induces a thermal stress σ_t given by:

$$\sigma_t = Y(z)\,\alpha(z)\,\Delta T \tag{4.19}$$

where Y is the Young's modulus and α is the thermal expansion coefficient. About a plane of zero-strain at $z=z_0$, the strain of the film can be split into a mean strain ε_0 and a bending strain that depends on z:

$$\varepsilon(z) = \varepsilon_0 + \frac{z - z_0}{r} \tag{4.20}$$

where the mean strain is:

$$\varepsilon_0 = \frac{<\sigma_t(z)>}{<Y>} = \frac{<Y(z)\,\alpha(z)>}{<Y>}\Delta T \tag{4.21}$$

where average of any quantity A is denoted by $<A>$ given by:

$$<A(z)> = \frac{\int_0^t \int_0^b A(z)\; dydz}{\int_0^t \int_0^b dydz} \tag{4.22}$$

where t is the thickness of the film and b is its width. To find z_0, we use the equilibrium conditions for the total forces: $<\sigma> = 0$, and for the bending moment: $<\sigma z>=0$, where the total stress in the film is: $\sigma(z) = Y\,\varepsilon(z) - \sigma_t(z)$. Combining all this together, we find an expression for the bending radius r:

$$r = \frac{<Y><Yz^2> - <Yz>^2}{<Y><Y\alpha z> - <Y\alpha><Yz>} \cdot \frac{1}{\Delta T} \tag{4.23}$$

where we have also used $z_0=<Yz>/<Y>$.

In the case of bimorph with uniform layers of thickness t_1 and t_2 and uniform widths of w_1 and w_2, it can be shown that r is given by:

$$r = \frac{1}{6}\,\frac{4(t_1+t_2)^2 - 2t_1t_2 + \dfrac{Y_1b_1t_1^3}{Y_2b_2t_2} + \dfrac{Y_2b_2t_2^3}{Y_1b_1t_1}}{(\alpha_1-\alpha_2)\,\Delta T(t_1+t_2)} \tag{4.24}$$

In the case of trimorph uniform layers of thicknesses t_i and uniform widths of w_i, with $i=1,2$ and 3, it can be shown that r is given by:

$$r = \left[-\frac{1}{2}\left(\sum_{i=1}^{3} Y_it_i^2\right)^2 + \frac{2}{3}\sum_{i=1}^{3}\sum_{k=1}^{3} Y_iY_kt_i^3t_k + 2\sum_{i=1}^{3}\sum_{k=i+1}^{3} Y_iY_kt_i^2t_k^2 + 2Y_1Y_3t_1t_2t_3t_{tot}\right] \times \left[\sum_{i=1}^{3}\sum_{k=1}^{3}(\alpha_k-\alpha_i)Y_iY_kt_it_k(t_i+t_k) + 2\,(\alpha_3-\alpha_1)Y_1Y_3t_1t_2t_3\right]^{-1}\left(\frac{1}{\Delta T}\right) \tag{4.25}$$

The thermal expansion coefficients for different materials are given in table 4.1. The Young's modulus of selected materials are given in table 1.2 in Chapter 1.

It can be shown that the most efficient method of producing displacements with cantilever beam bimorph actuators is to place the heater directly over the region where the beam joins the substrate. This is the high-stress area and bimorph expansion will result the largest possible deflections. Heating of the beam near its tip does not produce as much displacement and reduces the power efficiency of the actuator. Since, the thermal conduction is largest through the substrate, in poorly designed actuators the temperature is lowest in the clamping region of the cantilever beam compared to its tip. By reducing the size of the heater and confining it to the high-stress region, the design can be improved considerably.

Figure 4.13 shows a schematic of a microfabricated optical scanner . The mirror was attached to the device frame using two arms that acted as a bimorph (actually multimorph) thermal actuator that depending on the supplied power they tilted the mirror by different angles.

The bending beams were 40 μm long, 3.25 μm thick (total) and 6 μm wide. The polysilicon heating element is 1.2 μm wide and 0.3 μm thick and it is sandwitched between two SiO_2 layers (0.6 μm and 0.75 μm) that are capped by 0.6 μm aluminum (this aluminum layer is for mechanical purposes). A second aluminum layer (1μm thick) is used to contact the polysilicon heater. The top aluminum layer on the central plate acts as a highly reflecting mirror and it is capped by a protective Si_3N_4 layer (1 μm thick).

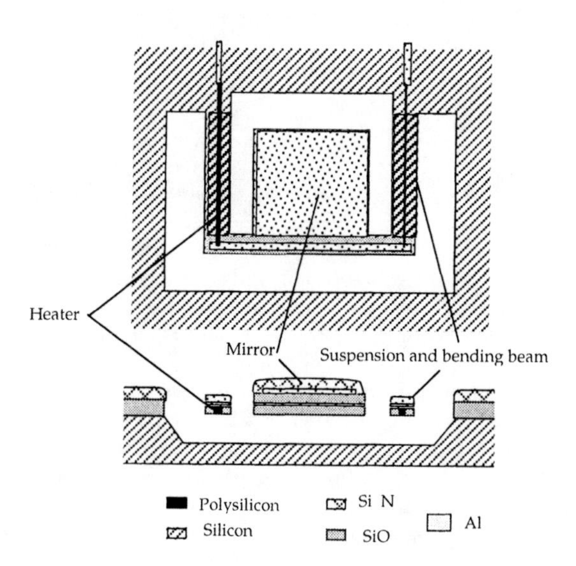

Figure 4.13 Schematic top view and cross-sectional views of the thermally-actuated micromirror.

The above micromirror deflected nearly linearly as a function of the heating power and at 4.6 mW its deflection was 4.6 °. Its response time was fast around 2 ms.

Another thermally-activated cantilever beam is reported in [8]. In this design, a 1.5 mm long silicon cantilever beam was used and deflections as large as 200 μm at 80 mW and 150 μm at 160 mW were observed for beam thicknesses of 6 and 10 μm respectively. The beam width was 250 μm. The silicon cantilever beams were covered with a SiO_2 insulation layer (0.6 μm thick) and had 6 μm thick aluminum layer. The heater was a diffused boron resistor with a typical value of 100 Ω. Operation voltage was 3-5 V at 20-30 mA. Strokes as high as 350 μm were obtained under pulsed power conditions.

Other thermal actuator structures that employ bimorphs are described in [11] where levers are used to amplify displacement. In some their structure 3 dB bandwidth as high as 1 kHz were obtained. They also discuss bimorph thermal actuators with both in-plane and out-of-plane displacements.

4.4 Phase Transformation Microengines and Actuators

Near the phase transformation temperature, some solids undergo a discontinuous change in their volume. This change can be much larger than the linear volume changes caused by the simple thermal expansion of the solid. TiNi alloys and some gels exhibit interesting phase transformation effects and their phase transformation temperature can be adjusted over a wide temperature range making them ideal materials for this types of actuators. Specially in TiNi shape memory alloys there is an interesting shape memory effect that can be used to produce very large forces and displacements.

4.4.1 Shape Memory Alloy

Shape memory alloys (SMAs), such as TiNi, exhibit thermoelastic martensitic transformation leading to interesting shape recovery characteristics upon heating. The shape recovery occurs at a transformation temperature called austenite temperature (T_a). These alloys are usually formed at high temperatures with a highly ordered crystalline structure called austenite phase. This phase changes to the martensite phase upon cooling the alloy below a composition dependent transition temperature.

In the martensite phase the crystal structure is usually plagued with twin planes that facilitate deformation of the alloy by propagating through the structure. When the alloy is re-heated above T_a, the alloy regains its highly ordered austenite crystalline structure. As a result, the SMA returns to its pre-deformation shape that it was originally formed. Thus, the original formation is of great importance in SMA actuator and it is called "training" of the SMA.

A simple experiment, schematically shown in figure 4.12, can be used to demonstrate the application of shape memory alloys in actuators. The spring in figure 4.14 has a Hooke's constant of 2.5 N/cm and it is slightly stretched at room temperature under the load of spring. Upon application of 5 volts (≈300 mA) to the trained TiNi wire (SMA), the wire heats up and contracts by $\Delta\ell/\ell \approx 4\%$ and attains its austenite phase.

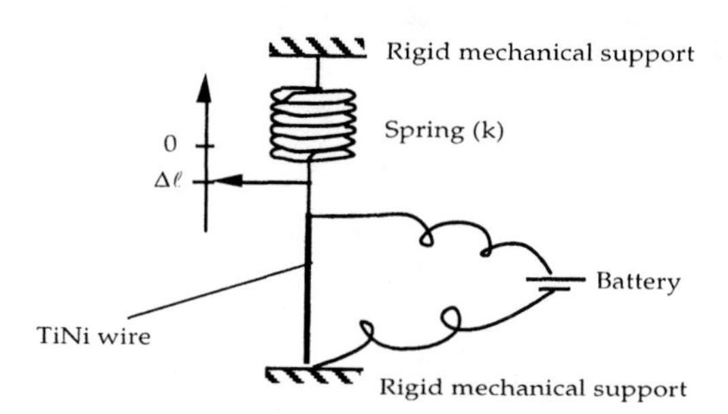

Figure 4.14 Schematic of a simple experiment to demonstrate the application of shape memory alloys in actuators.

Taking the length of the TiNi wire in figure 4.14 to be 2.5 cm, it contracts by 1 mm in its austenite phase. The spring applies 0.25 N force to the contracted wire. Upon cooling, the spring pulls back the wire to its initial position (martensite phase) and restores the equilibrium at room temperature. When the SMA contracts it exerts a force of approximately 2.5 N. Thus, the net work performed by the SMA against the spring is approximately (2.5-0.25)0.01=0.0225 Watts. The input power (VI) is 1.5 Watts and the efficiency (P_{out}/P_{in}) of this simple actuator is 0.015.

Figure 4.15 shows the dependence of the contraction of a TiNi wire on the applied electric field. In this experiment, a load of 1.8 N was applied that restored the length of wire to near its original value up on cooling. The hysteresis effect is quite interesting and present in all TiNi actuators. In this case, the hysterisis was quite large spanning over approximately 5 V/m.

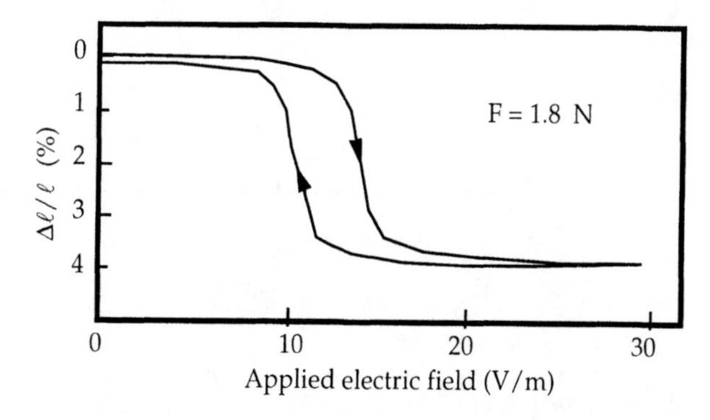

Figure 4.15 Contraction of a TiNi wire as a function of electric field at a load of 1.8 N.

Similar hysterisis effects are also found by monitoring the TiNi wire contraction as a function of temperature (figure 4.16). The hysteresis, in this case, occurred over nearly $\Delta T_h \approx 25$ °C while the transition occurred over a temperature range of nearly $\Delta T \approx 35$ °C.

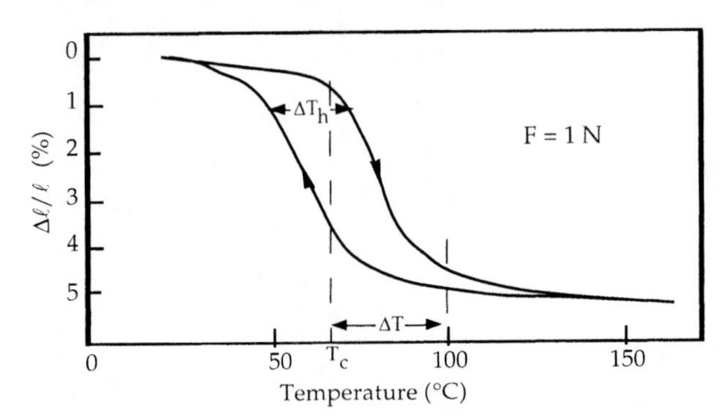

Figure 4.16 Contraction of a TiNi wire as a function of temperature a load of 1 N.

The resistance of the TiNi wire also shows interesting temperature dependence as shown in figure 4.17. When the wire is heated, it regains its highly ordered austenite phase. Higher order results in lower scattering rates reducing the resistivity. However as the temperature is further increased, the scattering rate increases due to phonons. At lower temperatures, the resistivity is low and it increase rapidly as the phase transformation temperature is approached.

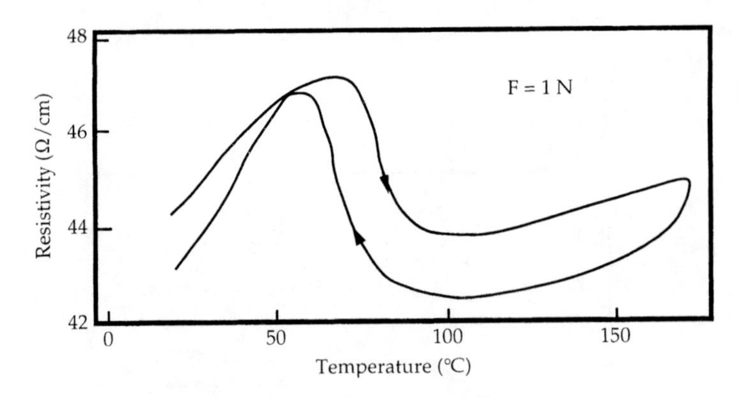

Figure 4.17 Resistance of a TiNi wire as a function of temperature a load of 1 N.

All the above parameters' behaviors depended on the rate of the application of the voltage or on the rate of heating/cooling. The variations are less pronounced at lower excitation rates.

TiNi thin films, more suitable for MEMS applications, are also studied and produced. According to [19], shape memory alloy actuators have a larger energy density of $6.0x10^6$ J/m^3 . As shown in table 4.4, the energy density of TiNi actuators is higher than most other microactuation methods.

Table 4.3 Energy density of various actuators [19,20].

Actuator Type	Energy density (J/m^3)
TiNi	$6.0x10^6$
Solid-liquid phase change	$4.7x10^6$
Bimetallic (bimorph)	$5x10^5$
Thermopneumatic	$1.2x10^6$
Thermal expansion	$4.6x10^5$
Electromagnetic (Experimental)	$4.0x10^5$ ($1.6x10^3$-$2.8x10^4$)
Electrostatic (Experimental)	$1.8x10^5$ ($1.6x10^3$-$2.8x10^4$)
Piezoelectric (Experimental)	$1.2x10^5$ ($1.8x10^2$)
Muscle	$1.8x10^4$
Microbubble	$3.4x10^2$

TiNi films with good characteristics were deposited using sputtering at ultra-high vacuum environment. Films were deposited over Si, SiO_2 or Si_3N_4 covered silicon substrates. Due to the difference in the sputtering rates of Ti and Ni, it was necessary to add Ti foil to the target surface to increase Ti content. Film crystallization was best achieved by heating the film between 515 and 525 °C in the UHV chamber immediately after the deposition without breaking the vacuum. It was also found that it was necessary to deposit a thin TiNi film over SiO_2 and anneal it at 700 °C before depositing the bulk of the film to achieve good adherence with SiO_2. Micromachining of the silicon was also done before the TiNi deposition to reduce the deterioration of the shape memory alloy film during processing.

The substrate force or some other means are used to "train" the TiNi films in MEMS structures. Interesting microrobots and microgrippers were designed and fabricated. Reference [26] covers applications of TiNi shape memory alloys in medicine and other fields. Interesting catheters and bone anchors are designed , fabricated and successfully tested.

Figure 4.18 shows the cross-section of micromachined silicon microgripper. $Ti_{50}Ni_{42}Cu_8$ shape memory alloy was used in this gripper and the sides of the gripper were attached together using selective eutectic (Au-Si) bonding. The interesting property of $Ti_{50}Ni_{42}Cu_8$ is that its transition temperature is just above body temperature (≥ 37 °C) making it suitable for medical implants. Its hysteresis is also narrower than the binary TiNi alloy as shown in figure 4.19.

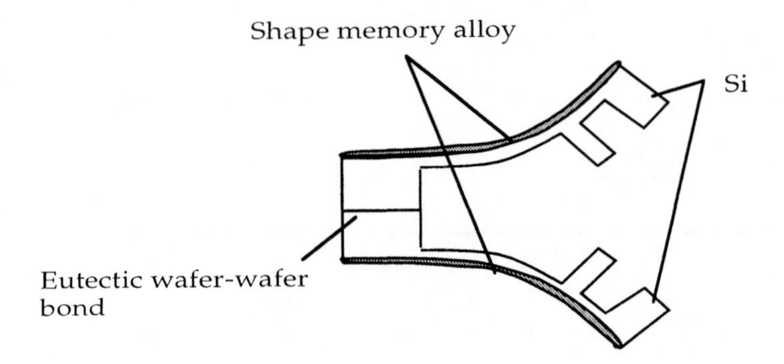

Figure 4.18 Cross-section of micromachined microgripper actuated by TiNiCu shape memory alloy.

Figure 4.19 shows the tensile stress in the TiNiCu thin film deposited over silicon. Actuation occurs by the recovery of tensile residual stress in the shape memory film. The film, which was deposited near 500 °C, developed a tensile thermal stress as it is cooled after the deposition. When cooled below 37 °C, the martensitic transformation starts and the thermal stress in the film relaxes by twin-related deformation. When re-heated above 37 °C, the film recovers this thermal stress. The silicon gripper acts as bias spring in this application and the gripper opens whenever heated above 37 °C and it closes because of the spring loading below 37 °C [24].

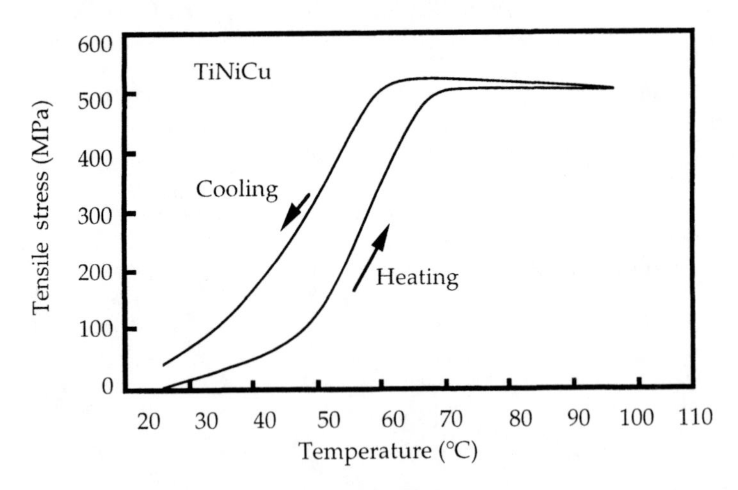

Figure 4.19 Tensile stress versus temperature plot of TiNiCu shape memory alloy thin film on Si used in the microgripper shown above.

4.4.2 Other Phase Transformation Actuators

As discussed in Chapter 7, the structural behavior of gels is intermediate between that of a solid and a liquid. Mostly fluidic, they are given form by a tangled network of polymer strands. A balance of forces maintain this state of affairs that can be disturbed by minute perturbations brought about by a variety of influences. Optical, thermal, electrical, and chemical methods have been shown to cause drastic reduction in the volume of certain gels.

The drastic changes that occur in the state of the gel, is understood to be caused by its phase transformation. When the temperature is lowered, the polymer network loses its elasticity and becomes compressible. Below a critical temperature, the elasticity becomes zero and compressibility becomes infinite resulting in the collapse of the gel. The volume of certain gels shrink or swell by a factor of as much as several hundred when the temperature is varied. Around the critical temperature, these changes can become discontinuous resulting in a very large volume change for an infinitesimal change in the temperature.

Figure 4.20 shows the temperature dependence of the volume of a gel (polyacrylamide) immersed in a water/acetone solution. As the temperature was deceased to below 22°C the gel collapsed and swell again as the temperature was raised above this critical temperature.

Although not yet reported in the literature, it is conceivable to incorporate the above polyacrylamide gel in a cavity like the one shown in figure 4.6 to generate large diaphragm displacements for a small change in the temperature ($\Delta T \approx 1°C$). According to preliminary estimation, such an actuator will have an efficiency surpassing most actuators discussed in this book. The only problem that such an actuator may have is the hysteresis associated with volume recovery as discussed in chapter 7.

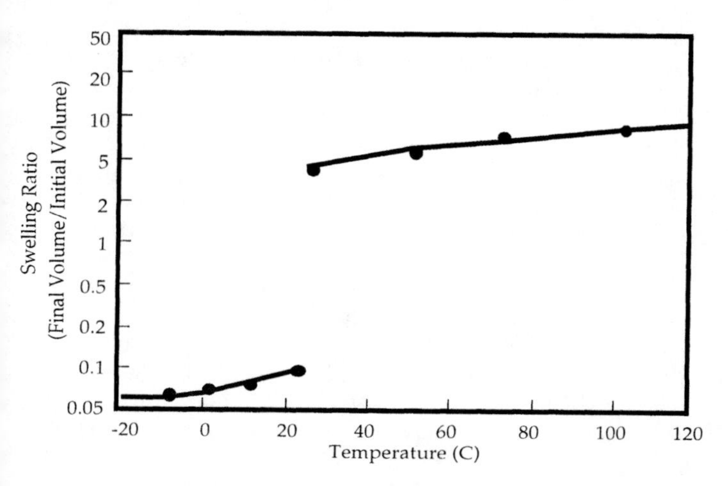

Figure 4.20 The swelling ratio of a hydrolyzed gel as a function of temperature immersed in a water (58%), acetone (42%) solution. At the transition temperature (22°C) the gel underwent a discontinuous but reversible change in its volume [27].

4.5 References

1. J. R. Welty, C. E. Wicks and R. E. Wilson, Fundamentals of Momentum, Heat, and Mass Transfer. 3rd Edition, John Wiley, NY, pp. 402-465 (1984).
2. J. N. Zemel, "Future Directions for Thermal Information Sensors." Sensors and Actuators A 56, pp. 57-62 (1996).
3. F. C. M. Van De Pol, H.T. G. Van Lintel, M. Elwenspoek and J. H. J. Fluitman, "A Thermopneumatic Micropump Based on Micro-Engineering Techniques." Sensors and Actuators A21-A23, pp. 198-202 (1990).
4. P. Krause, E. Obermeier and W. Wehl, "A Micromachined Single-Chip Inkjet Printhead." Sensors and Actuators A 53, pp. 405-409 (1996).
5. N. Nakajima, K. Ogawa and I. Fujimasa, "Study on Microengines: Miniaturizing Stirling Engines for Actuators." Sensors and Actuators, 20, pp. 75-82 (1989).
6. A. H. Epstein, et. al., "Micro-Heat Engines, Gas Turbines, and Rocket Engines - The MIT Microengine Project." 28th American Institute of Aeronautics and Astronautics Conference Proceedings, June 29- July 2 held in Snowmass Village, CO, pp. 1-12 (1997).
7. A. H. Epstein, et. al., "Power MEMS and Microengines." Transducers '97, pp. 753-756 (1997).
8. G. Greitmann and R. A. Buser, "Tactile Microgripper for Automated Handling of Microparts." Sensors and Actuators A, Vol. 53, pp. 410-415 (1996).
9. J. Buhler, J. Funk, O. Paul, F.-P. Steiner and H. Baltes, "Thermally Actuated CMOS Micromirrors." Sensors and Actuators A, Vol. 46-47, pp. 572-575 (1995).
10. H. -P Trah, H. Baumann, C. Doring, H. Goebel, T. Grauer and M. Mettner, "Micromachined Valve with Hydraulically Actuated Membrane Subsequent to a Thermoelectrically Controlled Bimorph Cantilever." Sensors and Actuators A, Vol. 39, pp. 169-176 (1993).
11. J. M. Noworolski, E. H. Klaassen, J. R. Logan, K. E. Petersen and N. I. Maluf, "Process for In-Plane and Out-of-Plane Single-Crystal-Silicon Thermal Microactuators." Sensors and Actuators A, Vol. 55, pp. 65-69 (1996).
12. M. Beragamasco, F. Salsedo and P. Dario, " Shape Memory Alloy Micromotors for Direct-Drive Actuation of Dexterous Artificial Hands." Sensors and Actuators, Vol. 17, pp. 115-119 (1989).
13. K. Kuribayashi, "Millimeter-sized Joint Actuator Using a Shape Memory Alloy." Sensors and Actuators, Vol. 20, pp. 57-64 (1989).
14. J. A. Walker, K. J. Gabriel and M. Mehregany, "Thin-Film Processing of TiNi Shape Memory Alloy." Sensors and Actuators, A21-A23, pp. 243-246 (1990).
15. R. Jebens, W. Trimmer and J. Walker, "Microactuators for Aligning Optical Fibers." Sensors and Actuators, Vol. 20, pp. 65-73 (1989).
16. E. Quandt, et. al., "Sputter Deposition of TiNi, TiNiPd and TiPd Films Displaying the Two-Way Shape-Memory Effect." Sensors and Actuators A 53, pp. 434-439 (1996).
17. M. Beragamasco, P. Dario and F. Salsedo, "Shape Memory Alloy Microactuators." Sensors and Actuators, A21-A23, pp. 253-257 (1990).

18. P. A. Neukomm, et. al., "Characteristics of Thin-Wire Shape Memory Actuators." Sensors and Actuators, A21-A23, pp. 247-252 (1990).
19. P. Krulevitch, et. al., "Thin Film Shape Memory Alloy Microactuators." Journal of Microelectromechanical Systems, Vol. 5 (4), pp. 270-281 (1996).
20. R. H. Wolf and A. H. Heuer, "TiNi (Shape Memory) Films on Silicon for MEMS Applications." Journal of Microelectromechanical Systems, Vol. 4 (4), pp. 206-212 (1996).
21. R. Mukherjee, T. F. Christian and R. A. Thiel, "An Actuation System for the Control of Multiple Shape Memory Alloy Actuators. " Sensors and Actuators A 55, pp. 185-192 (1996).
22. S. Miyazaki and A. Ishida, "Shape Memory Characteristics of Sputter-Deposited TiNi Thin Films." Materials Transactions, JIM, Vol. 35 (1), pp. 14-19 (1994).
23. A. P. Jardine, J. S. Maden and P. G. Mercado, "Characteristics of the Deposition and Materials Parameters of Thin-Film TiNi for Microactuators and Smart Materials." Materials Characterization Vol. 32, pp. 169-178 (1994).
24. A. P. Lee et. al., "A Practical Microgripper by Fine Alignment, Eutectic Bonding and SMA Actuation." Sensors and Actuators A 54, pp. 755-759 (1996).
25. L. Hou et. al., "Structure and Thermal Stability in Titanium-Nickel Thin Films Sputtered at Elevated Temperature on Inorganic and Polymeric Substrate." Mat. Res. Symp. Proc. Vol. 360, pp. 369-374 (1995).
26. T. W. Duerig, "Present and Future Applications of Shape Memory and Superelastic Materials." Mat. Res. Symp. Proc. Vol. 360, pp. 497-506 (1995).
27. T. Tanaka, "Gels." Scientific American, Vol. 244 (1), pp. 124-138 (1981).

Optical Microactuators

5.1 Introduction

Silicon micromachining and optical signal processing are promising technologies that have not yet been joined productively. All-optical systems utilize fiber optics (or other optical waveguides) to carry power as well as the control signals and information to their various parts. Optical signals transmitted through waveguides cannot be easily affected by external electromagnetic interferences and they are also extremely difficult to listen to. More importantly, different wavelengths of light can be used to transfer information in parallel through the same fiber optics. These two important characteristics of light, called electromagnetic interference immunity and inherent parallelism, are mainly responsible for the popularity of optical systems. Realization of these systems has its own numerous technological problems, though. An all-optical system requires optical microprocessors, all-optical sensors, and actuators. The optical actuators are probably the least developed of all these different parts. Both optical processors and sensors are readily available [1-17] (although they may not be cost effective in most cases). Here we discuss optical microactuators.

Light can be either directly or indirectly transformed to a mechanical deformation. Examples of a direct optical microactuator, discussed in section 5.2, are given in [18-34] where, for example, light is used to generate photo-electrons that, by shielding the electrostatic force, actuate a silicon microcantilever beam, among other methods. In section 5.3 we discuss an interesting method of using standing waves generated by contra-directed lasers to trap atoms [35,36]. This is an example of direct manipulation of matter by light and, although there has not been a report of an optical trap fabricated using MEMS technology, it is the author's belief that implementing these traps with MEMS will open interesting research areas in need of atomic-scale micropumps.

Indirect optical actuation methods utilize light in heating solids or gases that, upon expansion, cause actuation. Indirect optical methods are usually less involved and generate more actuation power than direct methods. A variety of photo-voltaic, photo-acoustic, and photo-thermal Si oscillators, sensors, and actuators have been reported during the past few years [21]. Indirect optical actuators are discussed in sections 5.4 and 5.5.

An interesting feature of the optical actuators, both direct and indirect, is that they can be directly interfaced with fiber optics or other forms of guided waves such as in integrated optics [27-32]. This feature makes optical actuators an ideal candidate for smart structures [37] (see also Chapter 8), where many actuators are multiplexed (in time or frequency) and located over large areas. All-optical systems are briefly discussed in section 5.6. This chapter is concluded in section 5.7 by a discussion of fabrication methods and section 5.8 lists the references.

5.2 Electron-mediated Direct Optical Actuators

Realization of the goal of integrated μ-systems will take place on at least three levels: 1) Fabrication procedures and techniques that allow combining mechanical, electronic, and optical components on a chip; 2) Novel system types, system architectures, and interfaces that allow dissimilar device types to communicate and to function together; and 3) Novel devices, device hybrids, and device controllers that bridge the separate realms of existing mechanical, electronic, and optical components, and that allow conversion of signals from one realm to another. This section is based on the author's past work regarding novel optical microactuators that address the third item in the above list.

Silicon microactuators can be excited directly by an optical signal using the four different methods discussed in this section. The first method employs photo-generated electrons to modify the electrostatic pressure on a cantilever beam that forms a parallel plate capacitor with a ground plane. The other methods use optically-generated electron-hole pair screening in a semi-insulating layer, permittivity modulation of a gas by light, and a solar cell that provides power for the actuators [18-20].

Several practical benefits distinguish direct optical methods from radiative-thermal processes. Direct optical processes can be much faster than indirect. They can take place at much lower power. They permit greater simplicity, versatility, and parallelism of the controller. They can be tailored to an application using the well-known techniques of semiconductor doping and etching.

5.2.1 Photoelectron Microactuator

The structure of the optical microactuator is schematically shown in figure 5.1. A Si cantilever forming the top plate of a capacitor deflects when a potential is applied. A cantilever 600 x 50 x 1 μm^3 with gap 12 μm used bias ≈ 6V and optical power <0.1 $mWcm^{-2}$ to actuate 4 μm in ≈ 0.1 ms. Since this actuation scheme involves a conduction current, a battery or other current source is needed if more than one cycle is to be performed. Continuously charging the capacitor with a current $I \leq I_{max}/2$, where I_{max} is determined by the battery circuit, allows light-controlled actuation in either direction. A continuous photon flux $\Phi < I/\eta$ short-circuits the capacitor more slowly than the battery charges it, causing a charge build-up which closes the plates. A photon flux $\Phi > I/\eta$ causes an opposing photocurrent greater than the charging current. The net charge then decreases, and the plates relax open [18-19].

As shown in figure 5.1, a thin p^+-Si cantilever is mounted on an insulating post overhanging an Au ground plane on an insulating glass substrate to form a parallel-plate capacitor ≈ $\varepsilon_0 \, Lb/d$, where ε_0 is the free-space permittivity, b and L are, respectively, the width and the length of the cantilever beam and d is the distance between the cantilever beam and the ground plane. A power supply (V_0) is connected across the capacitor through a resistor. The stored charge $Q=C_0V_0$ causes an electrostatic pressure that deforms the cantilever as discussed in Chapter 2.

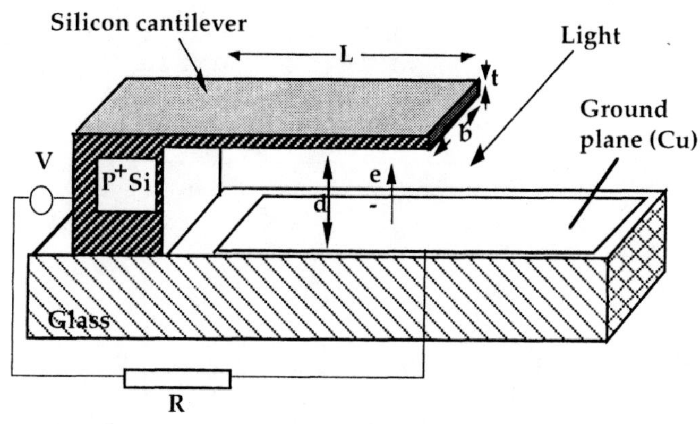

Figure 5.1 Side view of an optically controlled microactuator.

To achieve actuation of the cantilever beam, a monochromatic light is directed at the ground plane. At a quantum efficiency η and a work function W, a flux of photons Φ_γ of energy $h\nu$ striking the metal ground plane generates a flux of photoelectrons $\Phi_\varepsilon=\eta\Phi_\gamma$ of kinetic energy $T= h\nu - W$. Crossing the capacitor gap d to the top plate at potential $V_0>0$, each emitted photoelectron reduces the charge on the capacitor from $Q=\sigma Lb$ to $Q-q$. When the capacitor gap is a vacuum and electron transport is ballistic, an electron's transit time through the gap is much smaller than the mechanical time constant of the cantilever. This is also the case when the gap is permeated by a gas and transport is characterized by a carrier mobility. The simple relation

$$\frac{dQ}{dt} = - \eta q A \Phi_\gamma, \tag{5.1}$$

where A is the illuminated area, describes the current conduction through the cantilever beam assembly. The actuator-capacitor leaks at a rate controlled via the photon flux. Figure 5.2 presents the equivalent circuit of this simple device. The source current I_S divides among four routes:

i) Charging the capacitor through $I_C = dQ/dt = C_0 dV/dt$.

ii) Photoelectron current through the air gap. At low voltages (<10V) this path behaves as a light-controlled resistor, varying from dark to full illumination.

iii) Leakage at dark through the substrate glass, substrate surface, and the air. These we lump together into a dark leakage current I_d. This path has a resistance $R_d = 10^{10}\ \Omega$ at low voltages.

iv) A finite photoelectric current I_{ph} that flows through the air gap between the ground plane and the cantilever beam under illumination even at V_0=0 volts.

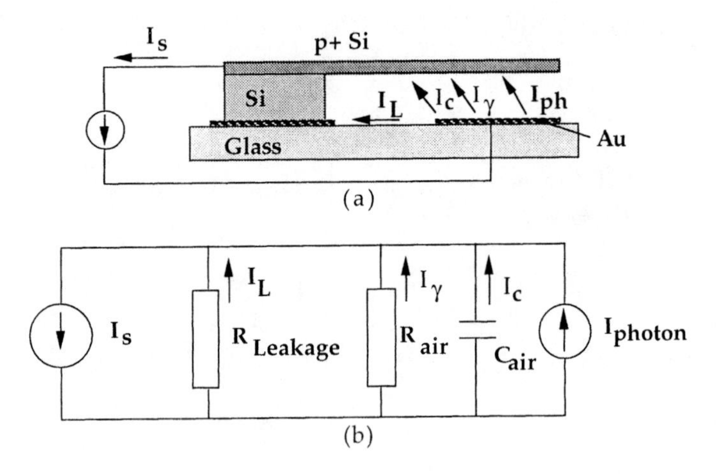

Figure 5.2 Equivalent circuit of photoelectric μ-actuator.

We relate the change in the charge Q on the capacitor to the source, dark leakage, and photo-leakage currents by

$$dQ/dt = I_S - I_L - I_{ph} - I_\gamma. \tag{5.2}$$

By selecting $I_L < I_S < (I_L + I_\gamma)$, the capacitor may be charged or discharged using an appropriate light. Integrating equation (5.2),

$$Q(t) = \int_0^t d\tau \, (I_S - I_L - I_\gamma - I_{ph})$$

$$Q(t) = Q_0 + (I_S - I_L)t - \int_0^t d\tau \left(\eta q b L \Phi_\gamma(\tau) + I_{ph}(\tau) \right) \tag{5.3}$$

Integrating over a "long" time period makes optical control of motion versatile, since the signal may be either analog or digital.

The stored charge Q(t) causes an electrostatic pressure that deforms the cantilever. Neglecting the change in capacitance as the cantilever deforms, the steady-state deflection of the cantilever end is in terms of the charge $Q = C_0 V_0$ on the capacitor

$$\delta(Q) = cQ^2 \tag{5.4}$$

with $c = 3L^2/(4\varepsilon Y t^3 b^2)$ where Y is the Young's modulus, and t is the thickness of the cantilever beam. The dynamic deflection in response to an optical control signal is given by the following convolution:

$$\zeta(t) = \delta(t) * F^{-1}[R(\omega)] = cQ^2 * F^{-1}[R(\omega)] \quad (5.5)$$

where $\delta(t)$ is given by equations (5.3) and (5.4), and $R(\omega)$ is given by

$$R(\omega)=[(\omega^2-\omega_0^2)^2+(b\omega)^2]^{-1/2}, \quad (5.6)$$

where ω_0 is the first mode resonance frequency and b is the damping coefficient.

Because of the non-linearity of [6], the cantilever can bend stably no further than the threshold of spontaneous collapse at

$$\delta = kL^4 / 8d^2 \quad (5.7)$$

where $k \approx 6\varepsilon V^2/Yt^3$. The fastest it can smoothly traverse this distance is roughly the period of the fundamental mode of free vibration T_0.

The fully stressed condition corresponds to a charge of some $Q_{th}C_oV_{th}$. The transport by photocurrent of a charge $-Q_{th}$ negates the electrostatic loading that stresses the cantilever to its maximum displacement. Since this transport need take no shorter a time than T_0, the photocurrent for maximum actuation need not exceed:

$$I_{max}Q_{th}/T_0. \quad (5.8)$$

Physically, this is simply the number of photo-electrons needed to completely "short out" the stored charge Q_{th} in the shortest time. The photon flux needed to generate this current is , with the corresponding energy flux or optical power density:

$$\rho_\gamma = h\nu\Phi_\gamma \approx \frac{1}{3}\frac{h\nu}{q\eta}Y\sqrt{\left(\frac{\varepsilon_0}{\rho_{si}}\right)}\frac{t^{5/2}d^{1/2}}{L^4} \quad (5.9)$$

where ρ_{si} is the density of silicon. When L, t, and d are expressed in micrometers, the numerical relation

$$\rho_\gamma = 1.4 \times 10^7 \frac{t^{5/2}d^{1/2}}{L^4} \; (W / cm^2) \quad (5.10)$$

is obtained for an Si air-gap actuator operated at quantum efficiency $\eta=0.1$ at $h\nu= 4.0$ eV. For our 600 x 50 x 2.5 μm Si cantilevers, $\rho_\gamma= 0.4$ mW cm^{-2} for a gap d=15 μm, and $\rho_\gamma=1.4$ mW cm^{-2} for a gap d=150 μm. With a fundamental frequency of free vibration near 8 kHz, the fastest spring-back actuation at these power densities is around 30 μs. Lower optical powers generate correspondingly slower motions than a spring-back.

We fabricated silicon cantilever beams using EPW anisotropic etch. Details of the fabrication are given in reference 3. Electrode areas were etched from an Au layer evaporated onto a glass wafer, and the cantilever assembly was bonded to the glass using the Si wafer's 250 μm thickness as a spacer. The resulting parallel-plate structure has a capacitance $C_0 \sim 10^{-14}$ F.

Static deflection measurement. A 0.5 mW He-Ne laser beam 0.5 mm-wide was reflected off a cluster of five identical cantilevers onto a screen 7m away. The actuator was charged with the ground plane negative to a potential of 105V by a current I_{source}=0.6 nA. When a UV lamp was switched on, illuminating the device with an optical flux ρ_γ= 1.5 mW cm^{-2} at 254 nm = 4.88 eV (for a gap d=15 μm), the five cantilever reflection spots on the screen moved absolute distances Δx=6.5±1.2 mm and Δy=4.1±1.3 mm from their positions at full charge in the dark. This corresponds to a cantilever tip deflection of 0.133 - 0.043 μm. This accorded with the maximum deflection predicted in equation (5.7) of 0.1μm, and constituted our first direct evidence of optically-controlled actuation. This experiment of several minutes duration did not provide any clue about the speed of actuation.

We noted that there was not any change in the bending of the cantilever beams due to the illumination by the He-Ne laser. The power densities of the UV source and the laser are comparable, and the bending of the cantilevers is due to the photoelectric effect which is sensitive to the photon energy.

Dynamic deflection measurement. The dynamic experiment apparatus comprised two elements in addition to the apparatus used in the static experiment. A shutter enclosed the UV pencil lamp within two concentric brass tubes. Highly polished, the smaller tube could rotate smoothly inside the larger, bringing apertures in each into coincidence. This design interrupted the UV quickly (<0.5s) and completely, isolated the lamp electrically, and did not obstruct the laser.

The clearest reflected laser spot fell onto a photoconductive cell ~ 1 cm in diameter. Fixed to the terminals of a digital multimeter connected to a pen plotter, the cell made a sensitive detector of spot motion. Together, the shutter, the DMM, and the plotter had a time resolution of roughly 0.5s.

Dynamic actuation runs were made at fixed source current Is and voltage limit V_{lim}, with the shutter opening and closing with a 60s or 120s period. Typically each run consisting of about four periods showed several clear mechanical transitions going between UV-on and UV-off. Two instances are shown in figure 5.3. In the first, the Au ground plane is biased negative by I_S=1nA, V_{lim}=105V. As the theory predicts, the cantilever's deflection decreased under illumination as photocurrent reduced the stored charge, and hence the electrical stress, on the cantilever. In the second, the polarity is reversed, biasing the cantilever negative. With I_S = -1.0 nA and V_{lim} = -105V, the traces from the detector were clearer, more uniform, and more reproducible than those obtained with normal bias. The amplitude was comparable or, in some cases, slightly larger.

Although the mechanical transients accompanying shutter motion obscured the sharpness of some of the transitions, clearly the detector system responded at its maximum speed to the actuation. The gross time resolution of the detection system was estimated to be 0.5s; thus, UV-controlled actuation took place in a time briefer than 0.5s. To determine exactly how much briefer would require a detection system of much greater speed. Future experiments using an electronic shutter and a digitizing oscilloscope will resolve easily, on a time scale ~0.1ms, the theoretical maximum speed of the actuator.

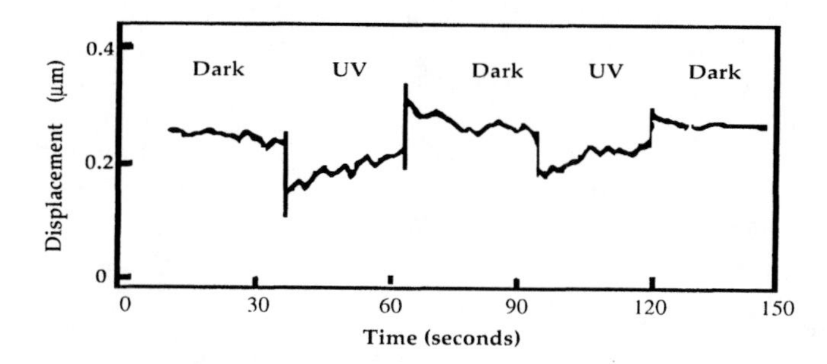

Figure 5.3 Displacement of the all-optical actuator as a function of time. The UV light was turned on and off at the different time intervals indicated in the graph [20].

5.2.2 Photoconductivity Modulation Actuator

Figure 5.4 shows a schematic diagram of a Si cantilever beam assembly with a semi-insulating layer (SIL) inserted between the cantilever beam and the ground plane. SIL can also be a part of the ground plane. The SIL can be formed by chemical phase vapor deposition (CVD) of a highly compensated semiconductor layer, or the ground plane can be chosen to be Cr-doped GaAs.

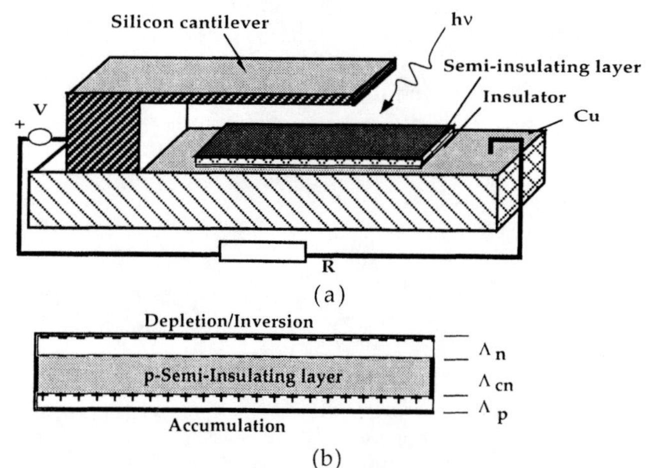

(b)

Figure 5.4 (a) Schematic of a electro-hole-pair screening controlled optical microactuator. (b) Different space charge regions in the semi-insulating substrate inserted between the cantilever beam and the ground plane.

A d.c. power supply is connected across the cantilever beam. In the SIL space charge regions will be formed in response to the external electric field that exists between the beam and the ground plane.

Since a semi-insulating semiconductor is usually either slightly n- or p-type, it is still valid to identify its carriers as being either majority or minority carriers. Therefore, space charge regions on both sides of the SIL, being of opposite polarity, will be formed by depletion of the majority carriers (or inversion under stronger electric fields) on one side and accumulation of the majority carriers on the other side. Thus, the SIL will be polarized (Fig. 5.4b). The SIL as a whole always remains charge neutral. Taking the SIL to be slightly p-type and applying positive d.c. voltage to the cantilever beam results in space charge regions that consist of a depletion/ inversion region (Λ_p width) in the upper surface and an accumulation region (Λ_n width) in the bottom surface of the SIL.

The width of space charge regions (Λ_p and Λ_n) is determined by the magnitude of the applied field and it is a function of the doping concentration, among other parameters. Furthermore, in the accumulation region the surface charge density is an exponential function of the potential, while in the depletion region it has a square root dependence on the potential. Therefore, L_p is in the few angstrom ranges while L_n is in the few micrometer ranges. These are very well known in metal-insulator-semiconductor (MOS) systems. Assuming that the thickness of the SIL is larger than $\Lambda_n + \Lambda_p$, there will be a middle region inside the SIL with a width of (Λ_{cn}) that will remain charge neutral.

The electric field inside the gap between the cantilever beam and the ground plane determines the electrostatic pressure exerted on the beam as discussed in the previous section. At dark, the amount of charge in the cantilever beam is determined by the applied voltage and by the capacitance of the structure. This capacitance is determined by the combined value of capacitors formed between the beam and the SIL (C_a), the depletion capacitance of the SIL (C_D), and the capacitance between the SIL and the ground plane (C_i):

$$C_a = A\varepsilon_0/d_a \tag{5.11}$$

$$C_D = A\varepsilon_{sc}/\Lambda_n \tag{5.12}$$

and

$$C_i = A\varepsilon_i/d_i \tag{5.13}$$

where A is the area, ε_0 is the permittivity of the free space (=8.854×10^{-14} F/cm) with a thickness of d_a, ε_{sc} is the permittivity of the semiconductor (=$11.9\varepsilon_0$ in silicon), and ε_i is the permittivity of the insulator between the SIL and the ground plane with a thickness of d_i.

When a light with photon energy larger than the semiconductor band gap is directed at the SIL, as shown in figure 9.a, electron-hole pairs inside the SIL are generated. Taking the optical generation rate to be g_{op} and carrier recombination lifetime to be t, the excess electron density (Δn) at the upper surface of the SIL is simply given by $\Delta n(0)=g_{op}$. In highly-compensated semiconductors, τ can be very short, in the nanosecond range. The excess holes will be swept into the bulk of the semiconductor and will eventually end up in the lower part of the SIL. Since they are majority carriers, their rearrangement takes place with a speed determined by the dielectric relaxation time constant inside the SIL. An increase in the minority excess carrier density in the depletion region changes the effective value of C_D to C_{DiII} and

$$C_{DiII}>C_D \tag{5.14}$$

C_i remains constant under illumination. Due to the increase in C_{DiII}, the total capacitance between the cantilever beam and the ground plane will increase. There are two different schemes that can be used here: i) constant charge scheme, and b) constant voltage scheme. In the constant charge scheme, as shown in figure 10.a, the electric field remains constant despite the fact that the capacitance becomes larger under illumination, as expected. In the constant voltage scheme, however, the charge density in the cantilever beam adjusts itself to respond to the change in the capacitance under illumination. This is shown in figure 10.b. Therefore, the electric field inside the gap increases, bending the cantilever beam further downward. The difference between the present and the previous (photoelectron) actuation mechanism is that here the capacitance is altered while keeping the voltage constant, and in the previous technique the charge is altered while assuming that the total capacitance remains constant.

The change in the electrostatic force exerted on the cantilever beam can be calculated as follows: The potential difference between the cantilever beam and SIL at dark is $V_a+\psi_s$ where V_a is the potential difference across the air and ψ_s is the surface potential of the SIL. Furthermore, $V_a=Q_s/C_a$, where Q_s is the charge on the cantilever beam (= charge in the SIL depletion layer Q_D). When the SIL is illuminated with a light with photon energy larger than the SIL energy band gap, the depletion layer collapses and C_D becomes very large. Since the potential across the cantilever beam and the ground plane is kept constant, the charge density in the cantilever beam increases in response to the increased capacitance of the cantilever beam assembly. The change in this charge density ΔQ is related to the change in the potential across SIL ($\Delta\psi$) as follows:

$$\Delta Q= -C_aC_i/(C_i+C_a)\Delta\psi \tag{5.15}$$

where the negative sign indicates that when Ψ decreases due to g_{op} (non-thermal equilibrium), because of smaller voltage drop across SIL, Q increases, as expected. The change in the electric field ΔE is:

$$\Delta E = - (1/d_a)C_i/(C_i+C_a)\Delta\psi \tag{5.16}$$

This change in the electric field corresponds to a change in the electrostatic pressure exerted on the cantilever beam (ΔP_E), which results in a change in the deflection of the cantilever beam $\Delta\delta$:

$$\Delta P_E = \varepsilon E \Delta E \tag{5.17}$$

and

$$\Delta\delta = (3\varepsilon L^4/2Yt^3)E\Delta E, \tag{5.18}$$

where E, the electric field, is given by:

$$E = (V_0 - \psi)C_i/d_a(C_i + C_a) \tag{5.19}$$

In cases where the SIL is part of the ground plane, C_i is infinite and ΔQ is simply ψC_a, and V_i in the above equations is zero, so we can set $\psi = \psi_s$ with ψ_s being the upper-surface potential of the SIL. Since setting $V_i = 0$ would result in a larger ΔP_E, in the remainder of this section we will assume that the SIL is part of the ground plane.

When the light is turned off, excess carriers recombine and the system moves toward thermal equilibrium. The decay of excess carriers, in this case, has an exponential form with a decay constant equal to the recombination lifetime. The excess electrons are spatially separated from excess holes, and the recombination lifetime might be slightly larger than usual. As long as the decay constant is shorter than the period of the free vibration of the cantilever, the response time is dominated by the later.

In addition to the above actuation mechanism, the photoelectron mechanism is also present here. However, a combination of photon energy, semiconductor band gap and work function, and d.c. bias amplitude and polarity can be selected to reduce the photocurrent contribution, if it is so desired.

5.2.3 Photodiode Microactuator

In this approach a p-n junction is connected in parallel with the cantilever beam-ground plane capacitor, as shown in figure 5.5. At dark, the leakage current through the p-n junction is on the order of a μA or so, and the voltage across the junction is $V_0 - I_0 R$. I_0 being very small, the voltage drop across the reverse biased p-n junction at dark is $\sim V_0$. When the junction is illuminated with photons having energies larger than the band-gap of the semiconductor, due to generation of electron-hole-pairs within a diffusion length of that junction, I_0 increases drastically. This reduces the voltage across the junction and the cantilever beam assembly, reducing the electrostatic pressure on the cantilever beam. The value of R is chosen to limit the current when the junction is illuminated.

The depletion capacitance of the p-n junction is usually higher than the capacitance of the cantilever beam, which is on the order of 1-0.01 pF. When the light is turned off, the excess carriers recombine and the voltage across the p-n junction increases. Since the p-n junction is reverse biased, the response time of the system is determined by the recombination/generation process in the semiconductor.

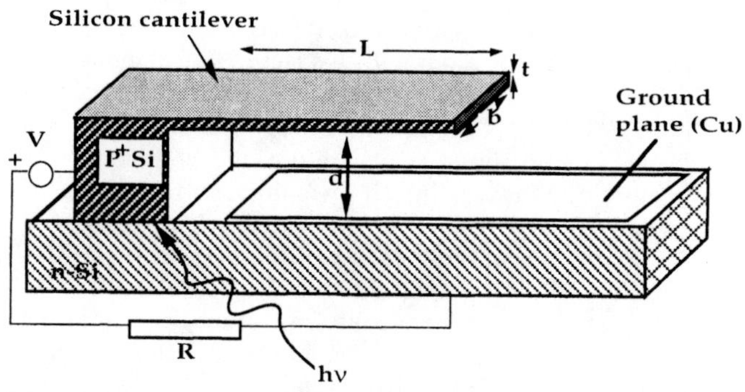

Figure 5.5 Schematic of an optical microactuator controlled by the voltage drop across a p-n junction that is modulated by light.

The amount of bias voltage that can be applied across the cantilever beam and the ground plane is essentially limited by the breakdown voltage of the reverse biased junction which is on the order of 10 volts or so. The value of R should be chosen to limit the current to no larger than a mA under the worst conditions, which means $R \sim 10^4\ \Omega$.

5.2.4 Optically-modulated Gas Permittivity Microactuator

The permittivity of a gas that fills the volume between the cantilever beam and the ground plane can be altered by illumination of an appropriate energy. One of the well-known mechanisms of this alteration is the transition of the gas molecules to higher excited states in the presence of the photon field. Gas molecules that are polar usually have larger electric dipole moments in their higher excited states. This results in larger permittivity of the gas. The relationship between the electric dipole of molecules and the permittivity is given by the Clausius-Mossotti (in the case of non-polar molecules) and Onsager (in the case of polar molecules) equations [19].

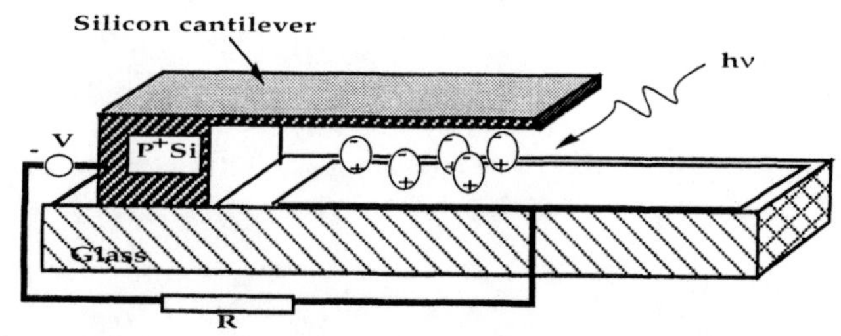

Figure 5.6 Schematic of an optical microactuator controlled by the permittivity of a gas modulated by light.

In the majority of cases, the displacements between the opposite charge centers are on the order of 10-100Å, and the change in displacement is at most 0.1 μm.

Modulation of the permittivity of gas molecules is expected to be small, but it is included here for completeness. NH_3, being a relatively well-studied gas, is a possible candidate. There are numerous of references that describe the relative position of the N atom with respect to the plane containing the H atoms. The N atom basically resides in a potential well with two minima. At zero electric field the probability of finding the N atom on both sides of the H-plane is nearly equal. Therefore, the electric dipole moment of the NH_3 molecule (N being slightly negative with respect to slightly positive H atoms) is very small, because of a slight difference in the potential of the minima from the rotation of the NH_3. But NH_3 has a very large polarizability and in the presence of an external electric field, the symmetry of the potential barrier breaks down leading to relatively large dielectric constant. It is well known that N atom has different energy levels inside the potential barrier. Due to the asymmetry of the potential barrier in the presence of external electric field, the expected value of the position operator at higher energy levels is further away from the H-plane. This leads to slightly larger electric dipole moment at higher energy levels. Thus, illumination with an appropriate energy photon can be used to excite the N-atoms and increase the dielectric constant of the NH_3.

5.2.5 Comparison of Direct Optical Microactuators

We defined a very elaborate figure of merit in Chapter 1. Here, to simplify our analysis, we only retain the essential parts of the figure of merit and reduce it to the following dimensionless form:

$$\mu = Fz / P\tau \tag{5.20}$$

where F and z are the force and displacement available from the actuator as it does mechanical work; P is the power of the signal that controls or initiates actuation; and τ is the time in which the actuator responds to its control signal, or its effective speed. Broadly, μ is the ratio of the mechanical work output during one actuation cycle to the energy required to control one cycle. The range z, force F, and response time of a photocurrent hybrid actuator are the same as for an electrostatic actuator of identical dimensions. The above figure of merit also provides for the evaluation of combinations of schemes and for the optimization of the design of a photoelectric actuator.

Photoelectron Microactuator. Using the peak optical power given in equation (5.9), μ for a photoelectric actuator becomes:

$$\mu = \frac{\left(\frac{3}{8}Lb\frac{\varepsilon V_{th}^2}{2d^2}\right)\left(\frac{d}{3}\right)}{\left(\frac{h\nu}{\eta}\frac{C_0 V_{th}}{q}\right)} = \frac{1}{16}\frac{\eta q}{h\nu}V_{th} \tag{5.21}$$

$$\approx \frac{1}{29}\frac{\eta q}{h\nu}\sqrt{\frac{Y}{\varepsilon}}\, t^{3/2} d^{3/2} L^{-2} \tag{5.22}$$

This is the ratio of the total photon energy required to liberate one photoelectron, on average, to the potential energy of an electron in the capacitor at maximum stress, times a geometric factor.

A 600 x 50 x 2.5 μm Si photoelectric actuator operated at $h\nu$=4.88eV, η=1%, with d=250 μm has as a figure of merit μ=0.34. By modifying the dimensions t, d, and L, and by selecting materials and techniques which permit operation at lower photon energies and higher quantum efficiencies, μ may be increased far above unity. This compares with μ=0.125 for a cantilever-type electrostatic actuator and with $\mu \approx 1.5 \times 10^{-5}$ for a bimetallic thermal actuator [20]. The figure of merit for a p-n junction switched actuator is $\mu \sim 10^{-6}$ when I^2R losses taken into account. When losses are ignored it can be shown that $\mu \approx 1$.

Photoconductivity Modulation Actuator. The force and displacement available from the actuation are $F=3Lb\varepsilon E^2/16$ and $z=\Delta\delta$, respectively. The power needed for actuation is $P=\delta nh\nu/\tau\eta$ and $dn=g_{op}$. Taking $Y=1.6\times10^{11}Nm^{-2}$, $\Delta\delta$=1 μm, b=100 μm, t=6.5 μm, and L=600 μm, the force is $F \sim 10^{-4}$ N. $\Delta\delta$=1 μm corresponds to an E field of 10^6 volts/m that can be achieved by V_0=100 volts and d_a=100 μm, and ΔE of 10^4 volts/m, which corresponds to $\Delta\psi$=1 volt. This change in the surface potential can be achieved by an illumination having photon energy of $h\nu$=1.5 eV, and $g_{op}=10^{22}$ e-h-p/cm^{3}sec., $\tau=10^{-6}$ sec., resulting in $\delta n=10^{15}$ cm^{-3}. Taking η=1 (this is reasonable as far as electron-hole pair generation is concerned when the photon energy is slightly larger than the SIL band-gap), $\Delta t \sim 10^{-3}$ sec., and the volume v=600 x 50 x 0.1 x 10^{-12} cm^3 we get the power that is needed for actuation: $P=vg_{op}h\nu/\eta=7.2\times10^{-6}$ W. Therefore, μ is around 1 for the present actuation scheme. This is an order of magnitude larger than the pure electrostatic actuator, three times larger than the photoelectric one, and five orders of magnitude larger than the bimetallic thermal actuator [20]. It should be emphasized that the structure that we have considered here is by no means an optimum one and the geometry as well as the SIL properties can be improved a great deal.

Photodiode Microactuator. The figure of merit can easily be calculated using the procedure used before. However, here we have to include the I^2R power loss as a part of the power needed for actuation. When the light was turned on, I^2R was 0.01W. The optical power was around 10 nano-W; therefore, it can be ignored. $\Delta\delta \sim 1$μm, $F \sim 10^{-5}$ N and $\Delta t \sim 10^{-3}$ sec. yields $\mu \sim 10^{-6}$. If we do not take the I^2R loss into account: $\mu \sim 1$. This is on the same order as μ for the previous actuation mechanism. However, with I^2R losses taken into account, it is six orders of magnitude less.

Optically-modulated Gas Permittivity Microactuator. If we accept a somewhat more realistic change of +1% in the permittivity of the NH_3, the change in the electrostatic pressure in the constant voltage scheme, being directly proportional to the permittivity, is also +1%. In this case, the change in the displacement of the tip of the cantilever beam is also +1% [19]. Assuming that the volume between the cantilever beam and the ground plane ($\approx$ 600 x 50 x 100 μm^3) is completely filled with NH_3 molecules with a mean free path of around 5 μm (at room temperature at one atmosphere), there are 24000 NH_3 molecules between the cantilever beam and the ground plane that can be excited. Each N atom, when excited, spends t second before it recombines. Taking t of one picosecond, an optical excitation rate of ε_{op} $\approx 2.4x10^{16}$ excited-NH_3/cm^3sec. is required to excite all the 24000 NH_3 molecules. The photon flux needed for a given excitation rate is $\varepsilon_{op}/\eta.\alpha$, where η is the quantum efficiency and a is the absorption coefficient. Taking $\eta \approx 100\%$ and $\alpha \approx 1$cm, the required photon flux for the above e_{op} is $\approx 2.4x10^{16}$ photons/cm^2sec. The optical power is $(h\nu\varepsilon_{op}/\eta.\alpha) \approx 4$ mW for $h\nu \approx 1$ eV. The figure of merit $(Fz/P\tau)$ for the present actuation scheme is, therefore, $\mu \approx 10^{-5}x10^{-6}/4x10\text{-}3x10^{-3} \approx 2.5x10^{-6}$. This is quite small compared with the other actuation schemes that we have discussed above.

5. 3 Direct Optical Manipulation: Optical Twizers and Traps

Atoms and molecules can be trapped by the stationary potentials of standing electromagnetic fields generated by lasers [35,36]. These traps, devised in the late eighties, are used to reach effective temperatures as low as a few microdegree Kelvin, and they are shown to be able to manipulate clusters of atoms as well as individual atoms. There has not been any report of integrating optical twizers with MEMS yet. However, lasers, cavities and positioning devices can all be realized in GaAs MEMS, so these direct optical methods of trapping and directly manipulating atoms are included here. Using these schemes, one can realize micropumps with atomic flow rates.

Figure 5.7 shows the schematic of a laser trapping and manipulation device. Up to six contra-directed lasers have been used to confine and manipulate atoms [36]. The basic idea is to generate a standing wave as shown in figure 5.7, and trap the atoms in the minima of the standing wave potential. There are mechanisms for the atoms having finite kinetic energies to "cool" down and settle at the bottom of the standing wave potential. It is beyond the scope of this book to review all these "cooling" mechanisms, so we discuss only some very elementary aspects of laser trapping of atoms here.

The motion of atoms in a standing light wave is governed mainly by the dipole force, the friction force and the momentum transfer. Under the conditions where both the friction and momentum transfer are negligible, the atoms see the standing wave potential given by:

$$U(z) = U_0 \cos^2 kz \tag{5.23}$$

where U_0 is the potential amplitude and k is the wavenumber ($=2\pi/\lambda$). Depending on the total energy of the atom, whether it is less than or larger than U_0, the atom will either escape the standing wave potential or become trapped in it and localize spatially. By generating standing waves in the three dimensions of x, y and z, these atoms can be completely localized. Then, using an electric field or another laser beam of appropriate energy, the trapped atoms can be moved around and manipulated.

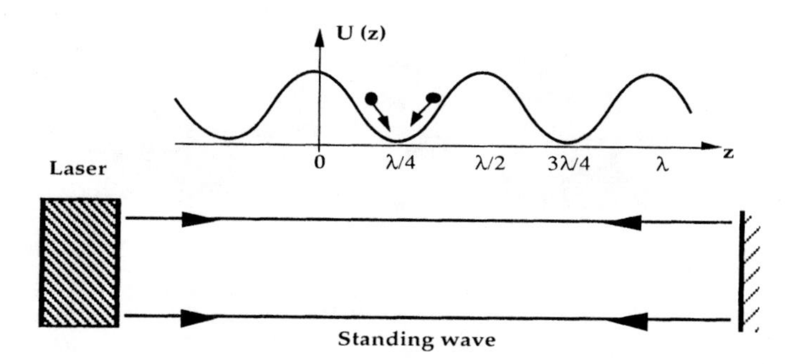

Figure 5.7 Schematic of an optical trap produced by standing optical fields [35,36].

There has been recent developments using the angular momentum of light to rotate molecules. One such experiment used nematic liquid crystals as the rotating molecule and used the light itself to sense the molecule's rotation [36.b]. At this stage of its development, these experiments are interesting academic exercises. But, coupled to the development of novel lasers, they may pave the way to the realization of optical manipulator with important applications in drug delivery, etc.

5.4 Opto-Thermal Actuators: Gas Expansion

Cavity-diaphragm Actuators. Optical heating and subsequent expansion of gases can be used in actuators, as schematically shown in figure 5.8. The deflected diaphragm can be used to control the flow of liquids or gases, as schematically shown in figure 5.9 [1,21-32].

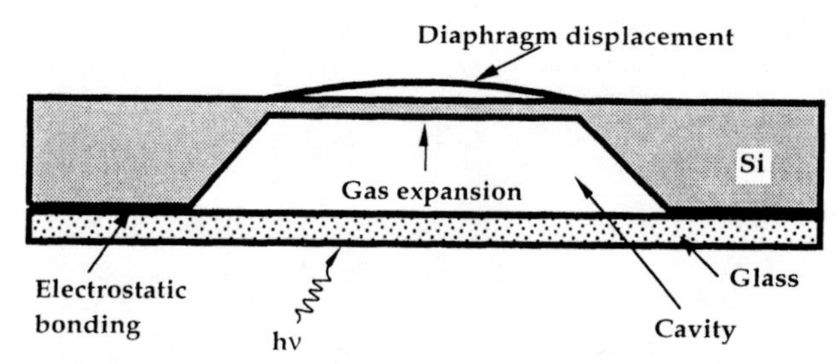

Figure 5.8 Schematic of an optically actuated silicon diaphragm.

Microfabricated devices that were capable of controlling both air and oils were reported. These flow controllers had speeds of 21 ms in air flow and 67 ms in oil flow, and showed respective sensitivities of 304 Pa/mW and 75 Pa/mW [21].

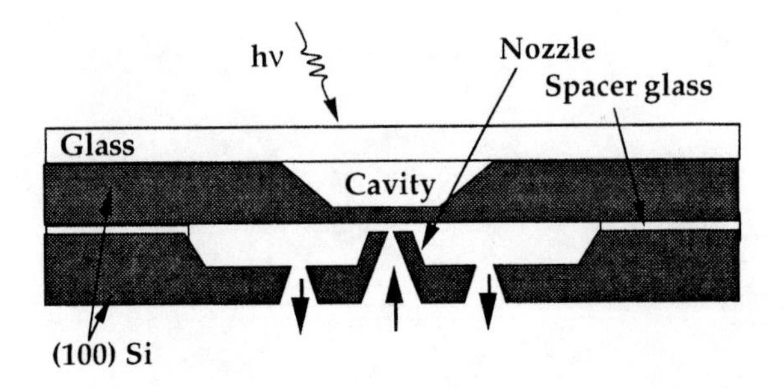

Figure 5.9 Schematic of an optically actuated silicon diaphragm used to control flow of gases and liquids [21].

Figure 5.10 shows an optically controlled micropump capable of pumping 30 nl/cycle using approximately 11 mW at 3 Hz operation frequency. To increase the efficiency of the gas heating by light, carbon wool fiber is incorporated inside the cavity.

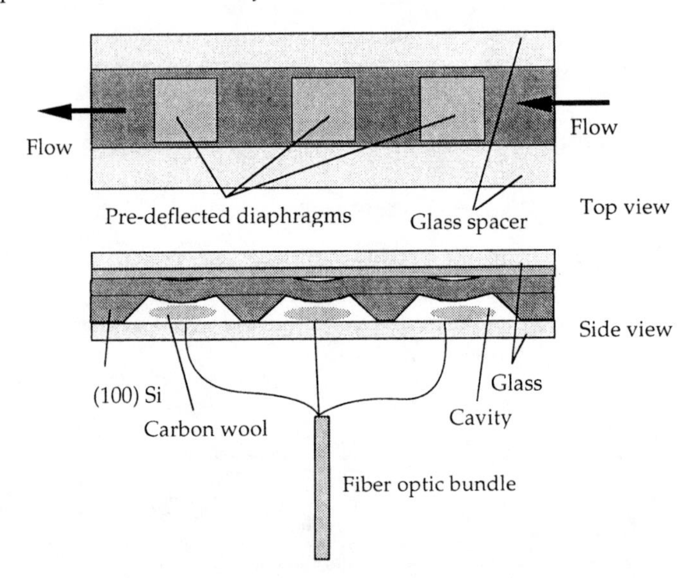

Figure 5.10 An optical micropump capable of pumping 30 nl/cycle [1,21].

<u>**IO Bridge resonators.**</u> Very similar to the laser trapping method is the excitation of a dielectric diaphragm with standing laser waves. The schematic of a device that uses a laser beam to set a Si_3N_4 diaphragm into oscillation is shown in figure 5.11. The principle of the oscillation of this device is very similar to that of the optical traps discussed in section 5.3. The laser passes through the Si_3N_4 bridge and reflects off the silicon substrate. The reflected and incident lasers set up standing waves. When the Si_3N_4 bridge is at the peak of the standing wave, it heats up and expands. As soon as it expands, it drops into the region where the standing wave amplitude is minimum and has a node. Being at the node of the standing

wave, the Si_3N_4 bridge cools off and contracts. The cycle repeats itself over a time period determined by the resonance of the bridge with an amplitude determined by the laser power, conductivity of the nitride (which depends on its stoichiometry, etc.) and heat conduction mechanisms, and varies with respect to a measurand (pressure, gases, etc.) [32].

The above device is self-oscillatory since the bridge is set into oscillation when illuminated with a CW light source. The self-oscillatory characteristic of the device makes it more practical because only one unmodulated laser source is required for both the excitation and interrogation of the microresonator.

The microbridge, shown in figure 5.11, can also be excited by intensity-modulated light incident on the structure. This incident light produces a thermal stress causing the structure to vibrate. When the incident light frequency matches the bridge's natural resonance frequency, the maximum output amplitude is achieved [22]. The bridge's fundamental undamped resonance frequency is expressed by $f_b = 1.028\frac{t}{L_b^2}\sqrt{\frac{kE}{\rho}\left(1+0.295\,L_b^2\,\frac{\varepsilon}{t^2}\right)}$, where t is the thickness of the beam, L_b is the bridge length, k is a coefficient related to the beam components, r is the beam density, E is the Young's modulus of the substrate, and ε is the longitudinal tensile strain of the bridge. This vibration amplitude is detected optically using an integrated optic interferometer. Measurand-induced changes in the strain of the device are detected as changes in the device resonant frequency.

Microresonators can be used to determine chemical components, small masses, temperature, pressure , and vibration. For a prototype pressure sensor [24] operating in 0-300 kPa range within ≈18 kHz span, the pressure sensitivity was ≈ 0.006 kHz/kPa. A vibration sensor can be used to detect Young's modulus and the built-in stress of thin deposited films.

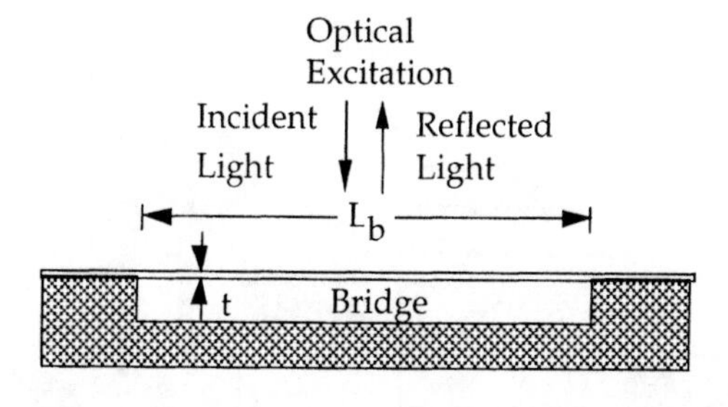

Figure 5.11 Side view of an optically excited bridge resonator [22].

Micromachined AlGaAs/GaAs Actuators/Sensors. Figure 5.12 shows an example of an AlGaAs/GaAs micromechanical device that uses a laser diode to thermally excite a GaAs cantilever beam and another laser diode and a photo detector to detect the vibrations of the cantilever [33,34]. The idea is to detect the cantilever beam's mass loading by tracking its resonance frequency.

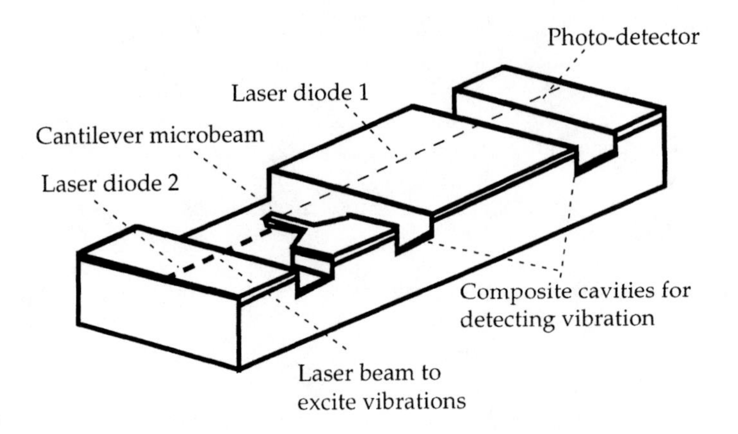

Figure 5.12 Schematic of a microbeam resonator driven photothermally by laser 2 and sensed by laser 1 and a photodetector all integrated on an AlGaAs/GaAs substrate [33,34].

To fabricate this device, three micromachining processes (figure 5.13) were carried out: a) an etch-stop layer of AlGaAs was formed in the laser diode structure; b) the shape of the microstructure was precisely defined by means of reactive dry-etching; and c) a wet-etch window was made with photo-resist, and the microbeam was undercut by selective etching of the sacrificial layer to leave the microbeam freely suspended. Table I compares several AlGaAs/GaAs micromachining schemes [33].

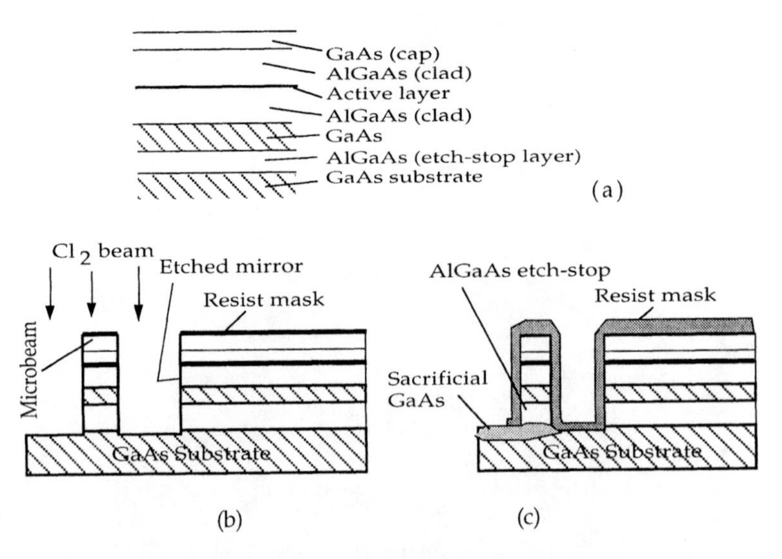

Figure 5.13 Process steps used to fabricate the AlGaAs/GaAs microbeam resonator structure shown in figure 5.12 [30].

Table 5.1 Comparison of several AlGaAs/GaAs micromachining schemes [33,34].

Sacrificial layer	Sacrificial layer etchant	Etch-stop layer	Etch-stop layer etch	Technology
AlGaAs	HF	GaAs	H_2O_2/ NH_4OH	Surface micro-machining
GaAs	Chlorine Plasma	GaAs	Cl_2 ion beam	
GaAs	H_2O_2/ NH_4OH	AlGaAs	Cl_2 fast ion beam	
$Al_yGa_{1-y}As$	HCl	GaAs	H_2SO_4/ H_2O_2/H_2O	
GaAs	H_2O_2/ NH_4OH	AlGaAs	H_2O_2/ NH_4OH	Bulk micro-machining
GaAs	H_3PO_4/ CH_3OH/ H_2O_2	GaAs	Br/CH_3OH	
AlAs	HF	GaAs/ AlGaAs	H_2O_2/ NH_4OH	Epitaxial lift-off

Optical Gels. As discussed in section 7.7 of Chapter 7, illumination by a UV light initiates an ionization reaction in photosensitive gels, creating internal osmotic pressure which induces swelling. This process is slow compared with the illumination with white light that causes the phase transformation by heating the gel. Very large volume changes occur in N-isopropylacrylamide/chlorophine copolymer gels as a function of optical power at 31.5 °C. This phenomena can be used in microactuators to generate large displacements.

5.5 All-Optical Systems

As we mentioned at the beginning of this Chapter, all-optical systems offer interesting features, such as parallel signal processing, possible immunity to electromagnetic interference, better sensitivity and simplicity. The basic elements of all-optical systems employed in smart structures are actuators, sensors, data links, and signal processors. A simple example of such a system is shown in figure 5.14. Bragg gratings can be used to leak the light out to power the actuator and provide it with a control signal. They can also be used to read the sensor output.

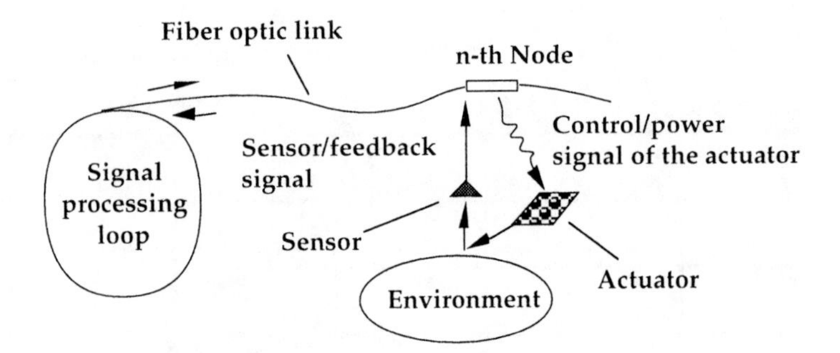

Figure 5.14 Schematic of an optical loop connecting a central signal processor to sensors and actuators located along a fiber optic. Both the power and control signal to and from the actuator are transmitted through the fiber optic. The state of the actuator/ environment is sensed by the sensor that outputs its signal into the fiber optic as well. The sensor may also need power that can be provided by the fiber optic. The signal processing loop keeps track of these different optical signals.

To address different sensors/actuators, a variety of multiplexing methods, including time-multiplexing and frequency (wavelength) multiplexing, can be used.

Bragg gratings can be used to leak the light out of the fiber optic to power the actuators and sensors and provide them with a control signal. Bragg grating can also be used to "read" the sensors' output. This is an active area of research that is undergoing rapid development. The important feature of such an all-optical system is compatibility of various sensors and actuators with fiber optics.

5.6 Fabrication Methods

In this section we briefly describe different fabrication steps used in integrated optics and other optical devices. Integrated optical microactuators require waveguides, optical couplers, and optical sources and detectors [2]. Co-fabrication of these devices on the same chip is a challenging task that remains to be addressed. (For a more detailed discussion of fabrication procedure refer to [2]. We have extensively used [2] in compiling this section.)

5.6.1 Waveguide Formation

Depending on the waveguide geometry, waveguide formation usually involves deposition of thin films, diffusion of impurities, or an ion-exchange process. As discussed in the previous section, the effective refractive index of the guide is higher than its surroundings; therefore, the waveguide formation involves defining or depositing a high refractive index region over a material with low refractive index.

Silicon Waveguides. Waveguides are usually formed by the diffusion of impurities into silicon [1-4]. Doping reduces the refractive index by increasing the local density of carriers. On the other hand, doping increases absorption of light in the semiconductor due to free carrier absorption. The relationship between the carrier density and the refractive index in semiconductors at photon energies below the band gap is given by:

$$n=n_f-\Delta n \tag{5.24}$$

where n_f is the refractive index of the intrinsic semiconductor. Δn, which is the change in the refractive index due to the free carriers, is given by:

$$\Delta n = \frac{N_c e^2 \lambda_0^2}{8\pi^2 \varepsilon_0 n_f c^2 m^*} \tag{5.25}$$

where N_c is the free carrier concentration, ε is the electronic charge, λ_o is the free space wavelength, ε_o is the free space permittivity, c is the speed of light in vacuum, and m* is the effective mass of the free carriers. Typical Δn are on the order of 10^{-5}-10^{-4}. Various dopants are used to fabricate waveguides in silicon. These include both acceptors, such as boron, and donors, such as phosphorus. Diffusion, ion-implantation, and epitaxial growth are all used in introducing these impurities.

To form a waveguide, a high refractive index silicon is sandwiched between two layers of lower index materials. The upper layer is usually a SiO_2 or Si_3N_4 layer which has a low absorption coefficient. The bottom layer, on the other hand, is either highly-doped silicon or another insulator, such as sapphire, glass, or silicon dioxide [2]. These later type layers have become available only recently as the silicon on insulator technologies has matured. The problem with the highly doped bottom layers is the excessive loss due to free carrier absorptions, which can be as high as 20-30 dB/cm. High doping concentrations in the waveguide regions are avoided to reduce free carrier absorptions.

Another method of forming silicon waveguides is to expose it to trap-generating particles such as electron or proton beams. These beams generate large densities of defects that, by trapping the free carriers, reduce the carrier densities locally. To reduce optical absorption and scattering, the irradiated film is annealed. Irradiation with electron beams or lasers may also be used to locally densify SiO_2. This process locally increases the refractive index of the SiO_2, directly forming the optical waveguide. Polysilicon is also used as an optical waveguide. Its absorption spectrum is different than that of the single crystal silicon and depends strongly on the grain size and grain boundaries.

Silicon/Germanium Waveguides. With the emergence of SiGe technology, silicon integrated optics will be enriched considerably by the introduction of Si/SiGe heterostructure waveguides that would allow optical waveguiding and light detection in planar structures. Silicon optical detectors, will also be improved both in speed and spectral response by incorporation of Ge. This technology is quite new and only Si/SiGe heterostructure bipolar transistors are reported. Noting that the refractive index is 3.5 for pure silicon and 4.0 for pure germanium at photon energies below the band gap, it is clear that optical waveguides with incorporation of a few percent Ge can easily be achieved. The interesting feature of this technology is that very high speed detectors and amplifiers can be integrated directly onto the optical waveguide by locally increasing the germanium content to form detectors that absorb the guided light and produce electrical signals.

Silicon Dioxide. Doping SiO_2 changes its refractive index. A common dopant of SiO_2 to form an optical waveguide is phosphorus. Figure 2 shows the refractive index of SiO_2 as a function of the doping concentration of various impurities. It is interesting to note that its refractive index can also be decreased by doping .

To dope SiO_2 with phosphorus, usually phosphine is added to the oxide formation process. In the plasma enhanced chemical vapor deposition (PECVD) method, phosphine is added to silane (SiH_4) and nitrous oxide (NO_2). High pressure thermal oxidation, flame hydrolysis deposition, and sputtering are also used to deposit SiO_2 over silicon. These methods are compatible with silicon microelectronics.

To form the waveguide, the phosphorus doped SiO_2 is sandwiched between two layers of undoped SiO_2. The bottom layer itself may contain two layers of SiO_2.. The layer immediately above the silicon wafer is usually thermally grown. The second layer is deposited using PECVD. PECVD has high deposition rate at relatively low temperatures. Thus, relatively thick films can be deposited in short periods of time. The buffer region is undoped and its total thickness is usually larger than ten times the wavelength of the guided light. For $\lambda \approx 0.633$ μm, the buffer layer thickness is 8-10 μm.

The waveguide layer is composed of a phosphorus-doped SiO_2. Depending on the phosphorus content, the wavelength of light, and the number of propagating modes, it has different thicknesses. At $\lambda = 0.633$ μm, a guide thickness of 4 μm with a phosphorus content of 3 weight percent produces a single mode optical waveguide.

The main advantages of doped SiO_2 waveguide technologies is that large cross-section waveguides can be fabricated. They are suitable for interfacing with fiber optics having core diameters of > 5μm. In that respect, PECVD is quite suitable since it has a fast rate of deposition at low temperatures, which makes it compatible with integrated circuit technology requirements. On the other hand, controlling the thickness and phosphorus content uniformity of different layers across the wafer, and from run-to-run, is quite difficult in PECVD.

Silicon Nitride. Silicon nitride, having a refractive index of 2.1 at l=0.633, has also been used in optical waveguides. A variety of methods, including sputtering CVD, PECVD, and low pressure CVD (LPCVD) has been used to deposit Si_3N_4 on silicon dioxide. The key parameter in determining the quality of Si_3N_4 waveguide layers is the stoichiometry of the film. Silicon-rich films tend to have large absorption, while nitrogen-rich films have a smaller refractive index.

When depositing Si_3N_4 over microstructures such as bridges, cantilevers and diaphragms, the stress generated by the nitride layer should also be carefully controlled. Otherwise, mechanical deformation may occur in the microstructure even in the absence of external forces. Table 5.1 shows some of the important parameters of silicon nitride films deposited using different methods or under different conditions.

Owing to its high refractive index compared to SiO_2, Si_3N_4 single-mode optical waveguides on SiO_2 tend to have very small dimensions. This causes problems, such as large coupling inefficiencies, when interfacing with fiber optics that are usually of larger diameters. Waveguide horns and other couplers are used to remedy this problem.

Table 5.1 Summary of Si_3N_4 Properties.

Deposition Method:	**CVD**	**PECVD**	**LPCVD**
Temperature (°C)	700	275-400	400-900
Typical dep. rate (Å/min)	100-200	200	<200
Etch rate in 49%HF (Å/min)	250-500	strongly reduced by hydrogen cont.	<250
Refractive index $\lambda = 0.63\ \mu m$	1.95-2.0	1.9-2.2	2.01
Optical losses TE_0 dB/cm	0.5-10	when contains Si: >20-30	<0.1
Film stress (dyn/cm^2)	tensile $\approx 5x10^9$	tensile to compressive*	highly tensile** $\approx 5x10^9$

* Depends on the Si/N ratio and deposition conditions.
** Can be made smaller and even compressive by making it Si rich.

Silicon Oxynitride. To remedy some of the shortcomings of the silicon nitride waveguides, silicon oxynitride (SiON) technology has been developed. By varying the nitrogen content of the SiO_2 layer, its refractive index can be varied from 1.45 (that of pure SiO_2) to 2.1 (that of the stoichiometric Si_3N_4), as shown in figures 3 and 4. The deposition methods used to form these layers are similar to that of Si_3N_4 except that in the PECVD method an agent, such as nitrous oxide (N_2O), is added to the gas mixture to produce $Si_xO_yN_z$. Reactive rf sputtering of $Si_xO_yN_z$ on silicon and SiO_2 has also been carried out by adding N_2 to the discharge gas mixture.

Polymers. Polymers are particularly interesting because of their ease of deposition and optical properties. In applications where waveguides with large elasto-optic and electro-optic coefficients are required, polymers will probably find many applications. Polymers are also becoming a permanent component of integrated circuits as a multilayer interconnect spacer and as conformal coatings. Their deposition can be accomplished using many different methods, including dip-coating, spin-coating, the Langmuir-Blodgett method, and spraying. Moreover, impurities and dopants, in some cases, can be easily incorporated by just adding them directly into the solvents, which may be acetone, Butyl-acetate or other organic solvents.

Among polymers that already have found applications in integrated optics are polyimide and PMMA. Polymers are also proposed as versatile materials for modulators, resonators and other optical components that may be used for global optical interconnects in very large-scale integrated circuits. Extensive research in this area is underway to increase the linear and non-linear electro-optic effects in polymers. In regards to integrated optic microsensors, polymers may find extensive application because of their ease of deposition and relatively large and easily controllable elasto-, and electro-optic coefficients.

Polyimides have relatively good environmental stability, thermal stability, and insulating and conformal properties. In the past, we experimented with Du Pont PI-2566, which also has good optical properties (such as a refractive index of around 1.55-1.62 at λ=0.633 Å and relatively low losses), and it is extremely easy to work with. The waveguide formation on polyimides is relatively straight forward. The substrate surface is first treated with an adhesion promoter, followed by spin coating with the premixed polyimide. After a brief initial partial curing of 30 min. at 80 °C, the polyimide is patterned using standard photolithography. Finally, the patterned film is completely cured at 150-330 °C for 30 min. The refractive index inversely depends on the curing temperature, and, in some cases, it also depends on the film thickness. The refractive index is usually between 1.55-1.62 (at λ=0.633 μm), respectively corresponding to curing temperatures between 330-150. Thus, the curing schedule as well as the film thickness across the wafer should be carefully controlled.

Other Dielectrics. In addition to the above dielectrics, a variety of other materials is also developed to form optical waveguides on silicon. Table 5.2 summarizes the optical properties of some of these other waveguide materials.

Table 5.2 Other optical waveguide materials.

Material	Structure	n	Loss at 633 nm (dB/cm)
Barium silicate	Amorphous	1.55	<1.0
SiO_2-TiO_2	Amorphous	1.6-1.9	<1.0
Ta_2O_5	Amorphous	2.20	0.5
Nb_2O_5	Amorphous	2.29	0.5
ZnO	Polycrystalline	1.98	0.1-1

The ion-exchange process can also be used to form optical waveguides. The process is very simple and the fabrication of waveguides using this method is quite straight forward. The idea is to bring portions of a glass substrate containing certain ions like Na^+ into contact with a molten salt containing ions that have a larger diameter than the ions in the glass, such as K^+. Then, ions in the melt and the glass may exchange their position, resulting in a local increase of the refractive index in the glass. The ion exchange is carried out at an elevated temperature >300°C. To define the waveguide, usually a metallic film that masks the glass substrate is used. Openings in this metallic mask allow the ion exchange to occur, forming waveguides and other integrated optic devices. A two-step ion-exchange process has also been used to make dual core waveguides. In this process, K^+ is exchanged first, which is then followed by the exchange of Ag^+. To increase the waveguide depth or to modify its profile, an electric field can be used. Very low-loss waveguides (<0.1dB/cm) with ion exchange technique have been produced .

The ion exchange technology is quite interesting because of its ease of application. The refractive index change that is produced by this process is usually small on the order of 10^{-4}. Thus, the SiO_2 waveguides that are produced using this method have large coupling efficiencies with fiber optic. On the other hand, this method requires initial ions to be present in the oxide, making it impractical for SiO_2/Si systems.

5.6.2 Anti-resonant Reflecting Optical Waveguides

Anti-resonant reflecting optical waveguides (ARROWs) have been proposed and implemented [39]. A schematic diagram of an ARROW is shown in figure 5.15.

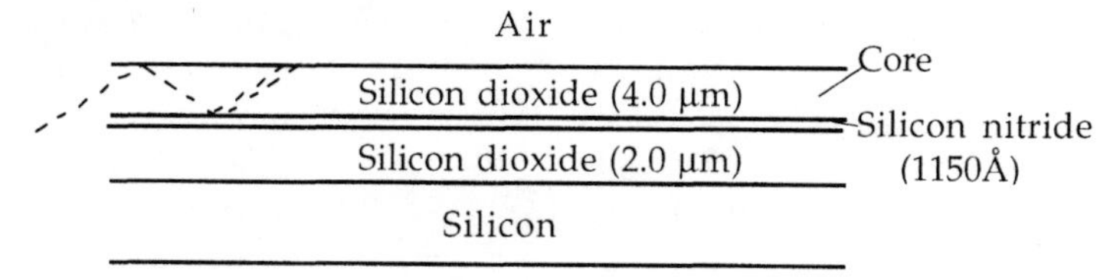

Figure 5.15 Schematic drawing of an anti-resonant reflecting optical waveguide on silicon. The dimensions are given for waveguiding at 0.63μm [39,40].

Light propagating in the core region undergoes total internal reflection at the core-air interface. At the core-cladding interface, the cladding thickness is carefully chosen to result in a very high reflectivity of the incident light. Although leaky in nature, ARROWs can be designed to have very low losses at the wavelength of interest. ARROWs allow efficient coupling to fiber optics since they can be designed to have large dimensions, and they guide light in any dielectric layer, including SiO_2 (which has nearly the same refractive index as the fiber cores). The refractive indices in ARROW are not dictated by the requirements of traditional resonant waveguides where the core layer must have a larger refractive index than the surrounding claddings.

Since ARROWs use dielectric layers similar to those of the conventional waveguides already discussed in this chapter, their fabrication also follows identical procedures. Due to its anti-resonant nature, fabrication of ARROWs is simpler and allows a large process latitude. ARROWs are polarization dependent and cannot be analyzed using scalar wave equations.

5.7 References

1. B. E. Jones and J. S. McKenzie, "A Review of Optical Actuators and the Impact of Micromachining." Sensors and Actuators A, 37-38, pp. 202-207 (1993).
2. M. Tabib-Azar, Integrated Optics and Microstructure Sensors. Kluwer Academic Publishings, Boston (1995).
3. O. Parriaux, "Integrated Optics Sensors." In: Advances in Integrated Optics. Eds: S. Martellucci, A.N. Chester, and M. Bertolotti, Plenum Press, New York, pp. 227-242 (1994).
4. P. V. Lambeck, "Integrated Opto-Chemical Sensors." Sensors and Actuators B, Vol. 8, pp. 103-116 (1992).
5. M. Tabib-Azar and D. Polla (Editors), Integrated Optics and Microstructures. Proceedings of International SPIE Conference (Pub. # 1793) (1992).
6. M. Tabib-Azar, K. K. Wang and D. Polla (Editors), Integrated Optics and Microstructures II. Proceedings of International SPIE Conference (Pub. # 2291) (1994).
7. M. Tabib-Azar (Editor), Integrated Optics and Microstructures III. Proceedings of International SPIE Conference (Pub. # 2686) (1996).
8. A. Garcia, and M. Tabib-Azar, "Sensing Means and Sensor Shells: A New Method of Comparative Study of Piezoelectric, Piezoresistive, Electrostatic, Magnetic, and Optical Sensors." Sensors and Actuators A. Physical Vol. 48 (2), pp. 87-100 (1995).
9. R.T. Kersten, "Integrated Optical Sensors." Proceedings of the NATO Advanced Study Institute on Optical Fiber Sensors, Kluwer Academic Publishers, Boston, Massachusetts, pp. 243-266 (1987).
10. J. Fluitman and Th. Popma, "Optical Waveguide Sensors." Sensors and Actuators 10, pp. 25-46 (1986).
11. C. Wagner, "Optical Pressure Sensor Based on a Mach-Zehnder Interferometer Integrated with a Lateral a-Si:H p-i-n Photodiode." IEEE Photon. Tech. Lett. 5 (10), pp. 1257-1259 (1993).
12. A. Garcia-Valenzuela, and M. Tabib-Azar, "Fiber Optic Force and Displacement Sensor based on Speckle Detection with 0.1 Nano-Newton and 0.1 Angstrom Resolution." Sensors and Actuators A. Physical, Vol. 36 (3), pp. 199-208 (1993).
13. P. Pliska and W. Lukosz, "Integrated-optical acoustical sensors." Sensors and Actuators, A41-A42, pp. 93-97 (1994).
14. A. A. Boiarski, J. R. Busch, B. S. Bhullar, R. W. Ridgway and L. S. Miller, "Integrated-optic Biosensor." SPIE Vol. 1886, pp. 15-26 (1993).
15. A. A. Boiarski, et al, "Integrated Optic Sensor With Macro-Flow Cell." SPIE Vol. 1793, pp. 199-213 (1992).

16. A. A. Boiarski, et al., "Integrated Optics Sensor for Measuring aflatoxin-B1 in Corn." Proceedings of International Integrated Optics and Microstructures III SPIE Conference Vol. 2686), pp. 45-53 (1996).
17. W. Lukosz, "Principles and Sensitivities of Integrated Optical and Surface-Plasmon Sensors for Direct Affinity and Immunosensing." In: Biosensors & Bioelectronics. Vol. 6, pp. 215-225 (1991).
18. M. Tabib-Azar and J. S. Leane, "Direct optical control for a silicon micro-actuator," Sensors and Actuators , A21-A23, pp. 229-235 (1990).
19. M. Tabib-Azar, "Optically Controlled Silicon Microactuators." Nanotechnology Vol. 1, pp. 81-92 (1990).
20. J. S. Leane, "The Photoelectric Micro-Actuator: A Novel Optically-Controlled Silicon Micro-Mechanism." M. S. Thesis, Case Western Reserve University, Cleveland, OH (1990).
21. J. S. McKenzie, K. F. Hale, and B. E. Jones, "Optical Actuators." in Advances in Actuators. Edited by A. P. Dorey, and J. H. Moore, Institute of Physics Publishing, Techno House, Redcliffe Way, Bristol BS1 6NX, UK (1995).
22. H. Uzeitig and H. Bartelt, "All-Optical Pressure Sensor with Temperature Compensation on Resonant PECVD Silicon Nitride Micro-structures." Electron. Lett., Vol. 28 (4), pp. 400-402 (1992).
23. M. V. Andres, K. W. H. Foulds, and M. J. Tudor, "Optical Activation of a Silicon Vibrating Sensor." El. Lett. Vol. 22 (21), pp. 1097-1099 (1986).
24. H. Wolfelschneider, et al., "Optically Excited and Interrogated Micromechanical Silicon Cantilever Structure." Fiber Optic Sensors II, Proc. SPIE, Vol. 798, pp. 61-66 (1987).
25. R. M. Langdon and D. L. Dowe, "Photoacoustic Oscillator Sensors." Fiber Optic Sensors II, Proc. SPIE Vol. 798, pp. 86-93 (1987).
26. B. Culshaw, "Silicon in Optics." Fiber Optic Sensors II, Proc. SPIE Vol. 798, pp. 346-353 (1987).
27. T. S. J. Lammerlink and S. J. Gerritse, "Fiber-Optic Sensors Based on Resonating Mechanical Structures." Fiber Optic Sensors II, Proc. SPIE Vol. 798, pp. 67-71 (1987).
28. D. Uttamachandani, K. E. B. Thornton, J. Nixon and B. Culshaw, "Optically Excited Resonant Diaphragm Pressure Sensor." Electron. Lett. Vol. 23, pp. 152-153 (1987).
29. S. Venkatesh and S. Novak, "Micromechanical Resonators in Fiber-Optic Systems." Opt. Lett. Vol. 12, pp. 129-131 (1987).
30. Y.J. Rao and B. Culshaw, "Comparison Between Optically Excited Vibrations of Silicon Cantilever and Bridge Microresonators." Sensors and Actuators A 30, pp. 203-208 (1992).
31. Y.J. Rao and B. Culshaw, "Continuously Stable Self-oscillation of Silicon Cantilever Microresonators." Electron. Lett., Vol. 27 (19), pp. 1697-1699 (1991).
32. J. Breguet, J. P. Pellaux, and N. Gisin, "Photoacoustical Detection of Trace Gases with an Optical Microphone." Proc. 10th Optical Fibre Sensors Conference, pp. 457-460 (1994).
33. Y. Uenishi, H. Tanaka, and H. Ukita, "AlGaAs/GaAs Micromachining for Monolithic Integration of Optical and Mechanical Components." Integrated Optics and Microstructures II, Proc. SPIE, Vol. 2291, pp. 82-91 (1994).
34. H. Ukita, Y. Uenishi, and H. Tanaka, "A Photomicrodynamic System

With a Mechanical Resonator Monolithically Integrated with Laser Diodes on Gallium Arsenide." Science, Vol. 260, pp. 786-789 (1993).

35. S. Haroche and D. Kleppner, "Cavity Quantum Electrodynamics." Physics Today, Vol. 42 (1), pp. 24-30 (1989).
36. (a) C. N. Cohen-Tannoudji and W. D. Phillips, "New Mechanisms for Laser Cooling." Physics Today, Vol. 43 (10), pp. 33-40 (1990).
    (b) T. Galstian, "Photon Angular Momentum Rotates Molecules." OE Reports (reported by: F. Su) No. 162, p. 1 June (1997).
37. Brian Culshaw, Smart Structures and Materials. Artech House, Boston MA (1996).
38. S. W. Smith, M. Mehregany, F. L. Merat, and D.A. Smith, "All-Silicon Waveguide and Bulk-Etched Alignment Structure on (110) Silicon for Integrated Micro-Opto-Mechanical Systems." Proceedings of International Integrated Optics and Microstructures III SPIE Conference Vol. # 2686, pp. 17-28 (1996).
39. M. A. Duguay, Y. Kokuban, T. L. Koch and L. Pfeiffer, "Antiresonant reflecting optical waveguides in SiO_2-Si multilayer structures." Appl. Phys. Lett. 49, pp. 13-15 (1986).
40. Y. Kokubun, T. Baba, T. Sakaki and K. Iga, "Low-loss antiresonant reflecting optical waveguide on Si substrate in visible-wavelength region." Electron Lett. 22, pp. 892-893 (1986).

Mechanical and Acoustic Microactuators and Micropumps

6.1 Introduction

As we discussed in Chapter 1, actuators can be viewed as devices that transform various forms of energy into mechanical work. In mechanical actuators, the starting energy itself is in the mechanical form and the actuator transforms one mechanical variable into another one [1-6]. Displacement, velocity and acceleration, and their equivalents in fluids [7-15], are the main variables of interest and importance in mechanical actuators. Examples of these actuators are solid links and levers that transform small displacement and large forces to large displacements with small forces and gears that perform similar tasks. Other interesting examples are mechanisms which can be used to "rectify" oscillatory motions and generate d.c. displacements [16-20].

Mechanical parts and devices are used in almost every actuator covered in the present book. This separate chapter is devoted to mechanical actuators since it is possible to design devices with a purely mechanical form of energy as their input and output. The fluidic amplifier of section 6.4 is one such example. A more interesting example of mechanical actuators is the bi-stable diaphragm discussed in section 6.2.4. These diaphragms are extensively used in calculator keyboards. Using a mechanical, load they can be biased just below their instability so that a very small mechanical perturbation causes a large change; possibly producing large displacements and forces. The following section discusses solid links, levers and gear boxes.

6.2 Solid Links and Other Mechanism

These microactuators were the first type to be fabricated using micromachining methods. Successful fabrication of free-moving three-degrees-of-freedom planar mechanisms have been reported by many researchers [1,2], where the three degrees of freedom are x and y motion, as well as rotation (Figure 6.1).

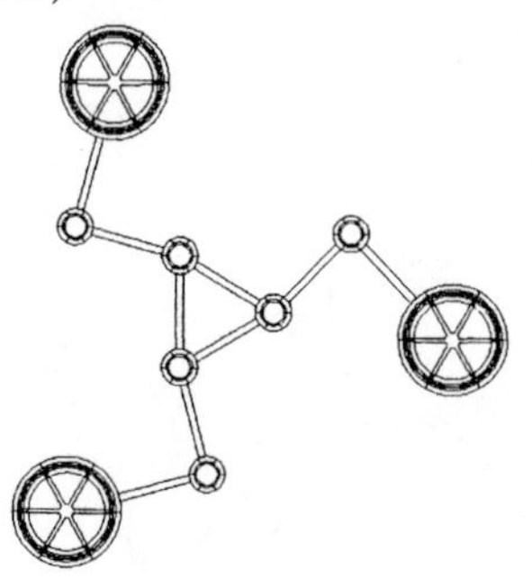

Figure 6.1 Schematic of a three-degrees-of-freedom mechanism [3].

This mechanism has been redesigned to incorporate three electrostatically actuated outer-rotor micromotors available for driving and controlling the three degrees of freedom. Figure 6.2 shows an SEM photo of a microfabricated joint that is used in the mechanism shown in figure 6.1.

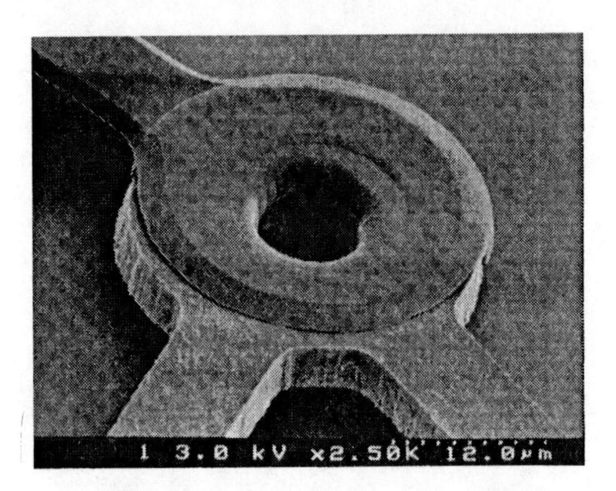

Figure 6.2 An SEM photo of a joint structure of a three-degree of freedom mechanism after release. Note that the nitride link is in the same plane as the polysilicon link and is free to rotate [3].

6.2.1 Deformable Links and Levers

Figure 6.3 shows a microfabricated deformable link coupled to a micromotor through a portion of a gear [1-5]. This particular structure was designed and fabricated to directly measure the torque that can be generated by the micromotor. However, deformable links can be used to store energy for high-speed actuation or other uses.

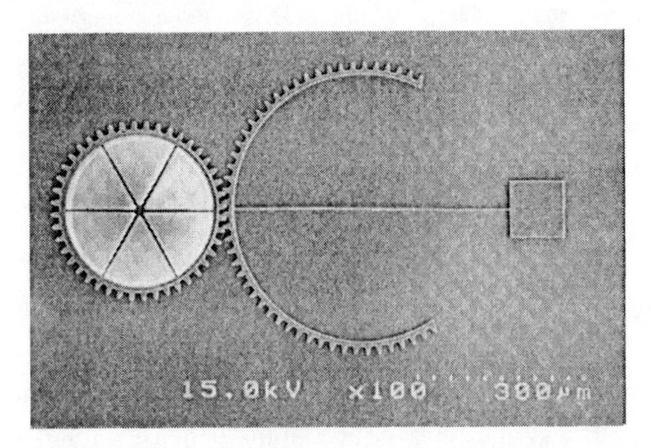

Figure 6.3 An SEM photo of a direct torque measurement device. A 500 μm-long beam with a partial rotor is coupled to a 150 μm radius outer-rotor micromotor [3].

Another design that utilizes links and joints for transmission of mechanical power is a piston. Figure 6.3 shows a simple piston-type crank, which allows oscillatory linear motion similar to that already achievable with lateral resonant devices. Here, the frequency of oscillation is equal to the number of rotations per second of the driving micromotor. Oscillation frequencies of a few Hertz are generated for a typical micromotor speed of about 50 RPM. The length of lateral deflection is dependent on the design of the link coupling and the size of the driving micromotor.

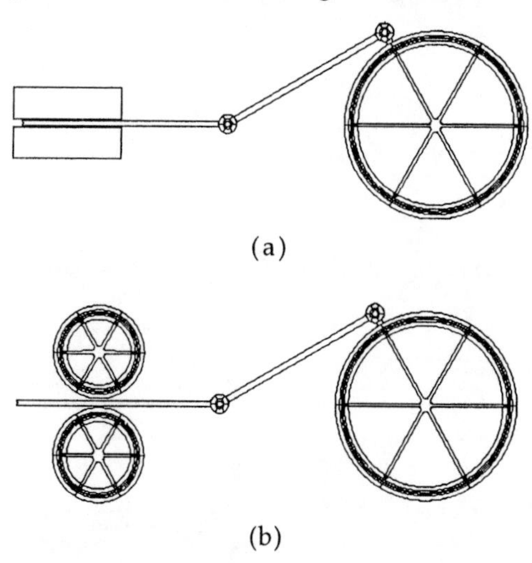

Figure 6.4 a) Schematic of a piston-type structure with guide blocks.
b) Schematic of a piston -type structure with rollers [3].

Figure 6.5 An SEM photo of a joint structure of a piston structure after release [3,5].

A key aspect of the surface micromachined link-based designs described above is that without planarization, full motion of the links is not possible. Figure 6.6 shows a cross-sectional diagram of links fabricated with and without planarization. The middle link is made from the same material as the micromotor bearing (in this case, silicon-rich silicon-nitride), and blocks the driving micromotor (figure 6.6(c)), so rotation of the rotor is limited. The planarization process used to solve these problems involves deposition of CVD low-temperature oxide (LTO) after patterning of the polysilicon layer, and subsequent planarization utilizing chemical-mechanical polishing techniques [3,4]. Once planarization is complete, fabrication can continue to create links and joints that are better suited for mechanical transmission of power.

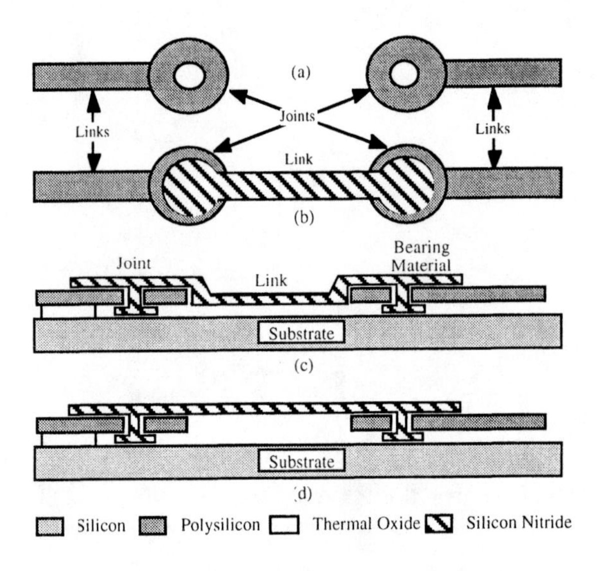

Figure 6.6 Cross-sections illustrating the fabrication process of a joint and link. (a) top view of polysilicon before deposition of the bearing material; (b) top view of the joints and links after deposition of the bearing material; (c) cross-section without planarization, side view; (d) cross-section of a device using planarization during fabrication, side view [5].

Figures 6.2 and 6.5 show the resultant joints and links which have been fabricated using this planarization technique. These devices have been released, and planarization has been successful as evidenced by the links, which are suspended above the substrate.

6.2.3 Gear Boxes

Gear boxes are familiar devices in mechanical systems. They are used to couple rotational motion among different parts of machines and can provide an effective method of increasing torque at the expense of reducing the rotational speed or vice versa. Figure 6.7 shows an SEM photo of two coupled gears that were fabricated using the surface micromachining methods descried in Chapters 1 and 2.

Figure 6.7 An SEM photo of coupled outer-rotor micromotors with involute teeth [3].

6.2.4 Bistable Diaphragms and Structures

Actuators that take advantage of large displacement/forces that can be generated with bistable structures are not reported in the literature. In principle, these actuators are quite easy to construct. To illustrate their operation, let us consider a diaphragm that is plastically deformed and has two stable configurations as shown in figure 6.8a. Residual stress can also be intentionally introduced in the diaphragm to make it buckle. The potential energy associated with the two stable configurations of the "buckled" diaphragm is shown in figure 6.8b.

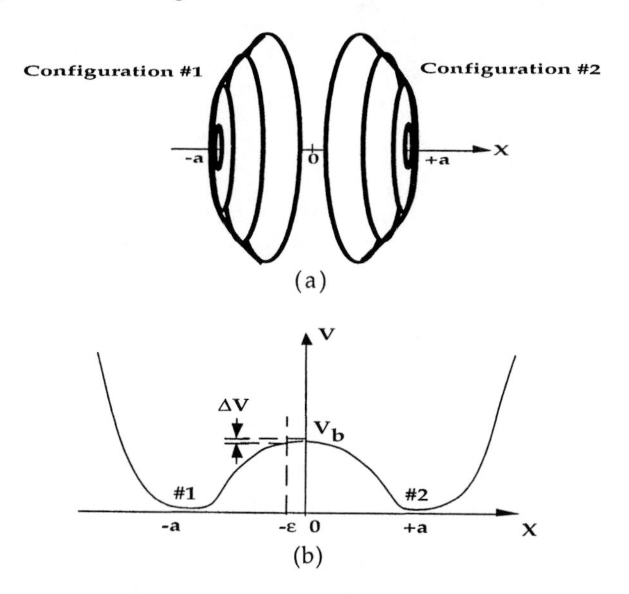

Figure 6.8 a) Schematic diagram of a "bistable" mechanical actuator. b) The potential energy corresponding to the transition of the bistable diaphragm from one stable configuration (#1) to another stable configuration (#2).

To take the diaphragm from configuration #1 to configuration #2 (figure 6.8b), a certain amount of energy, equal to the barrier energy denoted by V_b, should be delivered to the diaphragm. If the diaphragm is mechanically biased at $x=-\varepsilon$, the amount of energy necessary to cause the transformation becomes ΔV, which can be considerably less than V_b. Thus, the idea is to use another actuator that can be quite slow and bias the diaphragm at $x=-\varepsilon$. Then, using a small perturbation, generates possibly large displacement ($\Delta x \approx a+\varepsilon$) and force ($\approx V_b/\Delta x$) on a very short time scale.

6.3 Ultrasonic and Acoustic Microactuators

It is well known that energy can be transported using ultrasonic and acoustic fields [21-24]. In microactuators, the oscillatory ultrasonic and acoustic fields are either "rectified" or coupled directly to obtain a motion in solids, liquids or gases. The following sections discuss different microactuators that use ultrasound and acoustics to generate rotary and linear motions. Since these actuators usually use a piezoelectric actuator to generate the ultrasound or acoustics, some of them are discussed in Chapter 2 along with other piezoelectric actuators. Electromagnets can also be used to generate acoustics, but so far acoustic microactuators that are reported in the literature have employed piezoelectrics almost exclusively.

In the following section, an ultrasonically driven surgical cutter or knife is described. Section 6.3.2 describes bulk-acoustic wave actuators followed by section 6.3.3 that describes related devices used for inkjet printers. Section 6.3.4 describes an interesting "mist" generator device that uses surface acoustic waves (also called Raleigh waves).

6.3.1 Ultrasonic Microcutter

Microcutters for surgical applications can be inexpensively fabricated using micromachining of silicon. Silicon has mechanical properties that makes it an ideal candidate as cutting knife material. Moreover, neural activity sensors can be integrated on the tip of the silicon knife to enable monitoring of electrical activity in the tissue that is being cut [6].

To achieve the cutting action, the scalpel tip should be vibrated. Figure 6.9 schematically shows a cutting tool made out of silicon that uses a PZT actuator to vibrate the cutting edge.

The piezoelectric actuator should be placed on the knife so that maximum power can be transformed to the knife edge. Using a very simple design (figure 6.9), displacements as high as 9 μm at 15 $V_{p\text{-}p}$ in air have been demonstrated. In glycerin, the same cutter exhibited 3 μm displacement at 15 $V_{p\text{-}p}$ [6].

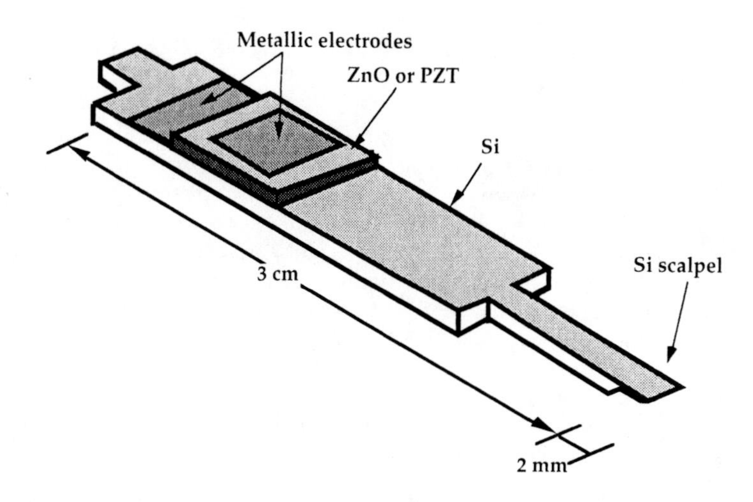

Figure 6.9 Schematic of a silicon ultrasonic surgical cutter [6].

6.3.2 Bulk Acoustic Wave Pumps

The application of ultrasound in industrial processing is quite well-known and documented. Cavitation, liquid streaming, cleaning, electrochemical machining, homogenization and emulsification, degassing of liquids, drying, fiber treatment, electrochemical etching, aerosol generator and atomizers are but a few examples of the applications of ultrasound to material processing [21-24].

One of the important applications of micromachining is in the area of fluidics (i.e., micropumps, valves, chemical reactors, DNA processors, etc.) [25] and ultrasonic methods can be readily used to realize compact, efficient and cost-effective micromachined fluidic devices. We discuss a few examples of ultrasonically actuated devices here.

Before discussing specific devices, however, we review the basics of ultrasonic/liquid interactions and resulting phenomena. When liquids come in contact with a vibrating surface, depending on the frequency and amplitude of vibration and the properties of the liquid (i.e., viscosity, density, chemical constituents, etc.) and its amount, different phenomena may occur. These are listed and described as follow [19,21-24].

Streaming. If the volume of the liquid is large and the vibration amplitude is small, streams may be set up inside the liquid due to the passage of the ultrasound. The process responsible for the generation of these streams is the difference in the liquid behavior when it is compressed compared to when it is strained. When compressed, the liquid acts more like a solid, while when strained, it acts more like a gas. Thus in the pressurizing cycle of the ultrasound, the liquid is pushed forward more than it is retrieved in the de-pressurizing part of the cycle. The net effect is the motion of the liquid in the direction of the ultrasound propagation [23]. The stream generation phenomena can be considered as the rectification of the oscillatory motion of the ultrasound by liquids. In references [19,23,24], the streaming theory is developed.

Cavitation. If the amplitude of vibration is large, ultrasound may cause cavitation in the liquid. Cavitation refers to the formation of a cavity which it occurs because when the vibrating surface is moving in the direction of decompressing and reducing the liquid pressure, any sub-microscopic bubbles tend to expand and grow. In the second part of the cycle, when the vibrating surface compresses the liquid, increasing its pressure, the expanding microbubbles collapse. The time scale of microbubble growth is around T/4, where T is the vibration period, while the time scale of the Microbubble collapse is usually a fraction of T/4. Thus, the collapse occurs much faster than the growth, causing large instantaneous pressures and temperatures to develop in the center of the bubble. Thus, it is the sudden collapse of growing bubbles that causes cavitation [19,23].

The microbubbles usually form around a nuclei, which may be impurities in the liquid etc. After the bubbles collapse, these nuclei usually remain intact and nucleate future bubbles. In ultrasonic cavitation, the sound only supplies energy at a slow rate to expand bubbles, and the sudden collapse of these expanding bubbles create the cavitation.

Cavitation can also be achieved by a sudden rise of local temperature and application of a localized electric field. Cavitation has many applications in industrial processing, including cleaning, electrochemical machining, homogenization and emulsification, degassing of liquids, drying, fiber treatment, and electrochemical etching.

Atomization. When the vibration amplitude is large, in addition to cavitation, aerosols or mists may be produced at the free surface of the liquid. The device that produces the aerosol is called an atomizer and it has a variety of applications in painting, fuel injection systems, humidifiers, catheters, etc. Atomization causes an increase in the surface-to-volume ratio of the liquid with the corresponding increase in its reactivity [24,23].

Atomization of a liquid droplet over an ultrasonic transducer occurs in the following manner. The vibrating transducer generates capillary waves in the droplet. When the amplitude of strain within the liquid droplet is sufficiently large to exceed the surface tension and tensile strength of the liquid, a free droplet forms. The size of this droplet is approximately half of the wavelength of the capillary wave. This wavelength can be related to the frequency of the ultrasound:

$$\lambda = (8\pi T / \rho f)^{1/3} \tag{6.1}$$

where T is the surface tension, r is the liquid density and f is the ultrasound frequency.

6.3.3 Inkjets

Inkjets use a variety of excitation methods, including thermal, electrostatics, rf and ultrasonics, to propel ink droplets [10,26-29]. High-frequency recording with electrostatically deflected inkjets is a very old-technique, so some of its operation principles go back to the time of Raleigh and Kelvin [10,26-29]. The modern inkjet printers use micromachined silicon to replace the costly ink manifold, nozzle and control plates and electronics.

Figure 6.10 schematically shows a micromachined silicon inkjet. An electrostatic field applied between the micromachined silicon electrode slot and the ink container (V≠0) charges the ink droplets, which are subsequently deflected by a high voltage deflector down-stream, preventing them from hitting the paper. When V=0, the electrostatic charging is absent and the neutral droplets hit the paper and print [10]. To be able to independently control the ink dots, each silicon slot must be electrically isolated. This device uses a binary voltage control as opposed to the other inkjet printers, which require analog control and are much more complex.

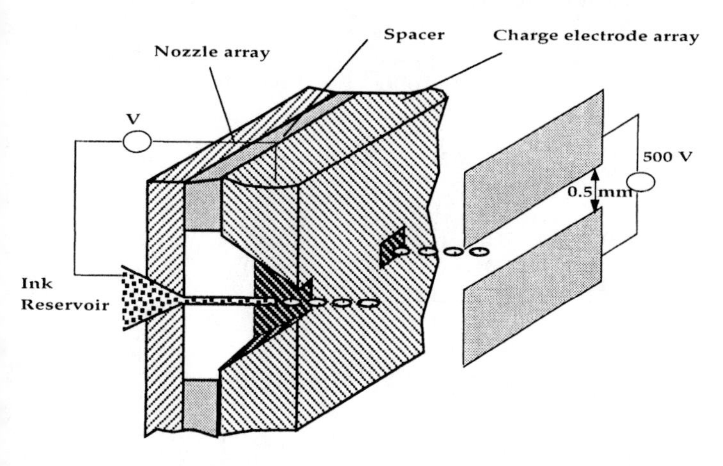

Figure 6.10 Schematic of a nozzle and a charge electrode used in inkjet array.

Ink droplets can be generated thermally by locating a heater in the chamber adjacent to the silicon slot. In this configuration, as the ink arrives in the chamber, the current to the heater is turned on and the heat vaporizes the ink generating a bubble and propelling the rest of the ink out of the silicon opening (these are also called bubble jets). Piezoelectric drivers are also used in inkjets. A piezoelectric patch over the substrate underneath a silicon nozzle can be used to propel the ink droplets. We discuss a similar device in the next section. In reference [29], a piezoelectrically-driven inkjet device is described, where the nozzle is formed on the edge of the silicon wafer (along the well-delineated <111> directions), as opposed to its surface.

Inkjet devices can be considered examples of devices that use atomization or the mist-generation process to achieve injection of droplets of ink fluid for printing purposes. One of the challenges of using silicon micromachining to fabricate inkjet arrays is the uniformity of the slot and nozzle openings. In reference [30], various methods are discussed that are used to achieve the desired uniformity.

6.3.4 Bulk Acoustic Wave Pumps

Figure 6.11a shows a schematic of a piezoelectric pump that generates streams in the liquid enclosed in its chamber to achieve pumping action. Usually in these devices, the input and output ports are not symmetric with respect to the flow process (figure 6.11b). Either a positive pressure on the input port guarantees the flow of the fluid in the positive direction or, in the case of micropumps, various methods are used to insure one-way flow of the liquids as the piezoelectric driver is excited sinusoidally. Figure 6.11b shows an input/output microdiffuser configuration that allows a dynamic and valve-less pumping action [8, 31]. Valves having a different input/output cross-sections have been shown to "dynamically" direct the flow [32,33] as shown in figure 6.11 .

Capillary action can also be used to bring the fluid into the chamber and the fluid is subsequently pushed out of the outlet port upon the excitation of the piezoelectric driver.

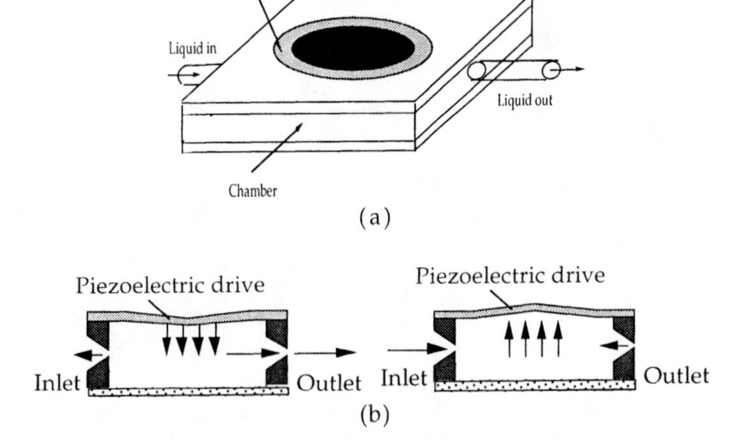

Figure 6.11 (a) Schematic of a piezoelectric micropump that can also be used to form ink or fluid droplets. (b) Schematic of a valve-less pump due to the asymmetry of the input and output microdiffuser configurations.

6.3.5 Flexural and Lamb Wave Actuators and Pumps

Symmetric and anti-symmetric Lamb waves are shown in figures 6.12a and 6.12b, respectively. In thin solid films, such as Si_3N_4 film, the anti-symmetric mode can be readily generated using a ZnO piezoelectric transducer deposited on its surface. Using a 2 μm thick Si_3N_4 diaphragm and a 1 μm-thick ZnO piezoelectric layer excited with interdigitated aluminum transducers, it was possible to generate anti-symmetric Lamb wave amplitudes as high as 14 Å at 14 V_{p-p} at 3.4 MHz in air [12]. The wave amplitudes were directly proportional to the excitation V_{p-p} and reach 60 Å at 16 V_{p-p} in water at 2.45 MHz (the change in the resonance frequency was due to the water loading.) The velocity of 2.3 μm polystyrene spheres that were introduced to water for flow visualization were also directly proportional to the square of excitation voltage (i.e., directly proportional to the excitation power) and reached 100 μm/s at 15 V_{p-p}.

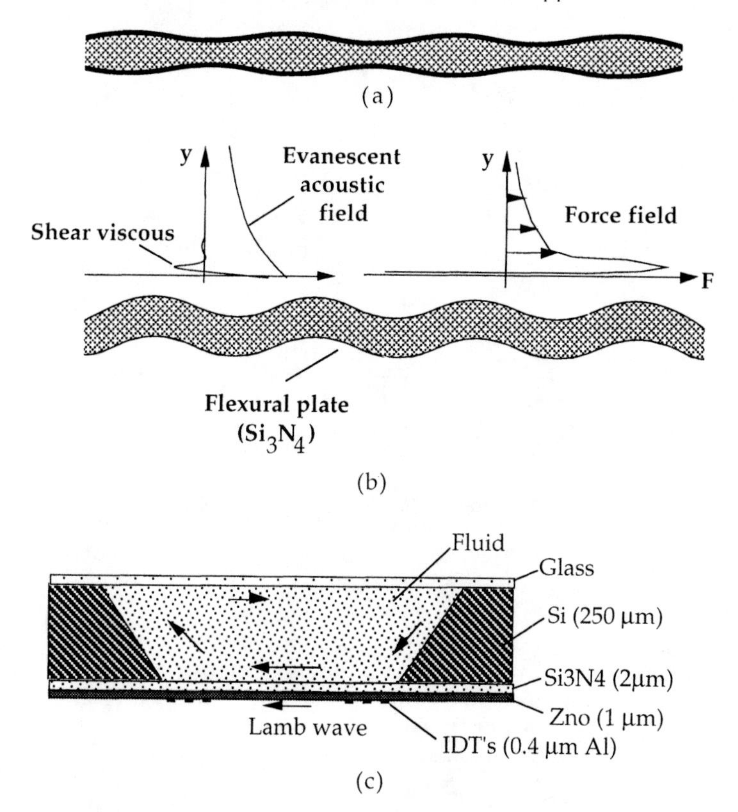

Figure 6.12 (a) Symmetric Lamb wave. (b) Anti-symmetric Lamb wave and the boundary layer profile of a fluid in contact with the diaphragm. (c) Schematic of a Lamb wave micropump [12].

6.3.6 Surface Acoustic Wave Actuators and Pumps

Surface acoustic waves (SAW) are extensively used in signal processing [34] and non-destructive evaluation of semiconductors [35]. A SAW device typically consists of a piezoelectric substrate, such as $LiNbO_3$, and a series of interdigital metallic transducers (IDT's). IDT's are used to apply an electric field pattern to the surface of the piezoelectric substrate and to generate the SAW. SAW devices are also called delay lines since they can produce a considerable time delay between input and output, on the order of 3.5 μs/cm [34,35].

Surface acoustic waves can be used to generate linear motion, fluidic micropumps, and atomizers [14,36]. Figure 6.13 shows the loci of the motion of material points on the surface of a solid substrate with a surface acoustic wave. Due to the requirement of boundary conditions, these material points have to execute an elliptical motion and, hence, their motions can be envisioned as having components parallel and perpendicular to the surface. The parallel motion can be used in linear actuators and in liquid micropumps. When the SAW power is sufficiently large, the elliptical motion of the surface material points can be used to set up capillary waves in liquids and atomize them.

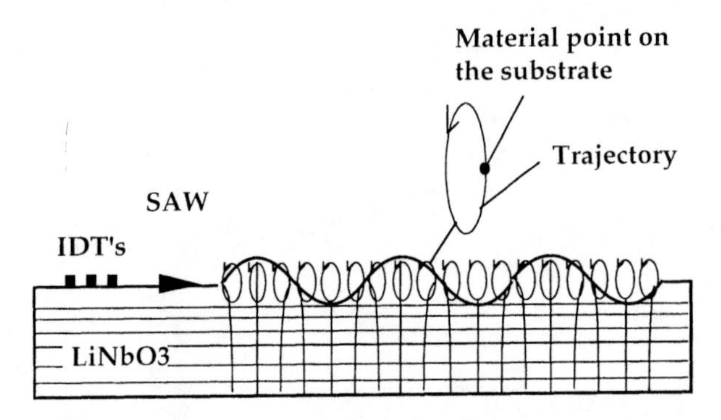

Figure 6. 13 A surface acoustic wave device or delay line with superimposed grids to visualize the surface deformation due to wave propagation. The material points at the surface execute elliptical motion. (Not to scale.)

SAW Linear Actuators. Using the elliptical motion of the material points on the surface of a SAW delay line, it should be possible to construct a linear motor, as schematically shown in figure 6.14. The vertical displacement of the material points on the SAW delay line are quite small - a fraction of the inter-atomic distances. This implies that the slider surface that is in contact with the SAW delay line should be extremely flat and smooth, which may not be practical. Using a very thin liquid film between the slider and the $LiNbO_3$, it may be possible to relax the flatness requirement.

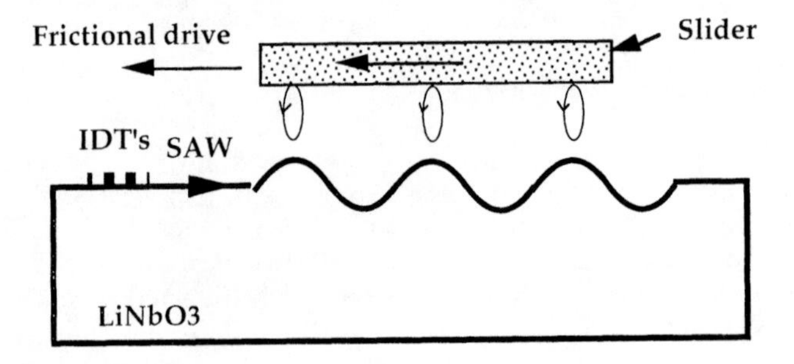

Figure 6. 14 Schematic of a SAW linear motor. (Not to scale.)

If the slider is semiconducting or metallic, it is also possible to take advantage of the acousto-electric interaction taking place between the electric field of the SAW and the free carriers to increase the slider force and speed.

SAW Micropumps. Two different regimes of interaction exist when a thin layer of fluid is situated above a SAW delay line. At low excitation amplitude the fluid is pumped in the direction that is opposite to the SAW propagation direction due to the frictional drive schematically shown in figure 6.15. At higher excitation powers and/or larger thickness of fluids, a radiation pressure drive which pumps the liquid in the SAW propagation direction, is observed [14].

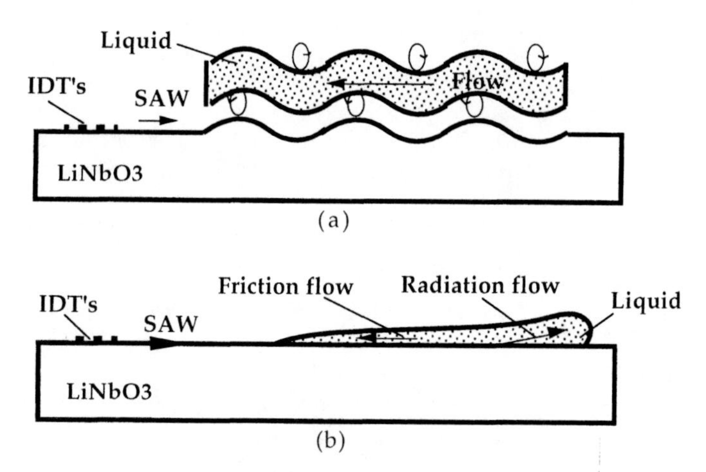

Figure 6.15 (a) The principle of SAW and liquid interaction causing friction flow. (b) Two different driving forces in SAW liquid pumps [14]. These figures are not drawn to scale.

Using a 10 MHz SAW generated on a 127.8° *Y*-rotated $LiNbO_3$ delay line, the edge of the water droplet could be driven at a speed of 0.1 mm/s at an excitation voltage of 80 $V_{p\text{-}p}$ [14].

SAW Atomizers. As SAW travels under the liquid film, figure 6.16a, it generates secondary waves inside the liquids. These secondary waves have a different velocity than the SAW, and to be phase matched, they have to travel at a slanted angle away from the SAW direction. These secondary waves generate the capillary waves needed in the atomization of the liquids.

The SAW velocity in $LiNbO_3$ depends on the crystallographic direction of wave propagation. In 127.8° *Y*-rotated $LiNbO_3$, the SAW velocity is 3800 m/s, resulting in a wavelength of 400 µm at 9.5 MHz. In distilled water at 25 °C, the sound velocity is approximately 1497 m/s resulting in a wavelength of 158 μm. By a simple argument based on the conservation of wave momentum (i.e.; phase matching), one can show that the sound inside water will travel at 67 ° angle with respect to the SAW propagation direction at 9.5 MHz [14].

The particle velocity at the $LiNbO_3$ surface (in the direction perpendicular to the SAW propagation plane) is an important parameter since it determines the liquid particle velocity. In piezoelectric materials, the surface particle velocity is directly proportional to the excitation voltage and to the velocity of the sound. In 127.8° *Y*-rotated $LiNbO_3$, the surface particle velocity was experimentally determined to have a linear dependence on the driving voltage and it reached 4.5 m/s at 240 V_{p-p}. The dependence of the particle size on the SAW wavelength was not as straight forward as in the case of bulk ultrasound atomizers because of the different angles of propagation of SAW, the sound in the liquid and the emerging mist.

The schematic of a SAW atomizer is shown in figure 6.12b. It consists of a 127.8° *Y*-rotated $LiNbO_3$ substrate with an overall dimensions of 2.2 cm X 5.2 cm X 0.2 cm. The interdigitated transducer was designed to have a center frequency of 10 MHz. At 100 V_{0-p} excitation voltage it atomized water containing a small amount of detergent at a rate of approximately 0.1 ml/min with a mean particle size of 19.2 μm and with flight velocities ranging between 1-2 m/s [14].

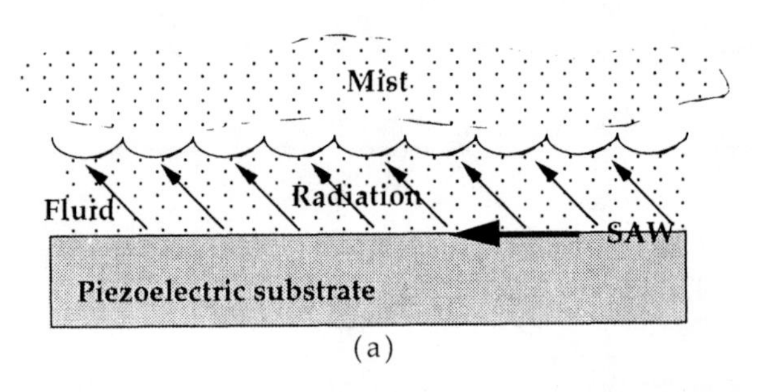

(a)

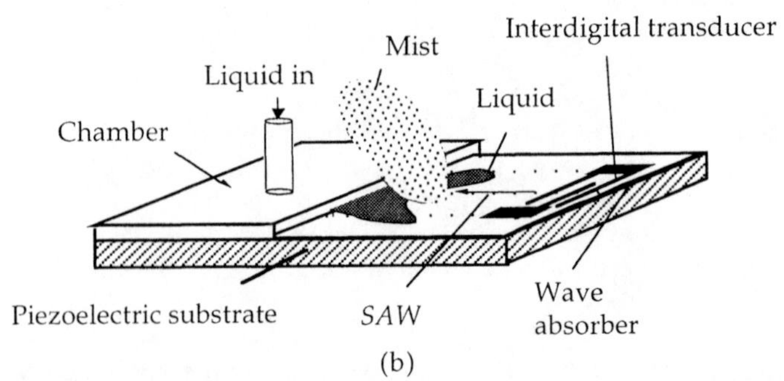

(b)

Figure 6.16 (b) Atomization with a capillary wave generated by SAW. (a) Schematic of a surface acoustic wave (SAW) atomizer [14].

In piezoelectric substrates, such as in $LiNbO_3$, SAW is accompanied by a traveling electric field. The presence of this traveling electric field may produce further interaction between the SAW and the liquid, very similar to the electro-hydrodynamic interactions discussed in Chapter 1. Due to this interaction, liquids with the same densities and viscosities but varying electrical conductivity or polarizabilities, will be pumped at different rates.

6.4 Fluidic Switches and Amplifiers

We have included fluidic switches and amplifiers in this chapter since they use a small fluidic signal to control a large fluid flow. Figure 6.17 schematically shows an example of such a device where the input fluid flow can be diverted between two outputs (#1 and #2) by passing a fluid through the control ports C1 and C2. When the input port is at a higher pressure than the output port ($P_{in}>P_{out}$), a laminar flow emanates and traverses the amplifier region and reaches the output ports and splits into two flows by the splitter. By applying a small differential control flow $\Delta F_c=F_{c2}-F_{c1}$, the laminar flow will be deflected toward one of the output ports, resulting in a differential output flow ΔF_{out} that is given by $A_F(\Delta F_c+F_{c0})$ for small differential control flow, where A_F is the flow gain of the device and F_{c0} is an offset flow that may be needed to set the output equal to zero when asymmetries in the amplifier structure unbalance the flow. Similar relationships also exist for differential pressure gain. Differential pressure gains as high as 3 and flow gains as high as 5 can be obtained.

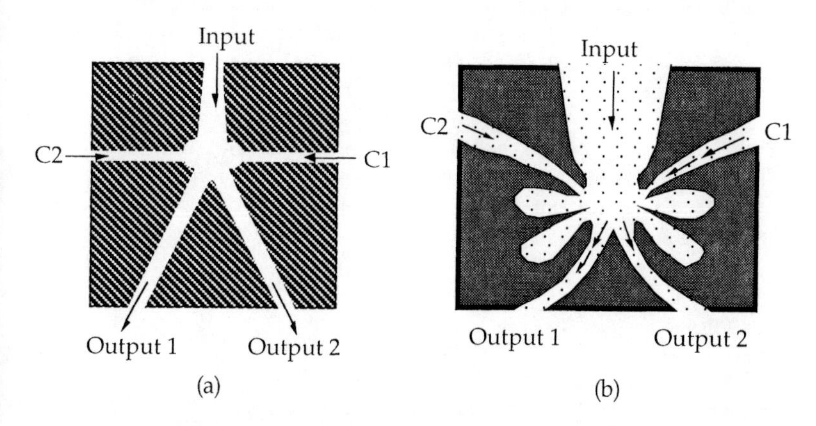

Figure 6.17 Schematic plan view of (a) a fluidic switch, and (b) a fluidic amplifier [9].

A deep reactive ion-etching method is used in etching deep channels (>35 μm) in silicon to define the channels in the fluidic amplifiers and switches. Amplifiers with very high-aspect ratio features are fabricated and reported. A laminar flow fluidic amplifier of 1 mm x 1 mm with 35 μm deep channels are fabricated and reported [9].

Hydraulically-actuated Valve. Figure 6.18 shows a schematic of a valve that is actuated by directing the input flow above or at a diaphragm that acts as the valve. The interesting feature of this design is that the valve opening is regulated by the input pressure. A thermally-actuated cantilever beam directs the flow in the desired direction. We have included this device in the present chapter since it uses the flow itself to actuate the valve. In principle, a small control flow can be used to actuate the cantilever beam.

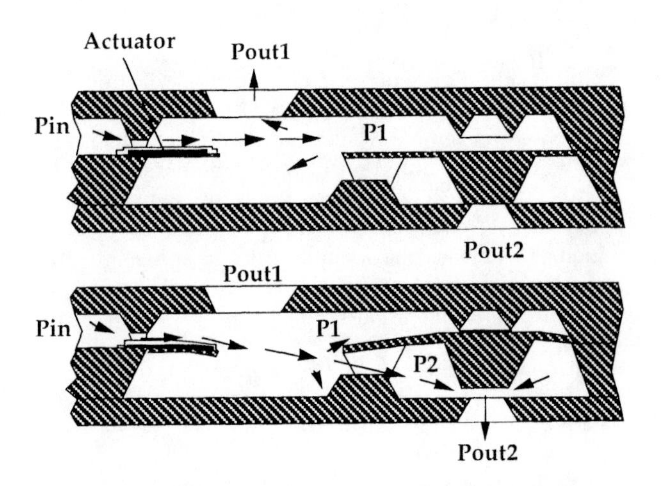

Figure 6.18 Schematic of a hydraulically actuated microvalve. When the flow is directed to the top of the diaphragm it is pressed down, blocking the flow. By directing the flow toward the edge of the diaphragm, it is opened [7].

Typical flow rates were 150 ml/min between input and output#1. When actuated, the flow rate through the output#2 was 10 ml/min. The input pressure was 7 bar and the thermal actuator consumed 8 W of power and actuated in 0.5 ms [7].

6.5 References

1. F. Behi, M. Mehregany and K. J. Gabriel, "A Microfabricated Three-Degree-of-Freedom Parallel Mechanisms." In Proceedings of the IEEE Micro Electro Mechanical Systems Workshop, pp. 159-165 (1990).
2. M. Mehregany and Y. C. Tai, "Surface Micromachined Mechanisms and Micromotors." Journal of Micromechanics and Microengineering, Vol. 1 (2), pp. 73-85 (1992).
3. K. C. Stark, Mechanical Coupling of Polysilicon Surface Micromachined Mechanisms. Ph.D. Dissertation, Case Western Reserve University (1997).
4. N. M. Mourlas, K. C. Stark, M. Mehregany and S. M. Phillips, "Exploring Polysilicon Micromotors for Data Storage Micro Disks." In

Proceedings of the IEEE Micro Electro Mechanical Systems Workshop, San Diego, CA, pp. 198-203 (1996).
5. K. C. Stark, M. Mehregany and S. M. Phillips, "Mechanical Coupling and Direct Torque Measurement of Outer-Rotor Polysilicon Micromotors." In Proceedings of the IEEE Micro Electro Mechanical Systems Workshop, Nagoya, Japan, pp. 221-226 (1997).
6. A. Lal, R. M. White, "Silicon Micromachined Ultrasonic Micro-Cutter." Proceedings of IEEE Ultrasonics Symposium, pp. 1907-1911 (1994).
7. H.-P Trah, H. Baumann, C. Doring, H. Goebel, T. Grauer, and M. Mettner, "Micromachined Valve with Hydraulically Actuated Membrane Subsequent to a Thermoelectrically Controlled Bimorph Cantilever." Sensors and Actuators A, Vol. 39, pp. 169-176 (1993).
8. (a) E. Stemme, and G. Stemme, "A Valveless Diffuser/Nozzle-Based Fluid Pump." Sensors and Actuators A, Vol. 39, pp. 159-167 (1993).
(b) E. Stemme and G. Stemme, "A Valveless Diffuser/Nozzle-Based Fluid Pump." Sensors and Actuators A, Vol. 39, pp. 159-167 (1993).
9. M. J. Zdeblick, P. P. Barth, and J. B. Angell, "A Microminiature Fluidic Amplifier." Sensors and Actuators A, Vol. 15, pp. 427-433 (1989).
10. P. Krause, E. Obermeier and W. Wehl, "A Micromachined Single-Chip Inkjet Printhead." Sensors and Actuators A 53, pp. 405-409 (1996).
11. a) C. E. Bradley and R. M. White, "Acoustically Driven Flow in Flexural Plate Wave Devices: Theory and Experiment." Proceedings of IEEE Ultrasonics Symposium, pp. 593-597 (1994).
b) K. Hashimoto, K. Ikekame and M. Yamaguchi, "Micro-Actuators Employing Acoustic Streaming Caused by High-Frequency Ultrasonic Waves." Transducers '97, pp. 805-808 (1997).
12. R. M. Moroney, R. M. White and R. T. Howe, "Fluid Motion Produced by Ultrasonic Lamb Waves." Proc. IEEE Ultrasonic Symposium, p. 355 (1990).
13. R. M. White, P. J. Wicher, S. W. Wenzel and E. T. Zellers, "Plate-Mode Ultrasonic Sensors." IEEE Trans. Ultrasonics, Ferroelectrics, Frequency Control, UFFC-34, p. 162 (1987).
14. M. Kurosawa, T. Watanabe, A. Futami, and T. Higuchi, "Surface Acoustic Wave Atomizer." Sensors and Actuators A, Vol. 50, pp. 69-74 (1995).
15. I. Edere, J. Grasegger, C. Tille, "Droplet Generator With Extraordinary High Flow Rate and Wide Operation Range." Transducers '97, pp. 809-812 (1997).
16. R. W. Brockett, "On the Rectification of Vibratory Motion." Sensors and Actuators A, Vol. 20, pp. 91-96 (1989).
17. A. P. Pisano, "Resonant-Structure Micromotors: Historical Perspective and Analysis." Sensors and Actuators A, Vol. 20, pp. 83-89 (1989).
18. T. Morita, M. Kurosawa, and T. Higuchi, "An Ultrasonic Micromotor Using a Bending Cylindrical Transducer Based on PZT Thin Film." Sensors and Actuators A, Vol. 50, pp. 75-80 (1995).
19. R. M. Moroney, R. M. White and R. T. Howe, "Ultrasonic Micromotors: Physics and Applications." IEEE Proceeding of Ultrasonics, Pub# CH832, pp. 182-187 (1990).
20. J. S. Danel, P. Charvet, Ph. Robert, P. Villard, "The Electrostatic Ultrasonic Micromotor." Transducers '97, pp. 53-56 (1997).
21. A. D. Pierce, Acoustics. McGraw-Hill Book Company, New York (1981).

22. P. M. Morse and K. U. Ingard, Theoretical Acoustics. Princeton University Press, New Jersey (1968).
23. J. R. Frederick, Ultrasonic Engineering. John Wiley & Sons, Inc., New York, (1965).
24. L. D. Rozenberg, Sources of High-Intensity Ultrasound. Plenum Press, New York, (1969).
25. (a) K. Petersen, "From Microsensors to Microinstruments." Sensors and Actuators A Vol. 56, pp. 143-149 (1996).
(b) J. Fluitman, "Microsystem Technology: Objectives." Sensors and Actuators A Vol. 56, pp. 151-166 (1996)..
26. R. G. Sweet, "High Frequency Recording with Electrostatically Deflected Ink Jets." The Review of Scientific Instruments, Vol. 36 (2), pp. 131-136 (1965).
27. L. Kuhn, E. Bassous and R. Lane, "Silicon Charge Electrode Array for Ink Jet Printing." IEEE Transaction on Electron Devices, Vol. ED-25 (10), pp. 1257-1260 (1978).
28. R. D. Carnahan and S. L. Hou, "Ink Jet Technology." IEEE Transactions on Industry Applications, Vol. IA-13 (1) pp. 95-105 (1977).
29. F. J. Kamphoefner, "Ink Jet Printing." IEEE Transaction on Electron Devices, Vol. ED-19 (4), pp. 584-593 (1972).
30. K. E. Petersen, "Fabrication of an Integrated, Planar Silicon Ink-Jet Structure." IEEE Transaction on Electron Devices, Vol. ED-26 (12), pp. 1918-1920 (1979).
31. M. Esashi, "Integrated Micro Flow Control Systems." Sensors and Actuators, A21-A23, pp. 161-167 (1990).
32. A. Richter, A. Plettner, K. A. Hofmann and H. Sandmaier, "A Micromachined Electrohydrodynamic (EHD) Pump." Sensors and Actuators A, Vol. 29, pp. 159-168 (1991).
33. A. Olsson, G. Stemme and Erik Stemme, "A Valve-Less Planar Fluid Pump with two Pump Chambers." Sensors and Actuators A, Vol. 46-47, pp. 549-556 (1995).
34. C. Campbell, Surface Acoustic Wave Devices and Their Signal Processing Applications. Academic Press, Inc., San Diego, CA, (1989).
35. M. Tabib-Azar, M. N. Abedin, Agostino Abbate, and P. Das, "Characterization of Semiconductor Materials and Devices Using Acousto-Electric Voltage Measurement." Journal of Vacuum Science and Techn. B, Vol. 9 (1), pp. 95-110 (1991).
36. M. N. Araghi, C. J. Kramer and P. Das, "Acousto-Optic Investigation of Layer Wave Properties." Proc. of the IEEE Ultrasonics Symposium, pp. 103-106 (1973).

Chemical and Biological Microactuators

7.1 Introduction

Both chemical and biological actuators are capable of producing very large forces and their energy densities and efficiencies are larger than other types of actuators. Mechanisms that are used in biological systems to extract useful work from chemical reactions are quite interesting and inspiring and can be quite beneficial in the design of novel "artificial" microactuators.

In the following sections we discuss how chemical reactions are used to produce useful work both in man-made and living organisms.

7.2 Conversion of Chemical Reactions to Electrical and Mechanical Energies

Chemical reactions are responsible for the operation of batteries. Batteries can be viewed as devices that use the stored potential energy of materials to generate electricity. A similar but much more sophisticated process takes place in living organisms to convert foodstuff to useful energy. In the following sections we discuss these different mechanisms both in living organisms and in man-made devices.

7.2.1 Operation of Batteries

Batteries are perhaps the simplest examples of devices that convert a chemical reaction to electrical energy. Although they themselves do not do mechanical work but they act as an intermediary step between chemical reactions and mechanical work since they can be used to power actuators that are discussed throughout this book.

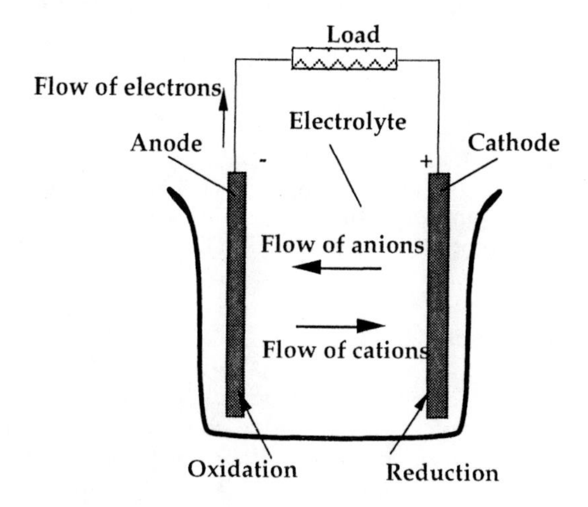

Figure 7.1 Structure of a battery.

In a typical battery operation [1,2], the anode becomes oxidized while the cathode is reduced. In the process, electrons flow from the anode to the cathode through the external circuit. Take for example the Zn-MnO_2 battery. The Zn anode in the process of oxidation gives up electrons to the external circuit while the Mn cathode received electrons from the external circuit as shown below:

$$2\,Zn + 2\,O^{--} \rightarrow 2\,ZnO + 4\,e^{-} \tag{7.1}$$

$$MnO_2 + 4\,e^{-} \rightarrow Mn + 2O^{--} \tag{7.2}$$

The four electrons produced in the first reaction provides the electrons needed in the second reaction. As the reaction proceeds, the cathode becomes reduced while the anode becomes oxidized. Depending on the oxidation mechanism (s) involved, after a thick oxide layer is grown over the anode, the reaction may or may not proceed further. Moreover, as the oxide thickness increases, the internal resistance of the battery may become excessively high and unacceptable. Thus, the battery becomes inoperable even before exhausting all its "fuel" (i.e.; the MnO_2 and Zn).

There are a variety of battery structures that are beyond the scope of this work. Fuel cells, for example, are other notable examples of power sources that are being extensively developed. Their operations are very similar to batteries except that one or more of their reactants are not permanently contained in the battery but supplied to it whenever power is needed.

There are a variety of oxides and materials that are used in batteries [handbook]. It can be shown that each gram of the active battery material (i.e.; Zn, MnO_2 etc.) approximately can produce 234 J of energy. This is approximately 702 J per cm^3 of the active material. A typical D-size dry-cell (Leclanche) battery produces approximately 25 kJ depending on the temperature, discharge conditions etc.

It is important to note that in the process of oxidation, the molecule loses electrons, while in the process of the reduction it gains electrons. This observation has become the basis of a more general definition of oxidation/reduction process based on the net change of the electrons in the starting molecule.

7.2.2 Chemical Reactions in Microactuators

In microactuators reversible electrochemical reactions can be used to generate pressure and perform useful work. When the voltage polarity of the actuator is reversed, however, these reactions should reverse themselves and reduce the pressure. The appealing aspects of electrochemical actuators is that when the excitation voltage is disconnected, the generated pressure may remain intact.

Consider the following reactions in an electrolytic cell with copper sulfate electrolyte, a Pt cathode, and a Cu anode. In the Pt anode we have [3]:

$$2H_2O\ (\ell)\ \text{--------}>\ O_2(g) + 4H^+\ (aq) + 4e^- \qquad (7.3)$$

In the Cu cathode, we have:

$$Cu^{+2}\ (aq) + 2e^-\ \text{------------}>\ Cu\ (s) \qquad (7.4)$$

The overall reversible cell reaction for $O_2(g)$ is:

$$2H_2O\ (\ell) + 2Cu^{+2}\ (aq)\ \text{--------}>\ O_2(g) + 4H^+\ (aq) + 2Cu\ (s) \qquad (7.5)$$

with an equilibrium potential of 0.89 V.

The above reaction can be reversed by short-circuiting the cell or by reversing the cell voltage. The reverse reaction reduces the generated oxygen to water and solid copper to copper ions.

The competing reactions at the copper electrode are:

$$2H^+\ (aq) + 2e^-\ \text{----------}>\ H_2\ (g) \qquad (7.6)$$

and

$$4Cu(s) + O_2\ (g)\ \text{----------}>\ 2\ Cu_2O\ (s) \qquad (7.7)$$

The first reaction requires 1.23 volts. The second reaction has an equilibrium potential of 0.76 V and it is not desirable since it deposits Cu_2O over the copper electrode reducing the actuators efficiency and ultimately reducing the lifetime of the copper electrode. To prevent the spontaneous Cu_2O formation on the copper electrode, a permselective membrane (Nafion^R) is used to shield the copper electrode from the oxygen that is produced in (7.5). Many positive and polar compounds can go through this membrane but negative ions and non-polar compounds are impermeable.

When the cell is energized (V_c<1.23 V), electrolysis takes place and oxygen gas evolves at the platinum electrode while copper ions deposit over the copper electrode. The time of the building up the pressure in the cell depends on surface area of the electrodes, ion and gas diffusion, electrolyte concentration, temperature, and cell voltage (current). When the cell current is set to zero, the cell maintains its pressure and only when the cell voltage is reversed or short-circuited, the cell pressure decreases.

A micromachined silicon cavity with Si_3N_4 diaphragm, shown in figure 7.2a, was used to construct an electrochemical cell capable of supporting the chemical reaction in equation (7.4) [3]. The diaphragm size was 1.2 mm X 1.2 mm x 1 µm and the cavity volume was 2 mm X 2 mm X 0.8 mm. The Si_3N_4 diaphragm deflected by 1.5 µm when the cell was excited with 1.5 Volts and the cell current was limited to 5 µA. At 1 mA cell current, the pressure built-up was severe and the diaphragm was burst open in a few seconds. When the cell reaction was reversed by reversing the cell potential, both the turn-on and turn-off times were around 2 s as shown in figure 7.2b. When the cell reaction was reversed by open-circuiting it, the turn off time was 15 s (figure 7.2b).

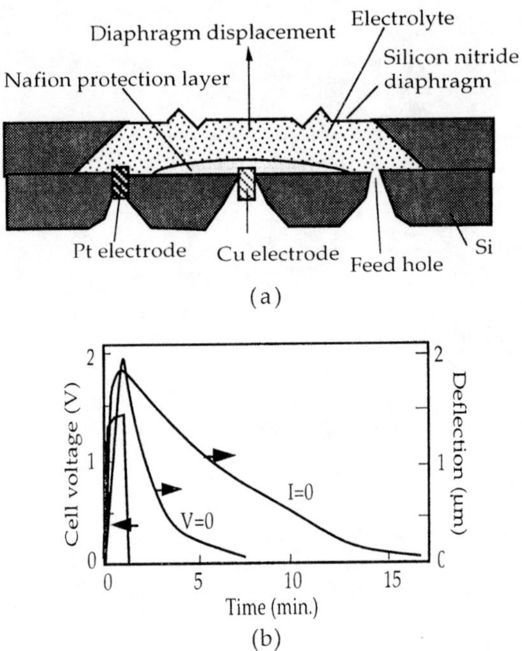

Figure 7.2 (a) Schematic cross-section of a mico-machined electrochemical actuator. (b) The deflection of Si_3N_4 diaphragm as a function of cell time under the different cell voltage and cell current conditions. At t=1 s, the cell voltage was turned off. It took longer (≈15 s) for the diaphragm to deflate when the cell was open circuited (I=0). At short circuit, it took around 2 s to deflate [3].

7.2.3 Chemical Reactions in Living Organisms

There are two sources that are used by living organisms to acquire enrage. If the energy source is light, the organism lifestyle is phototropic and the process of energy production is photosynthesis. If the source of energy is other materials in the environment (foodstuff) its lifestyle is chemotropic and the process of energy production is based on the hydrolyses of a molecule called ATP (Adenosine triphosphate). ATP is the "energy currency" of the living cell [4-10]. It is replenished in living organism either by photosynthesis or by consuming foodstuff that help replenish the ATP in the organism. Thus, essentially plants produce ATP by photosynthesis [7], animals acquire it from plants and humans acquire it from other animals and plants. Of course, eventually we also become the source of the ATP for other organisms.

Bioenergetics, or the thermodynamics of the biochemistry of living cells [4], is the study of energy changes accompanying biochemical reactions. In nonbiological systems, heat can be converted to mechanical or electrical energy. But biological systems are essentially isothermic and no direct use of heat generated by biological reactions are made to produce useful energy. Thus the usual endothermic and exothermic terms that are used in the thermodynamics of non-living systems are meaningless and are replaced by exergonic and endergonic terminology. Exergonic reactions result in the loss of free energy while endergonic reactions result in the production of the free energy.

Endergonic reactions produce chemical energy (free energy). To produce ATP an exergonic reaction is used that generates the high-energy compound at the expense of reducing the free energy. The ATP when broken down, or hydrolyzed, is an endergonic reaction that produces free energy. This cycle is schematically shown in figure 7.3.

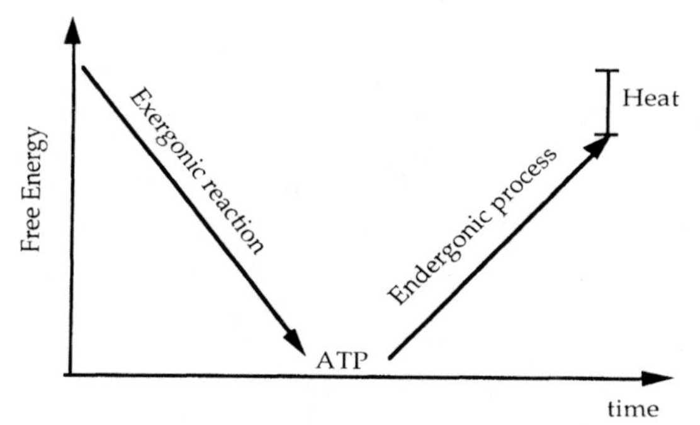

Figure 7.3 Transformation of free energy through an exergonic process that synthesizes a high potential energy compound, such as ATP, to free energy again through an endergonic process which also produces heat [4].

The process of storing energy in high potential energy molecules and their subsequent hydrolyses or oxidation is quite involved and beyond the scope of this article. However, we can understand some of the actuation mechanisms in living organisms by considering the evolution of free energy through ATP synthesis and breakdown [4]. When ATP is hydrolyzed it produces another molecule, which has lower potential energy, called ADP plus inorganic phosphate (Pi). The difference between the ATP and ADP molecules, hence, is that the ADP has a PO_3^- link less than ATP. In the process of the hydrolyses of ATP, 36.8 kJ/mole of free energy is also produced (i.e.; ΔG^0= -36.8 kJ). The hydrolysis of ATP is an oxidation process since it results in the loss of electrons.

In order to synthesize ATP molecules, all organisms must obtain supplies of free energy from their environment. In the case of plants and other autotrophic organisms, a simple exergonic process in the environment, namely the energy of the sunlight, is utilized. Heterothropic organisms, such as human beings, use the complex organic molecules in their environment as the source of free energy.

The efficiency of biological systems is around 50% while that of combustion engines is around 20% [4].

7.2.3 Chemical Reactions in Man-made Engines

Fossil fuel or man made organic molecules or in some cases hydrogen are used in engines. Oxidation of all these fuels occur spontaneously with the resulting loss of free-energy. Engines regulate the oxidation process to take advantage of the resulting heat and expanding gases. As will be discussed in section 7.4, there are ongoing efforts to design and fabricate power MEMS that use a combustion process to produce power and to generate electricity.

7.3 Biological Actuators: Muscle and Locomotion

Physical support and locomotion are the functions of animal skeletal-muscular systems [4-10]. The muscular system generates force, and the skeletal system is the mass against which the muscles contract to produce motion. In living organisms without skeleton, the locomotion is achieved with a variety of different "tricks". In the following section we briefly describe these mechanisms. Since this is not a book on the physiology and biochemistry of the mechanical work production, an in-depth coverage of all the aspects of locomotion is beyond its scope. We have only included the essence of skeletal mechanical work production mechanisms in this section.

7.3.1 Locomotion and Mechanical Work in Biological Systems

Mechanical work and locomotion are generated differently in different living systems. In the following subsections we describe some of the different mechanism used in different living organisms.

Unicells. Unicellular organisms do not have discrete skeletal-muscular systems. Protozoans and primitive algae move by the beating of cilia (plural of cilium: a microscopic hairlike process extending from the surface of a cell or unicellular organism) or flagella (plural of flagium which is a long, threadlike appendage, especially a whiplike extension of certain cells.) Each cilia or flagella contains a cylindrical stalk comprised of eleven microtubules, nine pairs arranged in a peripheral circle with two single microtubules in the center of the stalk (figure 7.4) [11]. Movement is affected by means of the power stroke, a thrusting movement, in which the microtubule cylinders slide past each other. Return of the cilium or flagellum to its original position is termed the recovery stroke.

Figure 7.4 The cilia or flagella containing a cylindrical stalk comprised of eleven microtubules, nine pairs arranged in a peripheral circle with two single microtubules in the center of the stalk [11].

Bacteria. Pseudopodia are utilized by sarcodines (amoebas) for locomotion; the advancing cell membrane extends, the cytoplasm liquefies and flows into the pseudopods as shown in figure 7.5 [11].

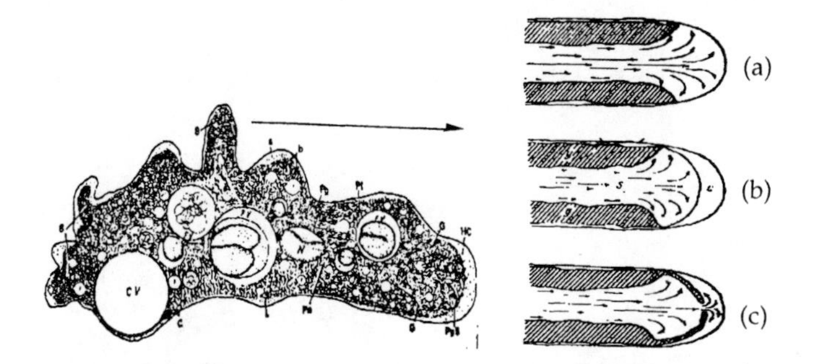

Figure 7.5 Left: Sketch of amoebae proteus showing its different regions. Right: Illustration of growth of a pseudopod. (a) The flowing endoplasm reaches the plasmalemma at the tip of the pseudopod. (b) A hyaline cap and (c) the hyaline cap is broken [11].

Worm. The muscles within the body wall of advanced flatworms such as planaria are arranged in two antagonistic layers, longitudinal and circular. The muscles contract against the incompressible fluid within the animal's tissues (this fluid is thus a hydrostatic skeleton.) Contraction of the circular layer of muscles causes the incompressible interstitial fluid to flow longitudinally, lengthening the animal. Conversely, contraction of the longitudinal layer of muscles shortens the animal. Annelids (e.g. earthworm), advance principally by the action of muscles on a hydrostatic skeleton. Bristles in the lower part of each segment called setae, anchor the earthworm temporarily in earth while muscles push the earthworm ahead.

7.3.2 Different Types of Muscle

Three types of muscle tissue are recognized in vertebrates: skeletal muscle, smooth muscle and cardiac muscle. Approximately 40 per cent of the body is skeletal muscle and almost another 10 per cent is smooth and cardiac muscle [5-11].

Skeletal muscle, also called voluntary or striated muscle, produces the movement of limbs, trunk, face , jaws, eyeballs etc. Each skeletal muscle cell or fiber, is roughly cylindrical, contains many nuclei and is crossed by alternating light and dark bands called striation as shown figure 7.6.

Smooth muscle, also called involuntary or visceral muscle, forms the muscle layers in the walls of the digestive tract, bladder, various ducts another internal organs. It is also the muscle present in the walls of arteries and veins. The individual smooth muscle cells are thin, elongate and pointed at their ends. Each has a single nucleus. The fibers are not striated. They interlace to form sheets of muscle tissue rather than bundles.

Cardiac muscle forms the heart tissue and its fibers show some characteristics of skeletal muscle fibers and some characteristics of smooth muscle fibers. They are striated like skeletal muscle fibers but are mononuclear and involuntary like smooth muscle fibers.

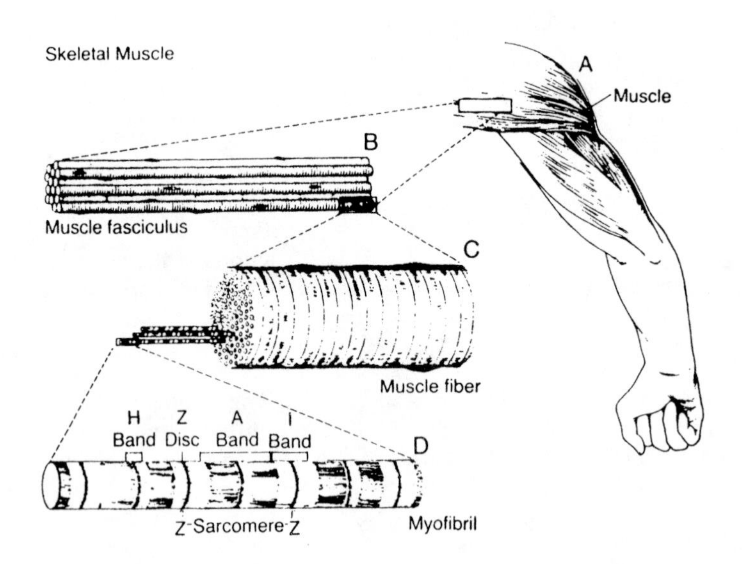

Figure 7.6 Schematic of skeletal muscle [5,6].

7.3.3 Skeletal Muscle Structure and Work Cycle

Skeletal muscle fibers are made of numerous fibers ranging between 10-80 microns in diameter. Most of the fibers extend the entire length of the muscle and except for 2 per cent of fibers, each is inverted by only one nerve ending, located near the middle of the fiber [5,6].

Sarcolemma is the cell membrane of the muscle fiber. At the end of the fiber, the surface layer of the sarcolemma fuses with a tendon fiber and the tendon fibers in turn collect into bundles to form the muscle tendons and thence insert into the bones. Each muscle fiber has thousands of myofibrils and each myofibril has about 1500 myosin filaments and 3000 actin filaments which are large polymerized proteins that are responsible for muscle contraction.

Myosin filaments are thick and actin filaments are thin (figure 7.7). Actin and myosin filaments partially interdigitate and thus cause the myofibrils to have alternate light and dark bands. Light bands contain only actin and are called I-band because they are mainly isotropic to polarized light. The dark bands, contain myosin and the end of actin filaments and are called A-band because they are anisotropic to polarized light. There are also small projections from the sides of the myosin filament which are called cross bridges. They protrude from the surfaces of the myosin filament along the entire extent of the filament except in the very center as shown in figure 7.7.

It is interaction between the cross bridges and the actin filament that causes contraction. Each actin filament is attached to a Z-disc and the filaments extend on either side of the Z-disc to interdigitate with the myosin filaments. Z-disc is composed of several filamentous proteins different from the actin and myosin filaments, also passes from myofibril to myofibril, attaching them to each other all the way across the muscle fiber. Therefore, the entire muscle fiber has dark and light bands as is also true of the individual myofibrils. These bands give skeletal and cardiac muscles their striated appearance.

The protein of myofibril or muscle fiber that lies between two successive Z-discs, is called a sarcomere. When the muscle fiber is at its normal, fully stretched resting length, the length of the sarcomere is 2.0 microns. At this length, the actin filaments completely overlap the myosin filaments and are just beginning to overlap each other and at this length, the sarcomere is capable of generating its greatest force of contraction. When a muscle fiber is stretched beyond its resting length, the ends of the actin filament pull apart, leaving a light area in the center of the A-band which is called H-zone. Normal sarcomere contraction occurs when the ends of the actin filament not only overlap the myosin filament but also overlap each other.

The myofibrils are suspended inside the muscle fiber in a matrix called sarcoplasm. The fluid of sarcoplasm contains large quantities of potassium, magnesium, phosphate and protein enzymes. Also present are tremendous number of mitochondria that lie between and parallel to the myofibrils, a condition which is indicative of the great need of the contracting myofibrils for large amounts of ATP formed by the mitochondria.

In the relaxed state, the ends of the actin filaments derived from two successive Z-discs, barely overlap each other while at the same time completely overlapping myosin filaments shown in figure 7.8. On the other hand, in the contracted state, the actin filaments have been pulled inward among the myosin filaments so that they now overlap each other to a major extent. Also the Z-discs have been pulled by the actin filaments up to the ends of myosin filaments. Actin filaments can be pulled together so tightly that the ends of the myosin filaments actually buckle during every intense contraction. Thus muscle contraction occurs by a sliding filament mechanism. But what causes the actin filaments to slide inward among the myosin filaments? This is caused by mechanical, chemical or electrostatic forces generated by the interaction of the cross-bridges of the myosin filaments with the actin filaments [5,6].

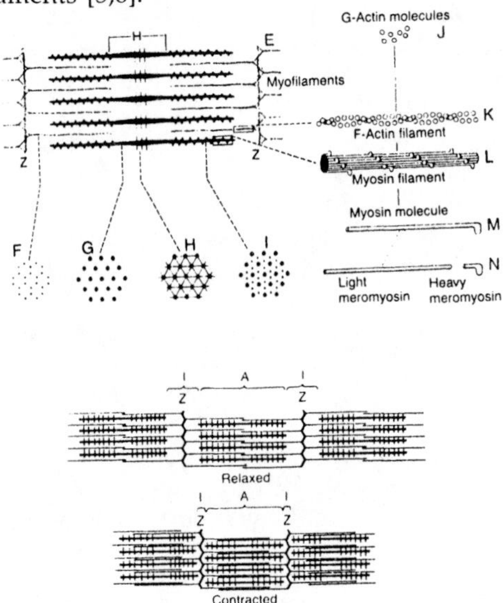

Figure 7.7 The sliding of the actin filament (thin) into the channels between the myosin filaments (thick) resulting in the relaxation (a) or contraction of the muscle [5,6].

Under resting conditions, the sliding forces between actin and myosin filaments are inhibited, but when an action potential travels over he muscle fiber membrane, this causes the release of large quantities of calcium ions into the sarcoplasm surrounding the myofibrils. these calcium ions activate the forces between the filaments and contraction begins. But energy is also needed for the contractile process to proceed. This energy is derived from the high energy bonds of ATP which is degraded to ADP to give the energy required [4].

The myosin filament is composed of multiple myosin molecules. The myosin molecule is composed of six polypeptide chains, two heavy chains each with a molecular weight of about 200000 and four light chains with molecular weight of about 20000. The two heavy chains coil around each other to form a double helix. However, one end of each of these chains is folded into a globular protein mass called the myosin head. Thus there are two free heads lying side by side at one end of the double helix myosin molecule as shown in figure 7.7. The other end of the coiled helix is called the tail. The four light chains are also parts of the myosin head, two to each head. These light chains help control the function of the head during the process of muscle contraction.

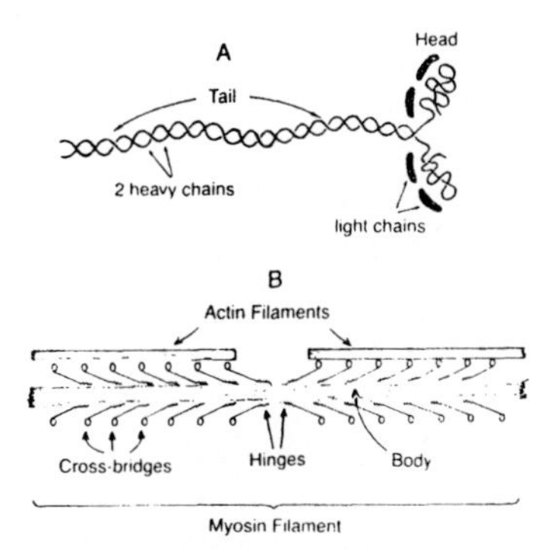

Figure 7.8 (a) An individual myosin molecule and (b) combination of many myosin molecules in a helical shape to form a filament. In (b), the cross-bridges between the myosin heads and active sites are also shown [5,6,8].

The myosin filament is made up of about 200 myosin molecules. The central portion of one of these filaments is shown in figure 7.8b, showing the tails of the myosin molecule bundled together to form the body of the filament while many heads of the molecules hang outward to the sides of the body. Also part of the helix portion of each myosin molecule extends to the side along with the head, thus providing an arm that extends the head outward from the body as shown in figure 7.8b. The protruding arms and heads together are called cross-bridges. Each of these cross-bridges are flexible at two points called hinges, one where the arm leaves the body of the myosin filament and the other where the two heads attach to the arm. The hinged heads are believed to participate in the actual contraction process. There are no cross-bridge heads in the very center of the myosin filament because the hinged arms extend toward both ends of the myosin filament away from the center. Therefore, in the center there are only tails of the myosin molecules and no heads.

Myosin head can function as an ATPase enzyme and this is essential to contraction. This property allows the head to cleave ATP and to use the energy for the contraction process [8].

The actin filament is composed of three different components: actin, tropomyosin and troponin. The back bone of the actin is a double stranded F-actin protein molecule. The two strands are wound in a helix in the same manner as the myosin molecule. Each strand of the double F-actin helix is composed of polymerized G-actin molecule. Attached to each one of the G-actin molecules is one molecule of ADP. These ADP molecules are the active sites on the actin filament with which the cross-bridges of the myosin filament interact to cause muscle contraction. The active sites on the two F-actin strands of the double helix are staggered, giving one active site on the overall actin filament approximately every 2.7 nanometers. Each actin filament is approximately 1 micron long. The bases of the actin filaments are inserted strongly into the Z-discs, while their other ends protrude in both directions into the adjacent sarcomeres to lie in the spaces between the myosin molecules [6-10].

The actin filament also contains two additional protein strands that are polymers of tropomyosin molecules, each molecule having a molecular weight of 70000 and extending a length of 40 nanometer. Each tropomyosin strand is loosely attached to an F-actin strand and that in the resting state it physically covers the active sites of the actin strands so that interaction can not occur between the actin and myosin to cause contraction.

Troponin is a complex molecule of three globular protein molecules which is attached approximately two thirds the distance along each tropomyosin molecule. One of the globular proteins (Troponin-I) has a strong affinity for actin, another (Troponin-T) for tropomyosin and the third one (troponin-C) for calcium ion. This complex attaches the tropomyosin to the actin. The strong affinity of the troponin for calcium ions is believed to initiate the contraction process.

A pure actin filament without the presence of the troponin-tropomyosin complex binds strongly with myosin molecules in the presence of magnesium ions and ATP, both of which are normally abundant in the myofibril. But if the troponin-tropomyosin complex is added to the actin filament, this binding dose not take place. Therefore, it is believed that the active sites on the normal actin filament of the relaxed muscle are inhibited by the troponon-tropomyosin complex. Consequently, they cannot interact with the myosin filaments to cause contraction. Before contraction can take place the inhibitory effect of the troponin-tropomyosin complex must itself be inhibited. In the presence of large amounts of calcium ions, the inhibitory effect of the troponin-tropomyosin on the actin filament is itself inhibited [6].

When calcium ions combine with troponin-C, each molecule of which can bind strongly with up to four calcium ions even when they are present in minute quantities, the troponin complex supposedly undergoes a conformational change that in some way tugs on the tropomyosin protein strand and supposedly moves the tropomyosin strand deeper into the groove between the two actin strand. This uncovers the active sites of the actin, thus allowing contraction to proceed. So the alternation of the normal relationship between the tropomyosin-troponin complex and actin by calcium ions, leads to contraction.

As soon as the actin filament becomes activated by the calcium ions, it is believed that the heads of the cross-bridges of the myosin filaments immediately become attracted to the active sites of the actin filament, and this in some way causes the contraction to occur. The walk-along theory of contraction describes precisely how the interaction between the cross-bridges and the actin leads to muscle contraction. Figure 7.9 illustrates the postulated walk-along mechanism for contraction. (There are some oppositions to this model [8].) this figures shows the heads of two cross-bridges attaching to and disengaging from the active sites of an actin filament [6].

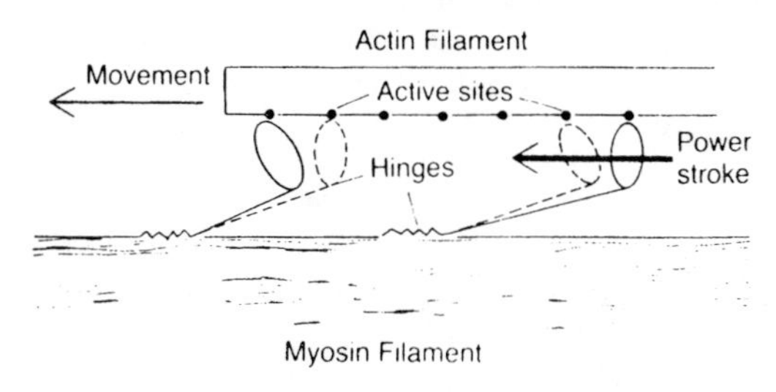

Figure 7.9 The walk-along mechanism, i.e., power stroke, resulting in the contraction of the muscle.

It is postulated that when the head attaches to an active site this attachment simultaneously causes profound changes in the intramolecular forces in the head and arm of the cross-bridges. The new alignment of forces causes the head to tilt toward the arm and to drag the actin filament to drag with it. This tilt of the head of the cross-bridge is called the power stroke. Then immediately after tilting, the head automatically breaks away from the active site and returns to its normal perpendicular direction. In this position it combines with an active site farther down along the actin filament; then a similar tilt takes place again to cause a new power stroke, and the actin filament moves another step. Thus, the heads of the cross-bridges bend back and forth and step by step walk along the actin filament, pulling the action toward the center of the myosin filament. Each one of the cross-bridges is believed to operate independently of all others, each attaching and pulling in a continuous but random cycle. Therefore, the greater the number of cross-bridges in contact with the actin filament at any given time, the greater, theoretically, is the force of contraction.

When a muscle contracts against a load, work is performed and energy is required. It is found that large amounts of ATP are cleaved to form ADP during the contraction process. Furthermore, the greater the amount of work performed by the muscle, the greater the amount of ATP that is cleaved, which is called the fenn effect. The following is a sequence of events that explains how ATP is used to provide energy for contraction:

1- Before contraction begins, the heads of the cross-bridges bind with ATP. The ATPase activity of the myosin head immediately cleaves the ATP but leaves the cleavage products, ADP plus Pi, bound to the head. In this state, the conformation of the head is such that it extends perpendicularly toward the actin filament but is not yet attached to the actin.

2- Next, when the inhibitory effect of the troponin-tropomyosin complex is itself inhibited by calcium ions, active sites on the actin filament are uncovered and the myosin heads do then bind with these, as illustrated in figure 7.10.

3- The bond between the head of the cross-bridge and the active site of the actin filament causes a conformational change in the head, causing the head to tilt backward toward the arm of the cross-bridge. This provides the power stroke for pulling the actin filament. The energy that activates the power stroke is the energy already stored in the head at the time of cleavage of the ATP.

4- Once the head of the cross-bridge is tilted, this allows release of the ADP and Pi and exposes a site on the head where new ATP can bind. Therefore, a new molecule of ATP binds, and this binding in turn causes detachment of the head from the actin.

5- After the head has split away from the actin, the new molecule of ATP is also cleaved, and the energy again " cocks " the head back to its perpendicular condition ready to begin a new power stoke cycle.

6- Then, when the cocked head, with its stored energy derived from the cleaved ATP, binds with a new active site on the actin filament, it becomes uncorked and once again provides the power stroke.

7- Thus, the process proceeds again and again until the actin filament pulls the Z membrane up against the ends of the myosin filaments or until the load on the muscle becomes too great for further pulling to occur.

Initiation of contraction in skeletal muscle begins with action potentials in the muscle fibers. These elicit electrical currents that spread to the interior of the fiber where they cause release of calcium ions from the sarcoplasmic reticulum. It is the calcium ions that in turn initiate the chemical events of the contractile process. This overall process for controlling muscle contraction is called excitation-contraction coupling.

There are some excellent textbooks and references that explain the physiology of muscle and its activation process in detail [5,6,10,11]. The above discussion was given in this section to illustrate how nature has solved one of the most important problems of living organisms, namely its mobility and ability to perform useful mechanical work.

7.4 Chemical Jet in Beetles

Beetles are a quite distinct group of insects, the main characteristic of which is the possession of hard wing covers, or "elytra." The name of the group comes from the Old English *bitan*, to bite, and refers to their tendency to chew and gnaw. The unimaginable variety in their forms has made the beetles the insect order with by far the greatest number of species. So far nearly 290,000 species have been described, which corresponds to more than two-fifth of all the insects which have been named. The number of beetle species is about six times as large as that of all vertebrate species.

The upper side of the body is usually more strongly arched than the underside, the latter usually being a smooth flat surface. The body is unmistakably divided into head, thorax and abdomen but when viewed from above only the dorsal plate of the prothorax (the pronotum) and part of the second thoracic segment (the elytra, which attach to the mesothorax) are visible, figure 7.10 [12].

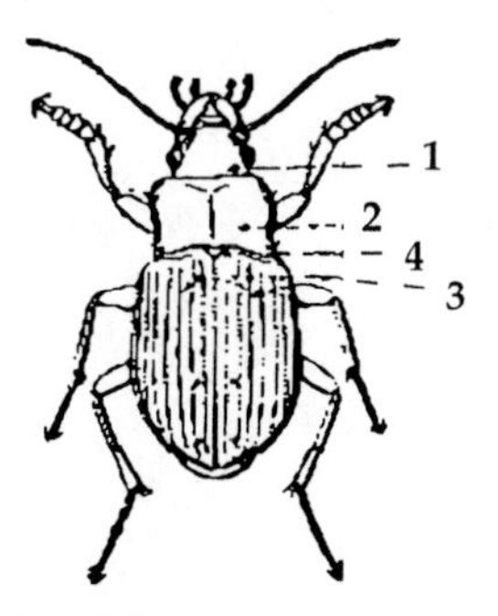

Figure 7.10 Dorsal aspect of a ground beetle: 1-head, 2-pronotum, 3-elytra 4- scutellum [12].

In many beetles another part of the mesothorax is also visible; this is the scutellum, a triangular plate behind the pronotum. The elytra are the strongly sclerotized forewings, no longer used for flight. The bodies of many beetles are characterized by ridges, humps or even hornlike formations. These are most often on the head and the protonum.

7.4.1 Squirting Action

A number of species like *lady beetles* are able to squirt out their blood at special places between femur and tibia when they are disturbed. Colored droplets then appear at these knee joints. In such species blood is charged with defensive substances which smell or taste objectionable to an attacker, or are poisonous or corrosive. This reflex bleeding thus has the same significance as other beetles' defensive spraying from the pygidial glands. Many beetles are protected to a certain extent by their bad-tasting or corrosive body fluids even though they do not exude them from the body. They bear "warning coloration" which is said to repel would-be attackers which have already had unpleasant experiences with identically colored, bad-tasting insects. Probably the most effective poison in beetle blood is cantharidin. Remarkably, this substance dose not harm the insectivorous vertebrates such as frogs, hedgehogs, bats and birds. But a dose of as little as 0.03 g of cantharidin taken internally is fatal to humans [12].

Occasionally their compound eyes are markedly kidney-shaped in outline, and in the *Whirligig Beetles* they are actually divided into two, each upper half adapted for vision on or above the water surface, and the lower half, for underwater vision figure 7.13.

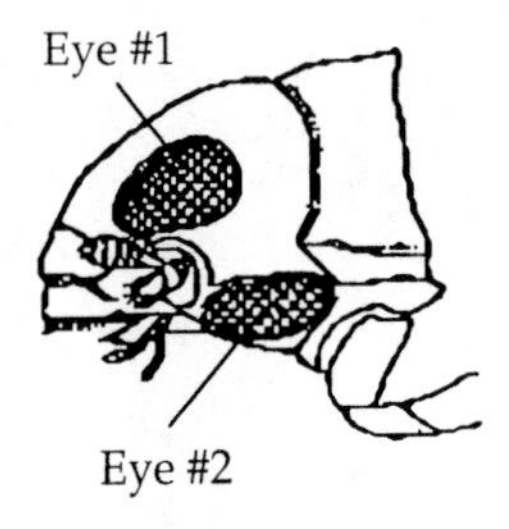

Figure 7.13 The *Whirligig Beetles* have an eye separated into two parts [12].

7.4.2 Surfactant Swimming

A quite different and peculiar method of locomotion is "surfactant swimming. " This is done by the large-eyed species of *Rove Beetles*, using glands at the end of the abdomen; the glands secrete a substance which lowers the surface tension of the water. Once the glandular secretion has touched the water, there is an area of reduced surface tension around the end of the beetle's body; the higher surface tensions at the other points where the beetle touches, act to pull the animal forward, and the direction of movement is controlled by bending the abdomen [12].

An amazing transformation is exhibited by the labium of certain species of the *Rove Beetle*. It can be shot far out to capture prey in a way reminiscent of the tongue-flick of chameleons, particularly since there is a sticky cushion at the tip in both cases (figure 7.14). The apparatus is erected by an increase in the internal blood pressure, and drawn back again by contraction of four pairs of muscle bundles.

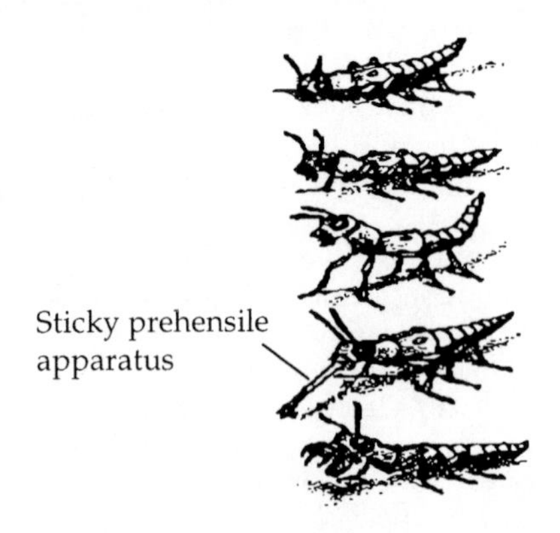

Figure 7.14 *Rove Beetles* catch their prey with a sticky prehensile apparatus which can be shot forward [12].

7.4.3 Explosive Jets

Some beetles like *Ground Beetles* and *Carrion Beetles* have a pair of glands (pygidial glands) in the posterior abdomen, opening at either side of the anus. The special function of these is to secrete substances, repellent in taste or smell which serve as defense mechanism when they are squirted at enemies. In some species like *Violin Beetles* these secretions also have a corrosive action.

Such protective devices are developed to an astonishing degree in the *Bombardier Beetles*. Here the pygidial glands produce two different chemical substances, which are stored in separate chambers closed off by muscular valves. The gland cells of one chamber secrete two different hydroquinone compounds, and those of the other, hydrogen peroxide at a concentration of twenty three percent. When the *Bombardier Beetle* is attacked, it opens the valves of the storage chambers, causing the contents to flow into an especially thick-walled space which serves as a reaction chamber. The instance the two substances come together there, an enzyme catalyzes an explosive reaction involving the decomposition of the hydrogen peroxide. The oxygen thus released supplies the gas pressure which forces out the quinone solution. The fine spray of this solution has such a deterrent effect on most pursuers that they immediately give up their attack. Even reptiles and birds spit out a Bombardier Beetle as soon as this corrosive substance comes into contact with the mouth. Under favorable conditions, the " explosion " of the larger *Bombardier Beetles* can actually be audible. There are occasional reports that a small veil like cloud is formed simultaneously [12].

The beetle is in a sense the most successful of all animal types. It has managed to invade essentially all habitats that the land can offer. At least two characteristics seem to help them arrive at this peak of success. One of these is the preservation of simple biting-chewing mouthparts, not modified into complicated proboscises for piercing, sucking or sponging. The second special property is the transformation of the anterior pair of wings into solid elytra; this gives the abdomen greater protection [12].

7.5 Combustion Engine: MEMS Microengines

As discussed before, spontaneous oxidation of fossil fuel and highly reactive gases such as H_2 generates heat and produces other expanding gases that can be used to generate mechanical and electrical power. It is well know that oxidation of one gram of H_2 at 25°C and at 1 atm produces 143 kJ of energy. Hydrocarbons, such as C_8H_{18}, produce approximately 48 kJ per gram. A macro-scale gas turbine with an air intake opening of one meter, for example, generates power on the order of 10^8 W. Scaling down such an engine to a millimeter scale power production on the order of 10W can be expected provided that the power per unit of intake air volume is maintained [13,14]. In section 4.2.4 of Chapter 4 we discussed these unique and interesting microactuators.

7.6 Conducting Polymer Actuators

In polymers oxidation and reduction is usually referred to as doping and it is well known that doping can result in an increase in the volume of the polymer. Such an increase in the volume or any increase in the linear dimensions of solid can be used to perform useful work.

Conducting polymer actuators are quite simple to construct. Operations of unimorphs consisting of either polypyrrole/Au/Poyethylene, poly (3-octyl-thiophene)/polypyrrole, and poly (3-octy-thiophene)/polyethylene sandwiches were reported [15]. Liquid electrolytes in these actuators were used, together with a spatially separated, mechanically inactive counter electrode. Devices' response were slow, measured in minutes, and actuator failure eventually occurred by delamination. Similar actuator structure laminated with insulating films have shown better performance and were capable of lifting gram weights [15-17].

Bimorph actuator devices using polyaniline on opposite sides of an adhesive polymer tape were also fabricated. Similar bimorph actuators with a wetted paper with aqueous HCl or a solid-state electrolyte inserted between the two polyaniline layers have shown promising characteristics. Polyaniline has two oxidation stages in electrochemical process, that is, there are three stable states as leucoemeraldine salt (LS), emeraldine salt (ES) and pernigraniline salt (PS). Polyaniline deforms upon electrochemical doping and its deformation is faster than polypyrrole due to proton migration on polyaniline.

Polymer gel actuators are also studied. Poly(vinyl alcohol) [15] and its copolymer with poly(acrylic acid) [16] have bean used in these actuators and excitation was achieved by either electrochemical means or exposure to acid/alkali. Conducting polymer gel fibers made of polythiophene [17] were also studied and showed promising results of fast response time. Electro-chemo-mechanical polymer actuators [17] were fabricated by using the electro-deposited polyaniline film.

7.7 Gels

Structural behavior of gels is intermediate between that of a solid and a liquid [15-22]. Mostly fluidic, they are given form by a tangled network of polymer strands. A balance of forces maintain this state of affairs that can be disturbed by minute perturbations brought about by a variety of influences. Optical, thermal, electrical, and chemical methods have been shown to cause drastic reduction in the volume of certain gels. These changes are the bases of the application of gels in actuators and are reviewed in this section.

The drastic changes that occur in the state of the gel, is understood to be caused by its phase transformation. When the temperature is lowered, the polymer network loses its elasticity and becomes compressible. Below a critical temperature, the elasticity becomes zero and compressibility becomes infinite resulting in the collapse of the gel. The volume of certain gels shrink or swell by a factor of as much as several hundred when the temperature is varied. Around the critical temperature, these changes can become discontinuous resulting in a very large volume change for an infinitesimal change in the temperature. Similar changes occur due to the optical, electrical and chemical perturbations as discussed in the following sections.

Thermal Effects. These effects were discussed in section 4.4.2 of Chapter 4. For a properly hydrolyzed gel, the change in the volume below the critical temperature was large and discontinuous . Longer hydrolysis results in larger discontinuities in the volume below the gel's critical temperature. In acrylamide gel hydrolyzed for eight days, the swelling ratio was 40 and it was possible to increase this ratio by longer hydrolysis [18,21].

Optical Effects. Figure 7.15 shows the effect of visible light on N-isopropylacrylamide/chlorophine copolymer gel. The gel temperature was 31.5 °C and illumination by visible light resulted in nearly 50% change in the gels' diameter (under the light focused spot size of 20μm). The actuation speed was low around a few second for a gel diameter of around 200 μm [19,22].

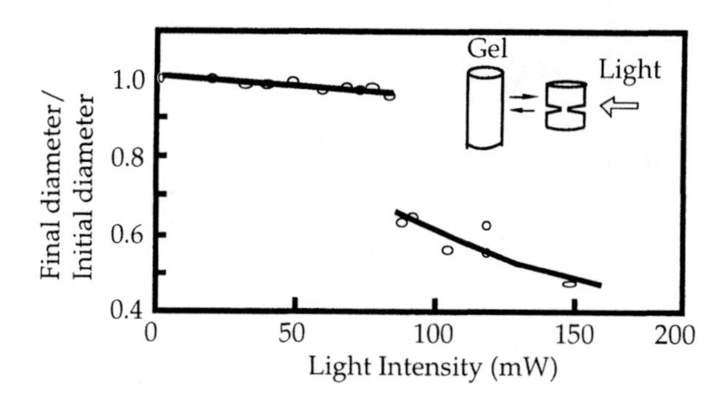

Figure 7.15 Diameter of N-isopropylacrylamide/chlorophine copolymer gel as a function of optical power at 31.5 °C [19].

Illumination by UV light initiates an ionization reaction in the gel, creating internal osmotic pressure which induces swelling. This process is slow compared with the illumination with white light that causes the phase transformation by heating the gel.

Electric Field Effects. Partially hydrolyzed acrylamide gels in a acetone-water mixture, is shown to undergo a discrete and reversible change in its volume by application of an electric field as schematically shown in figure 7.16a. The gel cylinder used in these experiments were 3 cm long and 4 mm diameter [20]. When the voltage across the cell is turned on, an electrostatic force is generated on the negatively charged acrylic acid groups in the polymer network inside the gel. This electrostatic force pulls the gel towards the positive electrode. This electrostatic force generates a uniaxial stress along the gel axis deforming the gel. Figure 7.16b shows the length of gel as a function of the voltage applied across the cell. The process is reversible and when the voltage is turned off, the gel becomes swollen and regains its original shape. At 1.25 volts 20% of the gel adjacent to the positive electrode collapsed resulting in 200 % reduction in its volume.

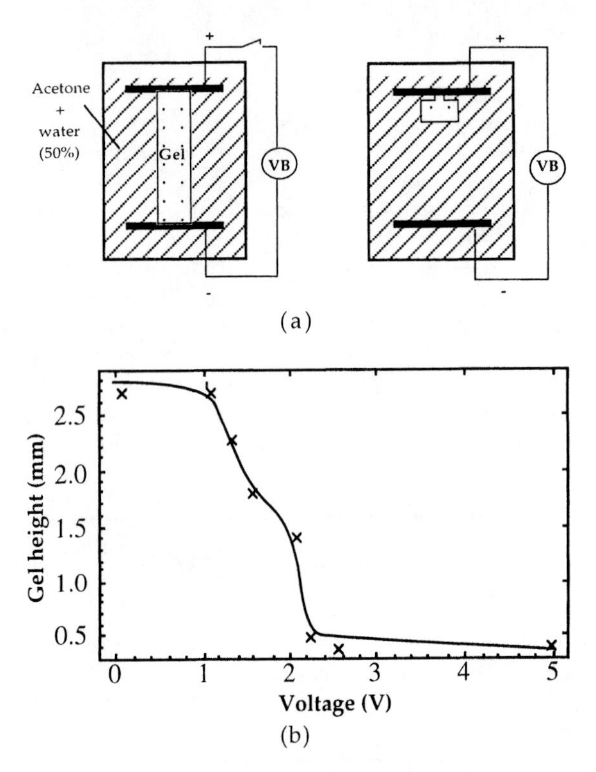

Figure 7.16 (a) The electrolytic cell used in applying an electric field across a partially hydrolyzed acrylamide gel. (b) The change in the gel height as a function of applied cell voltage at room temperature. Above 2.15 V the entire gel collapsed [20].

Chemical Effects. In an acrylamide gel in water and acetone solution, the polymer strand has larger affinity to other polymer strands than its affinity for the acetone molecule [18]. As a result, as the acetone concentration in the solution is increased, the gel collapses (in the absence of other forces) (figure 7.17).

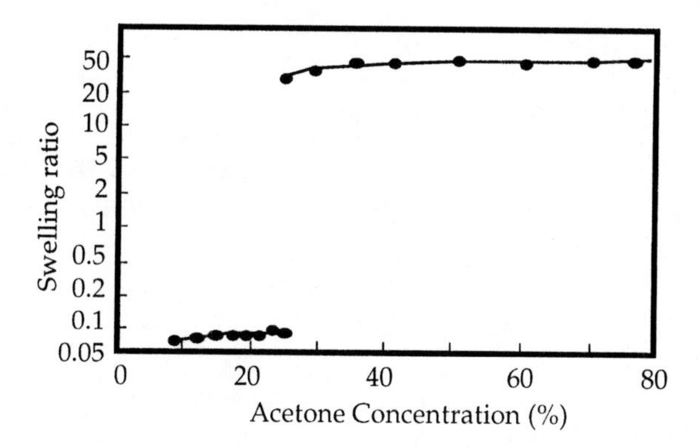

Figure 7.17 Increase in the acrylamide gel's volume as a function of acetone concentration in water [1].

Hysteresis Effects. Figure 7.18 shows the hysterisis effects observed in the N-isopropylacrylamide/chlorophine/sodium acrylate copolymer [22]. Hysterisis in the gel behavior as a function of external excitations (i.e., temperature, pH, and light intensity) is an indication that within the range that the hysterisis occur different phases of the gel coexist. The hysteresis effect can be used advantageously in optical and thermal switches since in these applications hysteresis is needed to achieve a stable operation.

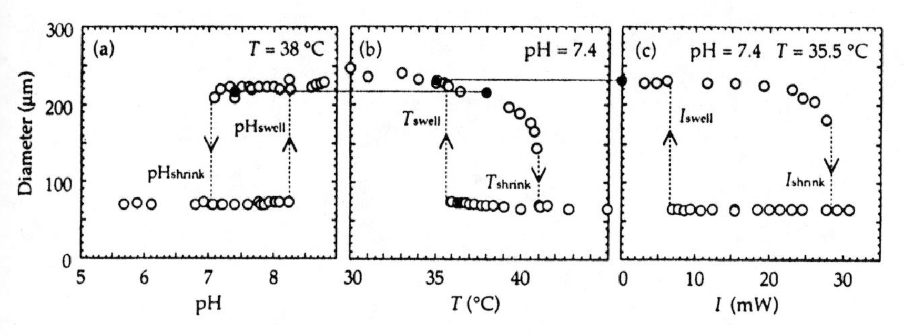

Figure 7.18 Equilibrium diameter of the gel as a function of (a) pH, (b) temperature, and (c) of visible light [22].

7.8 References

1. D. Linden, "Batteries and Fuel Cells." In: Electronics Engineers' Handbook." Edited by: D. G. Fink and A. A. McKenzie, McGraw-Hill Book Co., NY, pp. 7-66 - 7-56 (1975).
2. I. Buchmann, "Batteries." In: The Electronics Handbook. CRC and IEEE Press, NJ, pp. 1055-1061 (1996).
3. M. W. Hamberg, et. al., "An Electrochemical Micro-Actuator." Proceedings of IEEE Micro Electro Mechanical Systems, Amsterdam, the Netherlands, Pub.# 95CH35754, pp. 106-110 (1995).

4. P. A. Mayes, "Bioenergetics." In: Harper's Review of Biochemsitry." 20 ed., Edited by: D. W. Martin, Jr., P. A. Mayes, V. W. Rodwell, and D. K. Granner, Lange Medical Publications, Los Altos, CA, pp. 65-71 (1985).
5. D. W. Martin, "Contractiles & Structural Proteins." In: Harper's Review of Biochemsitry." 20 ed., Edited by: D. W. Martin, Jr., P. A. Mayes, V. W. Rodwell, and D. K. Granner, Lange Medical Publications, Los Altos, CA, pp. 480-496 (1985).
6. A. C. Guyton, Medical Physiology. 8th ed. W. B. Saunders Company, Philadelphia, pp. 67-79 (1991).
7. R. P. F. Gregory, Biochemistry of Photosynthesis. 3rd ed., John Wiley & Sons, New York (1989).
8. G. H. Pollack, "Contemporary Problems in Biology: Contractile Materials." In: Nanofabrication and Biosystems. Edited By: H. C. Hoch, L. W. Jelinski and H. G. Craighead, Cambridge University Press, Cambridge, UK, pp. 381-395 (1996).
9. T. A. McMahon, Muscles, Reflexes, and Locomotion. Princeton University Press, Princeton, New Jersey (1984).
10. Y. C. Fung, Biomechanics. 2nd ed., Springer-Verlag, New York, Inc., (1993).
11. H. Davson, "General Physiology." 4th Ed. The William and Wilkins Co., Baltimore, pp. 1-65 (1970).
12. R. zur Strassen, "The Beetles, Weevils, and Stylopids. In: Grzimek's Animal Life Encyclopedia. Editor-in-Chief: B. Grzimek, Van Norstand Reinhold Company, New York, pp. 231-297 (1970).
13. A. H. Epstein et. al., "Micro-Heat Engines Gas Turbines, and Rocket Engines-The MIT Microengine Project." Proceedings of 28th AIAA Fluid Dynamics Conference and 4th AIAA Shear Flow Control Conference, June 29 - July 2, 1997 Snowmass Village, Colorado.
14. A. H. Epstein, et. al., "Power MEMS and Microengines." Transducers '97, pp. 753-756 (1997).
15. N. E. Agbor, M. C. Petty and A. P. Monkman, Sensors and Actuators, B28, 173-179 (1995).
16. W. Schuhmann, R. Lammert, B. Uhe and H.L. Schmidt, Sensors and Actuators, B1,537-541 (1990).
17. K. Ramanathan, M. K. Ram, B. D. Malhotra and A. S. N. Murthy, Materials Science and Engineering: C3, 159-163 (1995)
T. Hirai et al., J. of Intelligent material systems and structures v4 n2 277-279 (1993).
18. T. Tanaka, "Gels." Scientific American, Vol. 244 (1), pp. 124-138 (1981).
19. A. Suzuki and T. Tanaka, "Phase Transition in Polymer Gels Induced by Visible Light." Nature, Vol. 346, pp. 345-347 (1990).
20. T. tanaka, I. Nishio, S.-T. Sun and S. Ueno-Nishio, " Collapse of Gels in an Electric Field." Science, Vol. 218, pp. 467-469 (1982).
21. D. K. Jackson, S. B. Leeb, A. H. Mitwalli, P. Narvaez, D. Fusco, and E. C. Lupton, Jr., "Power Electronic Drives for Magnetically Triggered Gels." IEEE Trans. on Industrial Electronics, Vol. 44 (2), pp. 217-224 (1997).
22. A. Suzuki, T. Ishii, and Y. Maruyama, "Optical Switching in Polymer Gels." J. Appl. Phys. Vol. 80 (1), pp. 131-136 (1996).

Smart Structures

8.1 Introduction

Smart materials change or modify their physical properties in response to the environment. Or in their most primitive form, they may report the effect of the environment on their different parts or attributes. Physical properties of materials may include color, physical dimensions, elasticity, or they may include other functions and properties, such as the ability of the material to "push" back in response to a force field. Smart materials may utilize a distributed sensing/signal processing/actuation scheme, or they may have a centralized signal processing unit along with distributed sensors and actuators. The sensors and actuators themselves can either be discrete or distributed. Figure 8.1 schematically shows a generic example of a smart material with embedded sensors and actuators. The goal or the "general command signal" may be fixed or may vary in response to its environment or according to an external command.

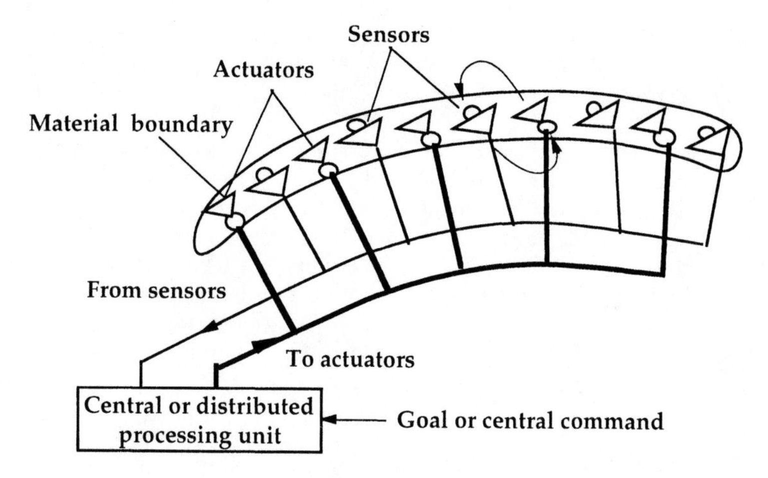

Figure 8.1 A schematic of a generic smart material with sensors and actuators.

In the literature, smart materials and structures are usually referred to as structures with mechanical tasks. Here, I use a broader definition and include optical, visual, acoustic and other tasks as well, and I treat the difference between smart materials and structures as a matter of differing complexity.

The most primitive example of a smart material is a material with distributed sensors of only one type. A practical example of such a material is a fabric with a liquid crystal incorporated in it so that it changes its color in response to a temperature change. In such a fabric, sensing, actuation and computation all occur at the molecular level. The computation is actually built into the fabric by the transition between different electronic or molecular levels.

Photochromic materials are another example. In these materials, the electron transitions caused by incoming light change the apparent color of the material. Thus, in intense daylight, photochromic eyeglasses become dark and enable the user to use the same pair of eyeglasses indoors and outdoors.

Under normal operation conditions, smartness implies gradual adaptability, purposefulness and ability to self-asses and self-repair [1-3]. In mathematical terms, smartness can be expressed as the ability of a system to minimize its errors. In this case, error can be defined as the difference between what is desired and what is achieved. "What is desired" can be an input given by human beings or it can be based on an award system promoting the well- being, usefulness and survivability of the smart system. It is outside the scope of this chapter to discuss artificial intelligence, which is the field that deals with some of these issues, but ultimately autonomous smart structures need to be intelligent at some level.

Sensors, actuators, and signal processors are the most obvious parts of a smart structure. I would like to add to this list another important, though less obvious, component: the physical links that are needed to connect the sensors and processors and actuators. These connections and links are probably the weakest part of large-scale smart structures and materials since they are most vulnerable to accidental or intentional "breakage," tapping or detection. They should also contribute minimally to the "characteristics" of the smart materials.

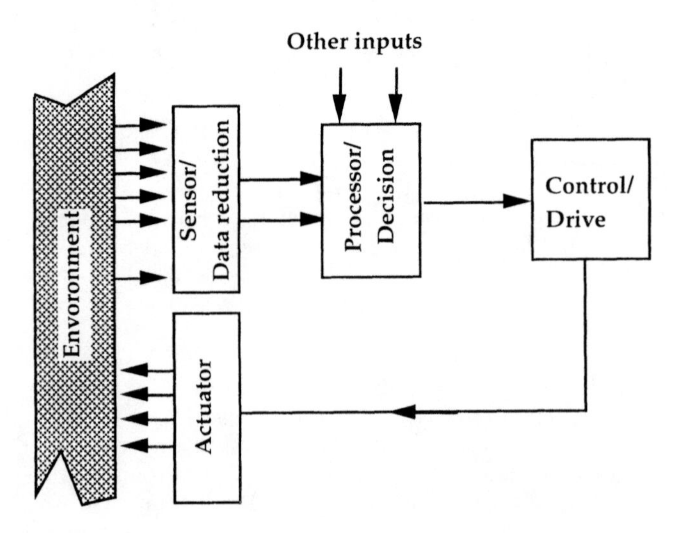

Figure 8.2 A schematic of a smart system interacting with environment.

Computation and signal processing can be performed explicitly or implicitly. By implicit signal processing, I mean a built-in physical characteristic that transforms the data from one form to another. A cantilever beam is a simple example of such implicit data processing since it transforms an applied force to displacement. Thus, it can be viewed as a computer performing the task given by $x=F/k$, where x (displacement) is the output, F (the applied force) is the input, and k (the effective Hooke's constant) is a constant related to the dimensions and the Young's modulus of the beam. By changing k, one can cover different ranges of displacements for a given range of applied forces. It is possible to dynamically change k by using piezoelectricity or electrostatic forces as discussed in previous chapters in this book, thus constructing a programmable force-to-displacement transformer. Examples of "implicit" signal processors are numerous and mechanical analog computers predating the digital computers were all made using mechanical devices, such as beams, gears, etc., that performed various computational functions [4].

Explicit signal processors have three main parts: an input terminal that is used to give the data in a numerical form; a processor with associated buffers, arithmetic units, memories, etc., that performs algebraic manipulations according to a program; and an output terminal that makes the result of the data processing available for viewing or other tasks. Microprocessors and specialized digital signal processors are all examples of explicit data processing systems.

Whether we use an implicit or explicit processor will be determined by the complexity of the tasks that our smart structure needs to perform. The implicit signal processing that takes place in photochromatic glasses is sufficient for the task of enabling people to use their eyeglasses indoors and outdoors. But such a natural response of the photochromatic glass alone may not be sufficient to enable using a very high sensitivity night-vision camera to function under varying light levels going from moon-lit to pitch dark areas. In this case, an elaborate feed-back mechanism may be needed to detect light intensities, perform some calculations and change the glass transmittance by applying an electric field across an appropriate glass or liquid-crystal light modulator.

The four essential parts in smart materials and structures: sensors, actuators, links and connection, and signal processors, are discussed in the following sections.

8.2 Sensors

Sensors are an important part of smart structures since they enable these systems to acquire information regarding themselves and their environment. There are many sensors and sensor technologies that can be readily used in smart structures and it is outside the scope of this chapter to discuss them all.

In general, sensors can be divided into physical and chemical sensors. Physical sensors are used in measuring physical quantities, such as displacement, velocity, and acceleration [5,6]. As the name implies, chemical sensors are used in sensing different gases and chemicals. At the heart of all these sensors is what condensed-matter physicist call "colombic" interactions. Most often, one can divide sensor structures into the two parts of a sensing method and a shell. The sensing method can be based on electrical measurements, such as in measuring the resistance of a strip of a heavily boron doped silicon used in strain gauges, or measuring the resistance of a polymer film in gas sensors. Sensing methods include optic, microwave, magnetic, acoustic/ultrasonic, thermal and mechanical [6]. These methods are directly employed in physical sensors as well as chemical sensors although in the latter case an intermediary step may be necessary.

The sensor shell is that part of the device that provides a transformation of the measurand to a parameter that can be sensed by the sensing method. A deformable diaphragm, for example, flexes when a pressure difference is applied across it. The amount of deformation can be easily quantified by a laser triangulation method, by a strip of piezoresistor or piezoelectric material, by capacitance method, etc. We call the deformable diaphragm the sensor shell and whatever method used to interrogate its deformation: the sensing mean.

In smart structures there is a need to acquire data and information from many points on the surface or inside the structure. Guided wave optics, microwave, and acoustic (both surface and bulk acoustic waves) offer some interesting characteristics in addressing some of these requirements. Of these three methods, optics and ultrasonics are extensively used.

8.3 Physical Sensors

In this chapter we present a comparative study of different sensing means that are used with comparable sensing shells to detect small displacement/force and accelerations. (The treatment presented here is based on the reference [6].) We consider the cantilever beam shown in figure 8.3 as the sensor shell commonly used by all the sensors discussed in the present chapter. The sensing mean translates the displacement of the cantilever beam to a signal. The sensor shell we have described is used in a variety of sensors including atomic force and scanning tunneling microscopes.

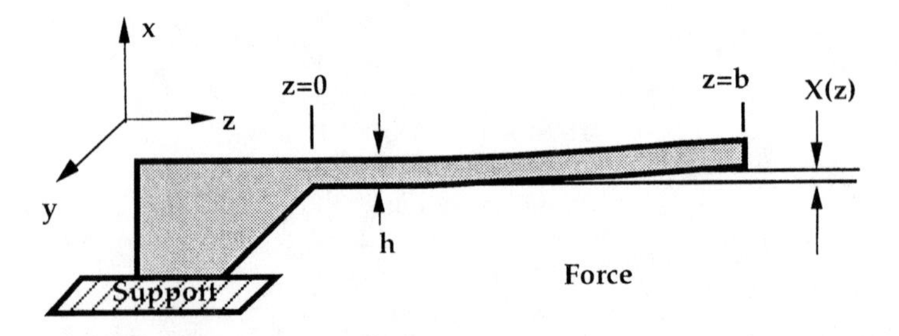

Figure 8.3 Sensor shell used in the present study to analyze different sensors.

The displacement of the cantilever beam along the x-direction in figure 8.3, when its weight is ignored and for small displacements, is given by the well-known relationship discussed in Chapter 1:

$$X(z)=\frac{F}{6YI_y}(3b-z)z^2\ , \tag{8.1}$$

where F is the applied force at the tip, Y is the Young's modulus, I_y is the moment of inertia of the beam's cross section with respect to the y-axis, and b is the beam length. At $z=b$, equation (8.1) can be used to find a relationship between the force and displacement:

$$F=kX, \tag{8.2}$$

where,

$$k=3YI_y/b_3, \tag{8.3}$$

and we have defined X(b)≡X. Typical dimensions for a silicon cantilever beam are: a (beam width) = 100 μm, b=5000 μm, and h=50 μm that result in a Hooke's constant of 4.2 N/m.

There are basically three categories of sensing means that may be used in <u>small</u> displacement sensors: a) electrical, b) magnetic, and c) optical. Electrical methods include: i) capacitive, ii) piezoelectric, iii) piezoresistive, iv) tunneling current, and v) microwave. Magnetic methods include: i) magnetization methods, ii) magneto-elastic methods, and iii) external magnetic field methods. Optical methods include two separate categories: a) free-space methods, and b) guided-wave methods. Guided-wave methods can be further divided into fiber-optic and integrated-optic devices. We will discuss the following optical methods: i) Michelson interferometry, ii) beam deflection method, iii) amplitude modulation method, iv) evanescent field method, and v) Mach-Zehnder interferometry. Acoustics, not discussed here, can also be used as sensing means.

8.3.1 Electrical Sensing Means

Capacitive Methods. Capacitive techniques are based on measuring the change in capacitance between two electrodes when one of them is displaced or deformed due to an applied force (figure 8.3). The size and geometry of the electrodes depend on the application, but usually the sensor is approximated by a parallel plate capacitor. There are several ways of measuring the induced change in a capacitance. For example, it can be measured with an impedance bridge, a capacitor-controlled oscillator or a charge measurement circuit. Some electronic components of these schemes should be considered as part of the sensor since they are a particular feature of the technique. Other components such as amplifiers, filters, etc., are common to all schemes.

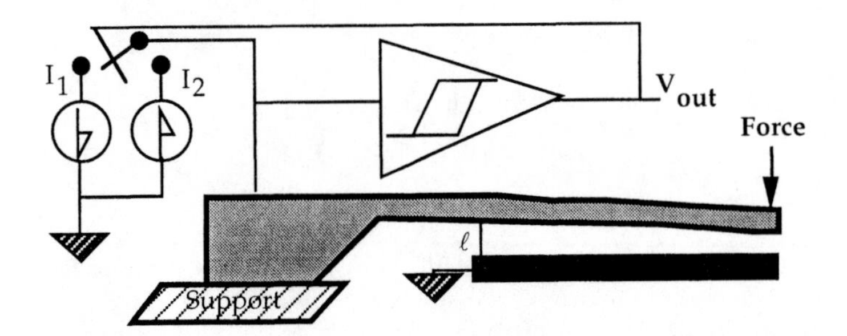

Figure 8.4 A schematic of a displacement sensor with capacitive sensing.

The simplest capacitive displacement sensor with a cantilever structure consists of a silicon beam fixed over an electrode as shown in figure 8.4. The cantilever beam variable capacitor is part of a read-out circuitry that translates the change in the capacitance into a readable electrical signal. The minimum detectable signal (MDS) is determined by the dominant noise that affects the output signal. A detailed analysis leading to the MDS of a capacitive sensor with different read-out circuits was presented in references 3-8. Here we follow their analysis, but apply it to the cantilever beam shell. The relaxation oscillator is probably the simplest choice for the read-out circuit, where the capacitance is determined by the oscillator period. Oscillator noise causes jitters and determines the MDS. The rms value of the change of either half-cycle time (σ_t) in the oscillator (jitters) is approximately V_{rms}/s, where V_{rms} is the rms noise voltage and s is the rate of change of the capacitor voltage.

We assume that bipolar devices are used for the source/switch combination and shot noise is the dominant noise. The capacitor voltage is approximately a linear ramp in time between the two comparator thresholds. The slope of the ramp is I/C_T, where I is the current (constant magnitude) and C_T is the total capacitance ($C_T = C_p + C_s$ where C_s is the sensing capacitance and C_p is any parasitic capacitance in parallel with C_s). The spectral density of the shot noise is 2qI, where q is the charge of an electron, and when the capacitor is allowed to swing the full available bias V_B, it is shown that the jitter in the oscillator period is:

$$\sigma_t = \sqrt{qC_TV_B}\,/\,I, \tag{8.4}$$

and the sensitivity of the capacitive displacement sensor is:

$$S_{cap} = \left|\frac{dC_S}{C_S dX}\right| = \varepsilon_0 \frac{A}{\ell 2}\left(\frac{\ell}{\varepsilon_0 A}\right), \tag{8.5}$$

$$S_{cap} = 1/\ell. \tag{8.6}$$

In the above equation, $C_s = \varepsilon_0 A/\ell$, where ℓ is the distance between the electrodes, A is their area, and e_o is the dielectric constant of air or vacuum.

The change in half-cycle time of the oscillator (Δt) due to a displacement dX of the cantilever beam is given by:

$$\Delta t = S_{cap} C_s V_B dX/I, \tag{8.7}$$

where V_B is the full available bias. The MDS (=dX) is found by equating Δt in (7) with σ_t in (4):

$$MDS = dX = \frac{1}{S_{cap}} \sqrt{\frac{q}{C_s V_B}\left(\frac{C_s + C_p}{C_s}\right)}. \tag{8.8}$$

As seen from (8.6) and (8.8), the sensitivity of the capacitor increases, and MDS decreases, as the separation between electrodes decreases. However, the attractive electrostatic force exerted between electrodes ($F=\varepsilon A V^2/2\ell^2$) also increases as ℓ is made smaller, and this limits the minimum separation between electrodes before the sensor collapses. This problem may be solved to some extent by using a three-plate capacitor. The minimum possible separation between electrodes is also limited by the roughness of the electrodes and by some unavoidable lack of parallelism of the electrodes.

The measurement of the capacitance is done assuming that its value is fixed during the oscillation periods of the relaxation oscillator. This will not be the case if the sensor is vibrating, as in atomic force microscopy (AFM). However, the frequency of the relaxation oscillator may be high enough for most applications of this sensor. A typical period for the relaxation oscillator is 20 ns (50 MHz oscillator). Hence, frequencies of vibration of around 10 kHz should be measurable with this sensor. This vibration frequency is high enough for most applications of this sensor.

The application of a capacitive sensor to an AFM has already been studied elsewhere [6], where two sensor configurations are discussed. The first one is an integrated silicon sensor, while the second type was implemented with a curved wire placed perpendicularly at a small distance from a ground plate. It is shown that the theoretical noise level due to the proposed electronic circuitry is equivalent to 0.028 Å, a good value for atomic resolution imaging. The principal limitation they found was that the separation between the electrodes that was needed to obtain the desired resolution was too small (~ 600 Å), making the roughness of the electrodes and contamination by dust particles critical problems that reduced their microscope's resolution appreciably.

Piezoelectric Methods. Piezoelectric materials have a stress-dependent polarizability and the displacement vector is related to the electric field and the stress, generally by a linear tensor equation. Applied stress on a piezoelectric slab induces surface charges, and the measurement of the magnitude and the polarity of the surface charge yields quantitative information regarding the applied stress and the deformation of the piezoelectric slab. A piezoelectric sensor using the cantilever beam structure is shown in figure 8.4. In its simplest form, the top and bottom face of the piezoelectric layer are metallized, forming a parallel plate capacitor. When a force is used to displace the cantilever tip, the stress induced on the piezoelectric layer induces charges at the piezoelectric-metal interface that are detected as a voltage across the capacitor. Such a sensor has already been fabricated and tested for AFM using a sputtered ZnO thin film as discussed in Chapter 1.

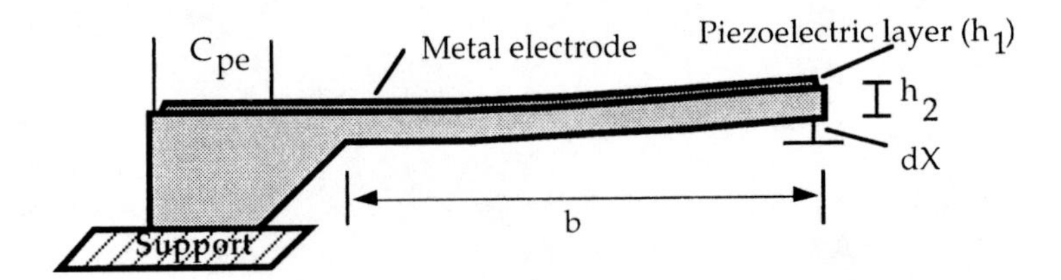

Figure 8.5 A displacement sensor with a piezoelectric sensing mean.

The charge induced on the electrodes in the piezoelectric sensor may be measured with the relaxation oscillator discussed before (see figure 3). If the noise generated at the piezoelectric layer or at the interface is neglected, the MDS of the sensor is set by the noise in the read-out circuit. We assume that the source/switch combination in the oscillator are bipolar devices so that shot noise is dominant. The change in the half-cycle time of the oscillator due to a displacement of the cantilever is dQ/I, where dQ is the stress-induced charge and I is the charging current:

$$dQ = S_{PE} Q_m dX, \tag{8.9}$$

where S_{PE} (= $dQ/Q_m dX$) is the sensor's sensitivity and Q_m is the charge on the electrodes at the comparator's threshold voltages when no stress is induced. The maximum charge Q_m is related to the threshold voltage V_B by $Q_m=V_B C_{pe}$, where C_{pe} is the stress-free capacitance between electrodes ($C_{pe}=\varepsilon A/h_1$, where ε is the dielectric constant of the piezoelectric). The induced charge on the electrodes is found from: $dQ = \int_0^b aD_x dz \approx d_{13} a \int_0^b \overline{T}_z dz$, where Δx is the x-component of the displacement vector, $\overline{T}_z$ is the average longitudinal stress across the piezoelectric layer and d_{13} is the appropriate piezoelectric constant. For $h_1 << h_2$ we have $\overline{T}_z = Yx_n/R_c$, where x_n is the distance to the neutral plane ($x_n \sim h_2/2$), Y is Young's modulus for the piezoelectric layer, and R_c is the local radius of curvature ($1/R_c \sim d^2X(z)/dz^2$ for $dX<<b$). Therefore, the sensitivity is:

$$S_{PE} = \frac{dQ}{Q_m dX} = \frac{3d_{13}Eah_2}{4bV_B C_{pe}}. \tag{8.10}$$

Taking $s_t = \sqrt{qC_T V_B}/I$ and using the above equations, we find the MDS:

$$MDS=(4b/3d_{13}Yah_2)\sqrt{q(C_{pe}+C_p)V_B}\ , \qquad (8.11)$$

where C_p is the parasitic capacitance.

Piezoresistive Methods. In this method, a piezoresistive material, like heavily-doped silicon (figure 8.6), whose resistivity is a function of internal stresses, is used. Piezoresistive sensors have been widely studied for pressure sensing. In our example of the silicon cantilever beam, the heavily-doped silicon beam itself is used as the piezoresistive sensor and as the beam deflects, its resistance changes because of the stresses induced inside the beam. A simple sensitive circuit, a Whetstone bridge, is used to measure changes in the piezoresistance (R_s) as shown in figure 8.6.

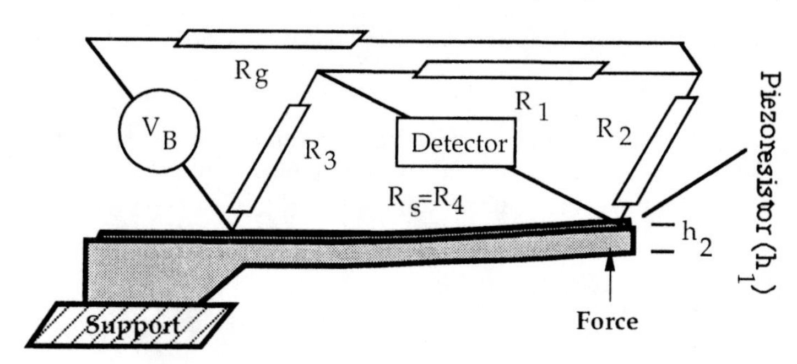

Figure 8.6 A schematic of a displacement sensor with a piezoresistive sensing mean.

The detector voltage (ΔV) near balance (R_1~R_s) is $(R_d V_B \delta u)/(R_1+R_2+R_3+R_s+AR_d+BR_g+DR_dR_g)$, where $A=2+R_1/R_3 + R_3/R_1$, $B=2+R_1/R_2 + R_2/R_1$, $D=1/R_1+1/R_2+1/R_3+1/R_s$, and R_s is the resistance of the piezoelectric layer which depends on the displacement of the cantilever, R_d is the resistance of the detector, R_g is the source resistance, and δu is the fractional bridge unbalance given by $R_s=(1+\delta u)R_2R_3/R_1$ (at balance: $\Delta V=0$, $R_s=R_2R_3/R_1$).

The longitudinal piezoresistive coefficient π_ℓ is defined through $\Delta\rho/\rho_0=\pi_{1s}$, where ρ is the resistivity of the material, ρ_0 is its resistivity with no induced stress, and σ is the longitudinal stress. We are assuming that the piezoresistive layer in our sensor is a diffused p-type thin ($h_1<<h_2$) layer on the back of the silicon cantilever. The longitudinal stress within the piezoresistive layer (σ) is approximately Yx_n/R_c, where x_n is the distance to the neutral plane, Y is Young's modulus for the layer, and R_c is the local radius of curvature. All other stresses are approximately zero. Since the layer's thickness is assumed to be very small, we may approximate x_n~$h_2/2$, and since dX is assumed to be very small compared to the length of the cantilever (dX<<b), we may approximate $1/R_c$~ $d^2X(z)/dz^2$, where X(z) is

given in (2). Assuming uniform doping across the diffused layer (when this is not the case, the following equations will be valid using average values), the resistance of the piezoresistor ($R_S = R_O + \Delta R$) is given by $\int_0^b (\rho_O + \Delta\rho)dz / ah_2$. The equilibrium resistivity is ρ_o=1/$qp\mu_p$, where p is the hole concentration, μ_p is the hole mobility (<<480 cm^2/V-s in heavily-doped materials). According to the above assumptions, $\Delta\rho$ is $\pi_l(b-z)3Yh_2\rho_0X/2b^3$. Thus we find: $\Delta R=3h_2Y\pi_l\rho_0X/4bah_1$. Depending on the relative values of the resistors, both the shot noise and Johnson noise can be important in this case. Assuming that R's are large, we can take the Johnson noise to be dominant. Taking $R_1=R_2=R_3=R_0$, and assuming that $\Delta R<<R_0$ and R_g~0, the sensitivity is:

$$S_{PR} = \frac{\Delta V}{V_B dX} = \frac{3h_2E\pi_l}{16b^2}\frac{R_d}{R_o + R_d}. \qquad (8.12)$$

The output noise is equivalent to the thermal noise generated at one of the four resistors in the Whetstone bridge. Assuming a single pole low pass filter response of the measuring circuit, the MDS is given as:

$$MDS = \frac{1}{S_{PR}V_B}\sqrt{2\pi kT R_s B_m}$$

$$= \frac{16b^2}{3h_2E\pi_l V_B}\left(\frac{R_o + R_d}{R_d}\right)\sqrt{2\pi kT R_s B_m}, \qquad (8.13)$$

where k is the Boltzmann constant, and T is the absolute temperature.

Tunneling Current. Figure 8.7 shows a schematic of a displacement sensor that uses tunneling current as the sensing mean. A pointed metallic tip is situated within the ~10Å of the cantilever beam that has a metallic coating. The resistance of the air-gap is $R_0e^{-2\kappa x}$, where R_0 ($\approx 10^7$ Ω)is the point contact resistance when the tip touches the cantilever beam.

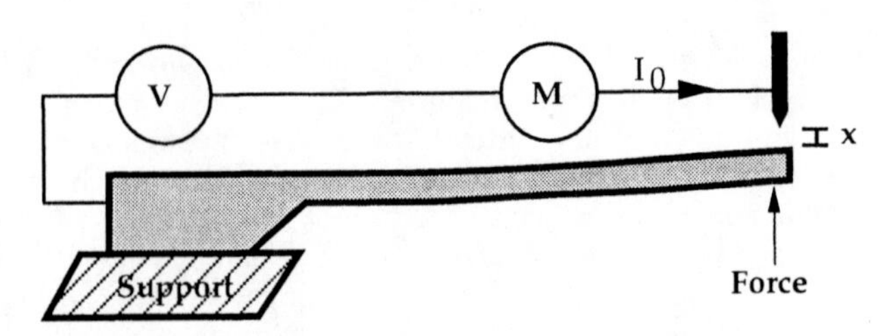

Figure 8.7 A schematic of a displacement sensor with a tunneling current sensing mean.

This sensing method belongs to the category of variable-resistance sensing means, such as piezoresistors and sliding contact potentiometers. With the exception that it has an exceptional sensitivity to displacement since its resistance exponentially depends on the distance between the tip and the beam with the scale factor ($\sim 1/\kappa$). One of its advantages over the capacitive sensing mean is that it has a very small tip-to-beam capacitance (on the order of 10^{-18} F or so) resulting in a very small capacitive back action force (the electrostatic force exerted on the cantilever beam by the charges on the tip). The current that flows through the tip is given by $I=I_0 e^{-2\kappa x}$, where I_0 is V/R_0 and κ is $\sqrt{2m_e\phi/\hbar}$, where ϕ is the work function of the probe, m_e is the electron mass, and h is the Planck's constant. Typical κ is 10^{10} m^{-1}.

To calculate MDS, we treat the tunneling current tip as a piezoresistor with $R \sim R_0 e^{-2\kappa x}$ and we note that the dominant noise at room temperature and for our cantilever beam, with the dimensions discussed in section two, is shot noise associated with the current I_0 that passes through the tip. The other noise sources that will not be considered here are: i) fluctuations in the back action force, which includes the image charge force (= QE, where Q is the total charge on the tunneling probe and E is the electric field in the gap) and the tunneling force associated with the electron momentum transfer (= Ip_e/q, where I is the tunneling current, and p_e is the electron momentum), and ii) the Brownian motion. Spectral density (S(f)) of the shot noise is $2qI_0$, and the apparent displacement corresponding to the shot noise is related to the fluctuation in the current (ΔI). The sensitivity of the tunneling current sensor (S_{tc}) is:

$$S_{tc} = \left| \frac{dI}{I_0\, dX} \right| = 2\kappa, \tag{8.14}$$

MDS is $\Delta I / I_0 S_{tc}$, where ΔI is $\sqrt{S(f).B_m}$ with B_m denoting the detection bandwidth:

$$\mathrm{MDS} = \frac{\sqrt{2qI_0B_m}}{I_0\, 2\kappa} = \frac{\sqrt{qB_m}}{\kappa\sqrt{2I_0}}. \tag{8.15}$$

From (8.15) it is seen that to minimize MDS, I_0 and k should be made as large as possible. Making I_0 large causes electro-migration of the cantilever surface atoms, deteriorating the stability of the sensor, and κ is a function of the electrode work function and temperature which are usually fixed.

Evanescent Microwave. A microstrip line microwave probe is shown in figure 8.8, where evanescent fields are produced at the probe that terminates in a quarter-wavelength resonator section. These fields interact with the cantilever and change the resonant frequency of the resonator. Microwave's main disadvantage is that its resolution limit is set by its wavelength (more accurately by $\lambda/2$) that is 30-0.3 cm (1-100 GHz frequency regime, and in free space). However, evanescent microwave is used to achieve resolutions better than $\lambda/1000$. This scheme is similar to the tunneling current sensor except that the scale factor can vary, depending on the frequency and the structure of the resonator that is used.

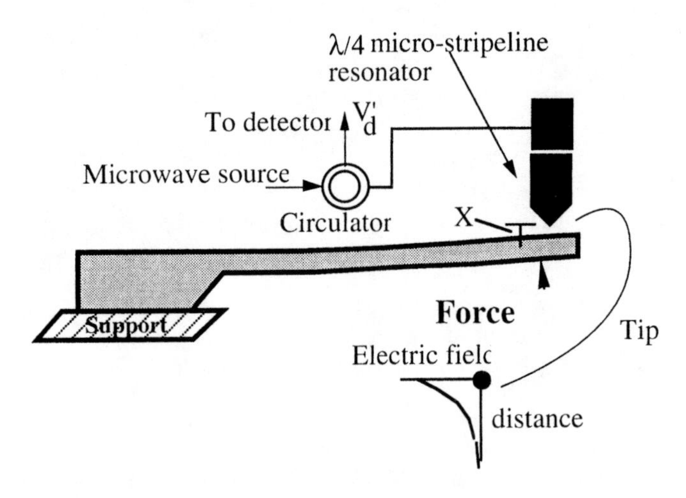

Figure 8.8 A schematic of a displacement sensor with a microwave sensing mean.

The detected voltage (V'_d) in the presence of the cantilever beam is approximately:

$$V'_d(f) = V_m e^{-(f-f'_c)^2/(\Delta f)^2}, \tag{8.16}$$

where f'_c is the resonant frequency and Δf is the bandwidth of the resonator. It is assumed that the cantilever beam weakly perturbs the quarter-wavelength resonator, slightly changing its center frequency, but not affecting its bandwidth or V_m. It is also assumed that "f" differs from f_c by a fraction of Δf, and that at f_c the resonator and everything else have a characteristic impedance of 50 Ω. Under these circumstances, it can be shown that the perturbed resonant frequency (f'_c) is slightly (δ) larger than its unperturbed value (f_c) and it can be written as $f'_c = f_c + \delta$ where
$\delta = C(f,\sigma)e^{-\alpha x}$ (σ is the conductivity of the cantilever beam at the probe frequency, C is a coefficient that is a function of σ and f, and α is the effective decay length of the evanescent field at the tip of the probe). We define a sensitivity for this sensor (S_{mw}) as:

$$S_{mw} = \frac{1}{V'_d} \left| \frac{dV'_d}{dX} \right| = \frac{1}{V'_d} \frac{dV'_d}{df'_c} \left| \frac{df'_c}{dX} \right| . \tag{8.17}$$

Using the above equations, we find $S_{mw} = 2|(f-f'_c)|\, \alpha\, \delta/(\Delta f)^2$. MDS is given by:

$$MDS \equiv \Delta X = \frac{\Delta V'_d}{V'_d S_{mw}}$$

$$= \frac{\Delta V'_d}{V'_d} \frac{\Delta f^2}{2|f-f'_c| \, \alpha C(f,\sigma)} \, . \tag{8.18}$$

Three stages contribute to the noise at the detector: the microwave source, the circulator, and the detector noise itself (we are ignoring the cantilever Brownian motion). The microwave source may have amplitude noise as well as jitter in its frequency and phase. Circulator noise is primarily Johnson noise. The detector noise is primarily shot noise with the spectral density of $2qI_d$ where I_d is the detector current induced by the incident microwave (amplitude of V'_d).

8.3.2 Magnetic Field Methods

Magnetic methods can be divided into three categories: i) magnetization methods, ii) magneto-elastic method, and iii) external magnetic field methods . In magnetization methods, a ferromagnet is usually used to change the inductance of a nearby coil (figure 8.9a) and the change in the inductance is measured to calculate the distance between the ferromagnet and the coil.

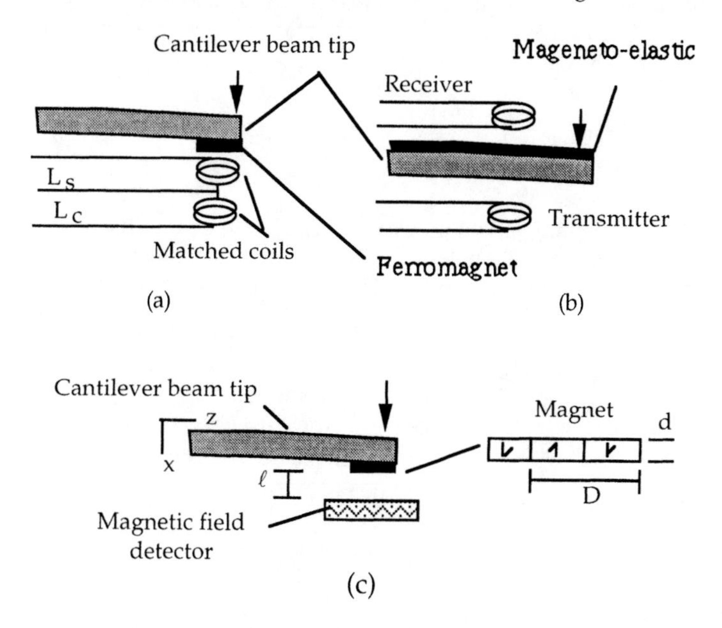

Figure 8.9 A schematics of displacement sensors with magnetic sensing means.

In the second method, a special group of materials, where their internal magnetization is a function of both external fields and their internal stresses, are used. These materials are either directly used as both the sensing shell and the sensing mean, or they can be incorporated with a separate sensing shell, such as a cantilever beam as depicted in figure 8.9b. In the latter scheme, as the cantilever beam bends, the magneto-elastic material is strained, and the change in its magnetization is detected using transmitter/receiver coils. In the last method, a magnetic field sensitive device, such as a Hall device or a superconducting quantum interference device, is used to measure the magnetic field. The magnetic field at the device is a function of the distance, and the device's output is measured to calculate the distance (figure 8.9).

Since the majority of the above methods require relatively large volumes, we do not discuss them in detail nor do we calculate their MDS's. However, among the above methods, the magnetic field method (method (c)) is probably the most promising technique for small displacement sensing. Progress in the deposition of thin-film magnetic rare-earth materials and advances in Hall sensors will make this method somewhat feasible. A room temperature, lateral bipolar junction magnetic field transistor (LBJMT) with 0.6 μT minimum magnetic field detection and voltage sensitivity of 1 V/T (current sensitivity of 0.1 mA/T at collector current of 3 mA) has been reported. Ferromagnetic films with remnant magnetization of ~300 mT can be readily manufactured and significant progress in depositing rare-earth thin films will soon result in yet higher remnant fields. Back action in the magnetic field scheme can be ignored and the dominant noise source in the LBJMT is the shot noise associated with its Hall current (I_H). We define the sensitivity of this scheme as:

$$S_{mf} = \frac{1}{I_H}\frac{dI_H}{dx} = \frac{1}{I_H}\frac{dI_H}{dH_x}\frac{dH_x}{dx} \tag{8.19}$$

Assuming that the magnetic film area is larger than the active area of the LBJMT and that it contains domains with 180∞ Bloch walls with magnetization perpendicular to LBJMT, it can be shown that the magnetic field of such a film is $H_x(x,z) \approx 8\ M_s\ (1-e^{-2\pi d/D})e^{-2\pi x/D}\cos(2\pi z/D)$ where M_s is the saturation magnetization of the film (~500-1000 mT) D is the periodicity of the magnetic domain and d is the film thickness. Moreover, I_H can be written as $I_l + C\ H_x(\ell,z)$, where C is the current sensitivity of the Hall sensor, I_ℓ is its leakage current, and ℓ is the distance between the cantilever and the Hall sensor. Therefore, the MDS is given by:

$$\mathrm{MDS} = \frac{1}{I_H}\frac{\Delta I_H}{S_{mf}} = \frac{D\sqrt{2q(I_l+CH_x(\ell,0)\ B_m}}{16\pi M_s C(1-e^{-2\pi d/D})e^{-2\pi\ell/D}}. \tag{8.20}$$

8.3.3 Optical Methods

Optics is widely used to measure displacement in biological as well as artificial systems. Optical measurements can be divided into interferometric and amplitude-sensitive methods. Interferometries are widely used because they often do not need calibration. Amplitude-sensitive methods, on the other hand, are simpler, but they require calibrations during their operation cycles.

Here we will be discussing the application of optical methods in sensing small displacement of a microfabricated silicon cantilever beam. It has also been shown that microfabrication methods can be used to fabricate optical components. These methods allow co-fabrication of optical and mechanical components which eliminates laborious alignment and calibration procedures. The title of the present conference, "Integrated Optics and Microstructures," refers to the integration of micromechanical and optical systems that is made possible by the microfabrication methods.

The fundamental limits of optical sensing means are set by diffraction and photon shot noise. In optical methods, different types of photo-detectors, such as photo-conductors, photo-transistors, PIN diodes, avalanche photo-diodes, and photo-multipliers, are employed. Usually, the detector current (I_d) has three components: the photo-current ($I_{pc}= q\eta P\, \Gamma_G/(\sqrt{2}h\nu)$, where η is the quantum efficiency, P is the incident optical power, G_G is the detector gain, $h\nu$ is the photon energy); the dark current (I_{dark}); and the background light current (I_B): $I_d= I_{pc}+ I_{dark} + I_B$. The rms shot noise current at the output of a photo-detector is :

$$\Delta I_d=\sqrt{2qI_d\Gamma_G FB_m}, \tag{8.21}$$

where F is the excess noise factor which can be large in photo-multipliers and avalanche photo-diodes. The sensitivity of the optical sensor is defined as:

$$S_{opt}= | dI_d/I_d dX |. \tag{8.22}$$

Taking I_d to be proportional to P, and taking P_0 to be the optical power reaching the photo-detector when the displacement is zero, we can also approximately write $S_{opt}{\approx}dP/P_0 dX|$. Furthermore, the detected power P can be expressed as $P_0T(X)$, where T(X) is a function of the sensor's displacement from the equilibrium position. S_{opt} then becomes $S_{opt}{\approx}dT(X)/dX$. The MDS is found by solving for minimum dX:

$$MDS = \left[\sqrt{2q\Gamma_G FB_m/I_d}\right]/S_{opt}\,. \tag{8.23}$$

Typical values of Γ_G are: 10^5-10^6 for a photo-multiplier, 10^2-10^4 for an avalanche photo diode, and ≈ 1 for a PIN diode. Depending on h and I_{ph}, I_d can vary between a few nA to a few mA in these detectors. *F* is usually high (~10~100) in photo-multipliers and avalanche photo-diodes. B_m is usually in the range 10 Hz-1 MHz.

In the following sections, we consider, and compare the following different schemes: Michelson interferometer, beam deflection method, amplitude modulation method, evanescent optical field method, and Mach-Zehnder interferometry.

Michelson Interferometry. Figure 8.10 shows a Michelson interferometer used to detect silicon cantilever beam deflections.

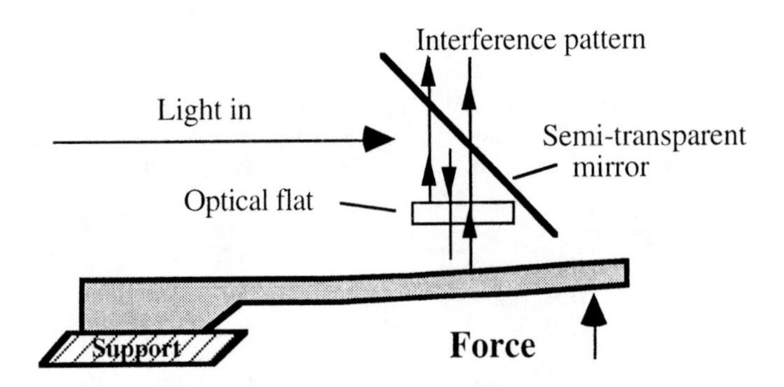

Figure 8.10 A schematic of a displacement sensor using Michelson interferometry.

Assuming fringe visibility of unity, the transmittance function (T(X)) and the sensitivity are:

$$T(X) = \frac{1}{2}\left[1+\mathrm{Cos}\left(\frac{2\pi}{\lambda}X\right)\right], \text{ and } S_M = \frac{\pi}{\lambda}\,\mathrm{Sin}\left(\frac{2\pi}{\lambda}X\right), \tag{8.24}$$

and MDS is $\lambda\left[\sqrt{2q\Gamma_G F B_m / I_d}\right]/\pi$, where we have taken S_M to be at maximum, i.e., $X=m\lambda + \lambda/4$ (with integer m).

Beam Deflection Method. Figure 8.11 shows a displacement sensor based on the beam deflection method that requires free space propagation. A collimated laser beam is reflected off the back of the silicon cantilever and it is detected using a pinhole photodetector. We assume that a collimated laser beam is reflected with an oblique angle from the back of the cantilever. A relatively large mirror (larger than the waist of the laser beam at the reflection) will be needed if the dimensions of the cantilever are small compared to the laser beam diameter in order to avoid undesirable diffraction effects.

We will assume that a pinhole photodetector is placed within a cross section of the laser beam. The intensity profile of the laser beam is $I_{ph} = I_0\exp(-8r^2/D_d^2)$, where D_d is the laser beam diameter at the photodetector ($R=R_d$) and I_0 is the laser beam intensity at r=0 at $R=R_d$. As the cantilever bends, the intensity of the light detected by the photodetector changes.

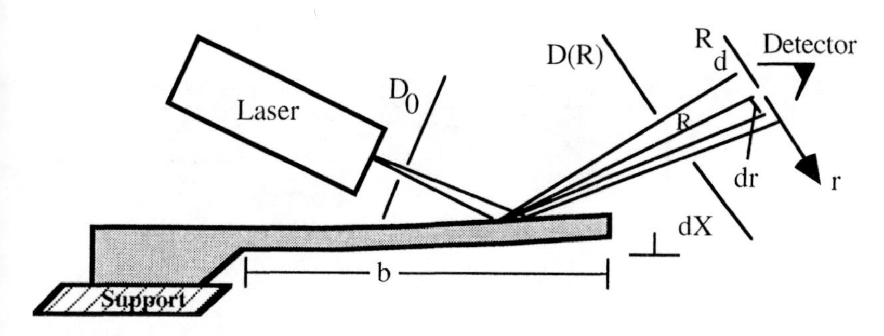

Figure 8.11 A schematic of a displacement sensor with beam deflection sensing mean.

The change in the relative position of the deflected laser light at the detector (δr) is related to the displacement of the cantilever (dX) by $\delta r = 2R_d dX/b$. The change in the detected power ($P_d = a_p I_{ph}(R_d, r_d)$) is given by $dP_d = a_p \frac{dI_{ph}}{dr} dr = a_p \frac{dI_{ph}}{dr}\frac{2R_d}{b} dX$, where a_p is the viewing area of the pinhole detector ($a_p << D_p$). Thus the sensitivity (S_{BD}) is $(32\, r\, R_d/bD_d^2)\exp(-8r^2/D_d^2)$. The diameter of the laser spot depends on R as: $D^2(R) = D_0^2(1 + (4\lambda R/(pD_0^2))^2)$. The maximum sensitivity is achieved if R is large compared to $1/D_0^2$. This situation may be reached using a lens instead of moving the photodetector away. When $\lambda R >> 1/\pi D_0^2$, we have $D_d \sim 4\lambda R_d/\pi D_0$. The maximum value for dP_d/dr is at $r = D_d/4$, yielding the maximum sensitivity:

$$S_{BD} = 2\frac{\pi D_o}{\lambda b} e^{-0.5}. \tag{8.25}$$

The MDS is:

$$\text{MDS} = b\lambda e^{0.5}\left[\sqrt{2q\Gamma_G F B_m / I_d}\right] / 2\pi D_0 . \tag{8.26}$$

It is interesting to note that MDS is inversely proportional to D_0, which is somewhat counter-intuitive.

Amplitude Modulation Method. Figure 8.12 shows a displacement sensor based on the amplitude modulation scheme. The input and output optical waveguides are fabricated using thin dielectric films of thickness h_1 (e.g., SiO_2/Si_3N_4).

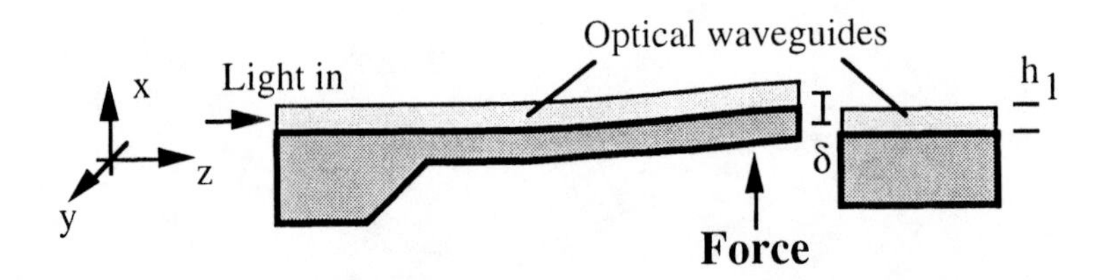

Figure 8.12 A displacement sensor with the amplitude modulation sensing mean.

The transmittance function T(d) may be found from the coupling integrals of the incident electric field and the different available modes of propagation at the entrance of the output waveguide. Let us assume that the only propagation modes involved in the process are the fundamental modes in each waveguide. To evaluate the coupling integral, we may approximate the guided electric (E(x,z) field as $E \sim E_o e^{-x^2/\alpha^2} e^{j\omega t - j\beta z}$. We further assume that the gap between waveguides is small compared to the radiation wavelength so that the incident field at the entrance of the output waveguide can also be given by the above E. The power coupling efficiency is then given by:

$$T(\delta) = \left(\int_{-\infty}^{\infty} e^{-\frac{x^2}{\alpha^2}} e^{-\frac{(x-\delta)^2}{\alpha^2}} dx \right)^2 \Big/ \left(\int_{-\infty}^{\infty} e^{-\frac{x^2}{\alpha^2}} dx \right)^2 = \frac{1}{2} e^{-\frac{\delta^2}{\alpha^2}}, \quad (8.27)$$

where d is the offset between the center of the input and output waveguides. The parameter a may be obtained exactly. A reasonable value for a is h_1/p which results in the transmittance function (T(d)) of $0.5\, exp(-\pi^2\delta^2/h_1^2)$ and sensitivity (S_{AM}) of $(\delta/h_1^2)\exp(-\pi^2\delta^2/h_1^2)$. Thus, the MDS is given by:

$$MDS = (h_1^2/\delta)\exp(\pi^2\delta^2/h_1^2)\sqrt{2q\Gamma_G F B_m / I_d}, \quad (8.28)$$

where δ is $\sqrt{2}h_1/2\pi$ for the best MDS. Clearly, to have the best MDS, h_1 should be made as small as possible and the beam should be dc biased slightly off so that $\delta \neq 0$.

Evanescent Optical Field Method. Above the optical waveguide, the electric field intensity decays exponentially as a function of the distance. This field is called evanescent field, and it enables objects outside the waveguide to interact with the optical fields inside the waveguide. If the outside object is another waveguide, like a piece of fiber optic, photons in the evanescent field may tunnel through the free space and propagate inside the fiber. The probability of the tunneling is scaled by the distance d, figure 8.13, between the waveguide and the fiber. Hence, one may monitor the optical field intensity inside the fiber optic to measure "d", and since the intensity inside the fiber is an exponential function of the distance, "d" can be measured with great accuracy.

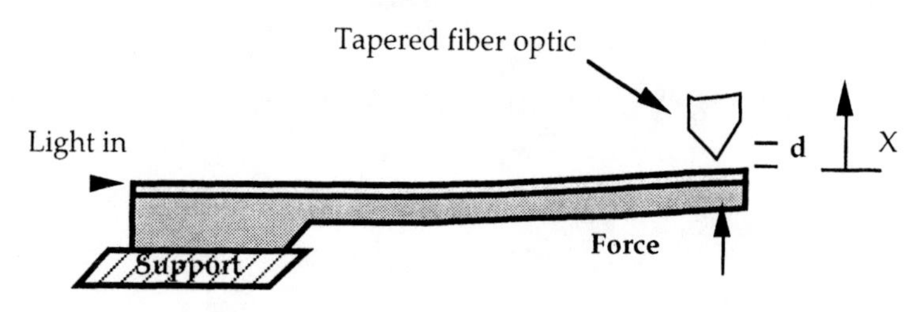

Figure 8.13 A schematic of a displacement sensor using an evanescent field sensing method.

In many ways this sensing mean is very similar to the tunneling current scheme with the exception that instead of electrons, photons tunnel through the free space. However, the spatial scale factor here ($\lambda_{photon}/10$-$\lambda_{photon}/100$) is larger than the tunneling electron's scale factor ($\sim\lambda_{DeBroglie}$), there is no appreciable back action when photons tunnel from the waveguide to the fiber, and the noise is dominated by the shot noise of the photo-multiplier that is used to detect the light intensity in the fiber. There is also no real reason to taper the fiber that is usually done in microscopy to increase the resolution and it is not needed here. The sensitivity of the tunneling photon sensor (S_{tp}) is:

$$S_{tp} = \left|\frac{dI_d}{I_d\, dX}\right| = \frac{dI_d}{I_d dI_{ph}} \left|\frac{dI_{ph}}{dX}\right|. \tag{8.29}$$

Since $I_{ph} = I_0 e^{-\alpha x}$, the last term in the above equation is αI_{ph} and, noting that $I_{ph}dI_d/I_d dI_{ph}=1$, we have $S_{tp}\approx\alpha$. The MDS is:

$$\text{MDS} = \left[\sqrt{2q\Gamma_G FB_m/I_d}\right]/S_{tp} = \left[\sqrt{2q\Gamma_G FB_m/I_d}\right]/\alpha. \tag{8.30}$$

Mach-Zehnder Interferometry. Figure 8.14 shows a Mach-Zehnder displacement sensor. The distance between the back of the cantilever beam and the interferometer arm (denoted as L in figure 12) is less than the decay length of the evanescent fields outside the waveguide.

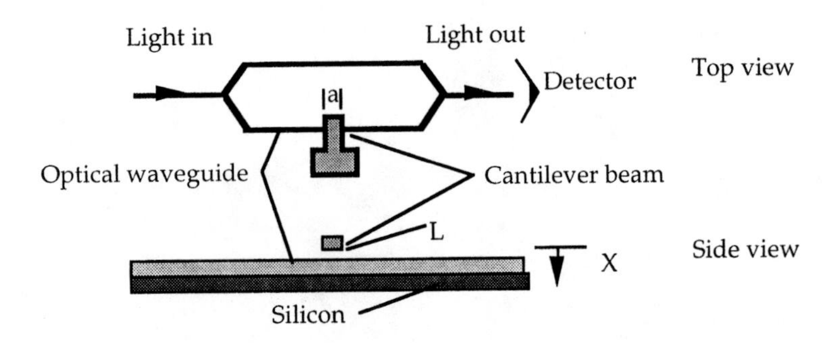

Figure 8.14 A Mach-Zehnder displacement sensor.

Under such circumstances, when the distance L varies, the effective index of refraction (neff) for the propagation modes inside the waveguide change. We assume TE_{00} mode propagation in the waveguide with unperturbed and perturbed refractive indices, respectively denoted by n_{00} and n_{eff}. The phase change due to the displacement dX of the cantilever is $\Delta\phi_{TE_{00}} = \frac{2\pi L}{\lambda}\frac{\partial n_{eff}}{\partial L}dX$. Lights that propagate through both arms of the interferometer are superimposed at the output and interfere at the detector's plane. The power at the photodetector (assuming fringe visibility of one) is $P_d=P_0T(x)=2P_0(1+Cos(2\pi a(n_{eff}-n_{00})/\lambda))$. The sensitivity is:

$$S_{MZ} = \left|\frac{dT(X)}{dX}\right| = \frac{4\pi a}{\lambda}\left|Sin(\frac{2\pi}{\lambda}a(n_{eff}-n_{oo}))\right|\frac{\partial n_{eff}}{\partial L} \,. \tag{8.31}$$

To obtain n_{eff}, we must solve the problem of a four-layer waveguide. However, because the cantilever beam interacts with the exponentially decaying evanescent fields, we can approximately write: $n_{eff} \approx n_{00}(1-\delta e^{-2\pi L/\lambda})$, where δ is very small. Thus, we have $S_{MZ} \approx \left[6\pi^3 a\delta^2 n_{00} e^{-4\pi L/\lambda}\right]/\lambda^3$, and the MDS is:

$$MDS = \lambda^3\left[\sqrt{2q\Gamma_G FB_m/I_d}\right]/(16\pi^3\delta^2 a\, n_{00}) \,. \tag{8.32}$$

$n_{eff}(L)$ can be calculated numerically and values for some particular cases are available in the literature. For example, dn_{eff}/dL is 0.4 μm^{-1} for a SiO_2/Si_3N_4/air thin film waveguide (L≈ t ≈ 0.1μm). The main disadvantage of this method is the requirement of small L and large a.

In table 8.1 we give the MDS's of different sensing means that are discussed above. A silicon cantilever beam (b = 5000 μm, a =100 μm, and h (=h_2) = 50 μm, Y for silicon {100}, <011> is 169×10^{-9} N/m^2) with spring constant of 4.2 N/m was used as a shell. The numerical values for the MDS given in table 8.1 are consistent with values in our own work and other published data. The MDS values may be improved in each case by changing one or more of the corresponding structural and other related parameters. However, this may not always be viable. For example, in the capacitive sensor, we may attempt to reduce the separation between electrodes ℓ; however, for a value of ℓ =1μm, the force between electrodes is 8.85 N, which will collapse the beam. In the piezoelectric scheme, we may try to reduce the value of the bias voltage V_B. This would require either decreasing the current or increasing the frequency of the oscillator which are limited by the properties of the bipolar devices used. Increasing the thickness of the piezoelectric layer may also be attempted, but the quality deteriorates as its thickness increases. For the piezoelectric sensor we have considered a PZT layer. Usually a higher sensitivity may be achieved with capacitive sensors than with piezoresistive ones; on the other hand, the circuit resolution is much better for piezoresistive sensors. For identical full-scale pressures and diaphragm areas, piezoresistive sensors have smaller MDS but capacitive sensors show better long term stability.

Tunneling current method has the smallest theoretical MDS ($\sim 10^{-3}$ Å) of all the sensing means studied here. Unfortunately, due to the migration of the surface atoms on the tip and on the cantilever beam, R_0 changes as a function of time. In the scanning tunneling microscope this is not a problem because the tip scans the surface. However, when the tip is stationary, as in our case, time dependent R_0 necessitates frequent re-calibration of the sensor. The microwave scheme is very similar to the tunneling current scheme except that its MDS is quite larger because of its much larger scale factor ($1/\alpha >> 1/2\kappa$). The MDS of the magnetic scheme (2Å) is quite impressive. However, increasing ℓ from 10 to 100 μm, which can be more realistically achieved, increases the MDS to ~1.5 μm. This example clearly shows the limitations of the displacement sensors that utilize external magnetic field sensing.

For the optical sensing means we may try to increase the power P_o (hence, I_d) that reaches the detector. While this may be done in the Michelson interferometer and in the beam deflection method, in the evanescent optical method this is not an easy task. For sufficiently large shells, the MDS for capacitive sensors may be considerably improved. The MDS for optical techniques, on the other hand, is generally insensitive to the size of the sensor shell (except for the beam deflection method). Optical sensing means have smaller MDS than capacitive, piezoelectric and piezoresistive means for a cantilever shell of the above dimensions. This result supports the main trend in recent years, where optical techniques are being utilized more widely in small displacement sensors.

Table 8.1 Minimum detectable signals of different sensors.

Sensing mean	MDS	Data	Value
Capacitive	$\ell\sqrt{\dfrac{q}{C_S V_B}\left(\dfrac{C_S+C_p}{C_S}\right)}$	$C_n\approx0$, $A=10^{-2}$ cm^2 $V_B=1V$, $\ell=10$ mm	43 Å
Piezo-electric	$\dfrac{4bV_B C_{pe}}{3Q_m d_{13} Eah_2}\sqrt{q(C_{pe}+C_p)V_B}$	$C_n\approx0$, $V_B=1V$ $d_{13}=-93.5$ pC/N $\varepsilon=260\varepsilon_o$, $h_1=1$mm	11 Å
Piezo-resistive	$\dfrac{16b^2}{h_2 E\pi_1 V_B}\left(\dfrac{R_o+R_d}{3R_d}\right)\sqrt{2\pi kT R_s B_m}$	$B_m=100$Hz, $h_1=10$mm $N_a=10^{21}cm^{-3}$, $R_d=1M\Omega$, $V_B=1V$ $p_1=25x10^{12}$ cm^2/dyn $\mu_n\approx120cm^2/V\text{-}s$,	2 Å
Tunneling Current	$\dfrac{\sqrt{qB_m}}{\kappa\sqrt{2I_0}}$	$I_0=10^{-7}$ amp. $k\approx10^{10}$ m^{-1}, $B_m\approx1$ MHz	10-3 Å
Evanescent Microwave	$\dfrac{\Delta V'_d}{V'_d}\quad\dfrac{\Delta f^2}{2\lvert f-f'_c\rvert\,\alpha\,C(f,\sigma)}$	$f_c\approx1GHz$, $\Delta f\approx f_c/100$, $f\text{-}f'_c\approx\Delta f/10$ $\alpha\approx300$ cm^{-1}, $C\approx1MHz$, $I_d\approx10\mu A$	104 Å
Magnetic	$\dfrac{D\sqrt{2q(I_1+CH_x(\ell,0)\,B_m}}{16\pi M_s C(1-e^{-2\pi d/D})e^{-2\pi\ell/D}}$	$C\sim0.1$ mA/T, $d\sim1$ µm, $M_s\sim500$ mT, $I_{11}\sim3mA$ $D\sim50\mu m$, $\ell\sim10$ µm, $B_m=1Hz$	2 Å
Michelson	$\lambda\left[\sqrt{2q\Gamma_G FB_m/I_d}\right]/\pi$	$\lambda\approx6000Å$, $I_d\approx1$ mA $F\Gamma_G B_m\approx100$ kHz	10-2 Å
Beam Defl.	$\dfrac{\lambda e^{0.5}\,b\left[\sqrt{2q\Gamma_G FB_m/I_d}\right]}{2\pi D_0}$	$D_0\approx400\mu m$, $\lambda\approx0.6\mu m$, $I_d\approx1$ mA, $F\Gamma_G B_m\approx100$ kHz	0.04 Å
Amplitude Mod.	$\dfrac{h_1^2}{\delta}e^{\frac{\pi^2\delta^2}{h_1^2}}\sqrt{2q\Gamma_G FB_m/I_d}$	$I_d\approx1$ mA, $F=1$ $\Gamma_G=10$, $B_m\approx1$ kHz $h_1\approx0.1$ µm	1.3 Å
Evanescent Optical	$\left[\sqrt{2q\Gamma_G FB_m/I_d}\right]/\alpha$	$\alpha\sim0.02$ $Å^{-1}$, $F=10^2$ $\Gamma_G=10^5$, $I_d\approx1$ mA, $B_m\approx10$ kHz	0.3 Å
Mach-Zehnder	$\dfrac{\lambda^3\left[\sqrt{2q\Gamma_G FB_m/I_d}\right]}{(16\pi^3\delta^2 a\,n_{00})}$	$\lambda\approx0.6\mu m$, $I_d\approx1$ mA, $\nu_{00}\approx1.4$ $F\Gamma_G B_m\approx100$ kHz $a=0.1$ mm, $L=10\mu m$, $d\approx10^{-3}$	0.18 Å

In conclusion, we have presented a consistent scheme to compare different sensors. The idea of decomposing sensors to sensing shells and sensing means were introduced for the first time. A comparison between different sensing means using a silicon cantilever shell was presented. Simple formulas were derived for eleven different sensing means. The results suggest that for cantilevers with small dimensions and a low spring constant (~1-2 N/m), tunneling current method has the smallest theoretical MDS (~10^{-3} Å) of all the sensing means studied here. Next to the tunneling current method are the optical methods with theoretical MDS on the order of 10^{-2}-1 Å, which are approximately two orders of magnitude smaller than the capacitive and piezoelectric methods. The evanescent microwave method had the largest MDS: 10^4 Å.

Temperature sensors are commonly used in control systems. We have not discussed them here since they readily available and are extensively discussed in the literature [7,8]. In the following section we discuss chemical sensors closely following reference [5].

8.4 Chemical Sensors

Due to the importance of chemical gases and substances in industry and the environment, chemical sensors have been among the important devices investigated during the past 30-35 years. Monitoring chemical substances and processes with electromagnetic radiation has been an important tool in chemical and related sciences. Probably the most common technique in chemical analysis involving electromagnetic radiation is spectroscopy. Spectroscopic techniques consist of irradiating a chemical compound with a well characterized radiation and analyzing the spectrum of the scattered radiation. The determination of the wavelength at which a substance absorbs or emits photons gives us quantitative information about the physical configuration of the atoms and molecules or of their interactions. In particular, optical radiation may interact with vibration modes of small molecules, chemical bonds of molecules, or some specific range of lattice vibrations in crystals among other interactions.

Chemical information using optics can also be obtained in an indirect manner, i.e., we may optically monitor a 'residue' of the chemical process. For example, if a chemical reaction is taking place and heat is being released, we may monitor the optical consequences on the surrounding medium or on a particular structure. Other possibilities may be that during some chemical process light is emitted. Then, we may collect the emitted photons and, by analyzing their spectrum and rate of emission, we may obtain information regarding the chemical state of the process. Optics being so versatile in the study of chemical substances, the interest in studying and developing optical chemical sensors is not surprising.

In environmental applications, it is important to monitor oxygen, carbon dioxide, ammonia, sulfur dioxide and hydrogen sulfide among other gases. Optical fibber sensors have been developed and are currently under study for these and other gases. Measurement of toxic emissions and substances, such as gases (NO_x, SO_2, NH_3, Cl_2), metal ions and hydrocarbons, is important in several applications, including agricultural, food industry, live stock, hygiene and bio-chemical. A number of methods are known in the literature for the measurement of these toxic substances such as reagent-based, infrared gas analyzers and electrochemical cells.

However, all of these methods have disadvantages: indicator-based methods are irreversible; infrared devices are bulky, expensive and have interference from other gases; electrochemical cells tend to have a relatively short useful life-time since they are difficult to microminiaturize and are prone to electromagnetic interferences. Therefore, a need for a reliable, reversible and inexpensive sensing scheme for monitoring industrial waste and emissions still persists. Increased "world wide" environmental awareness and ever-increasing strict legislations have pressed the scientific community to intensify their search for developing improved, new and novel methods of monitoring various toxic species of interest. It is preferable that these methods be adaptable in portable devices be easily calibrated.

In an attempt to overcome the problems associated with the conventional devices and to meet some of the many requirements of industry, optical fiber sensors are proposed. Aspects of fiber-optic chemical sensor technology have been discussed in detail in references [11-18].

Among the attractive features of fiber-optic chemical sensors is the possibility of remote sensing, which is of paramount importance in the case of toxic chemicals. Optical chemical sensors may be superior to their electrostatic counterpart due to the inherent properties of optical sensing means which include: a) the elimination of the possibility of an explosion in sensing flammable chemicals because no "spark" or appreciable heating is present in optical sensors and due to the low photon flux densities photo-synthesis is also minimal, b) the possibility of detecting more than one chemical by using multiple wavelength channels, and c) the immunity to electromagnetic interferences, among others. It is in part the objective of the present chapter to demonstrate that most of the fiber-optics chemical sensors can be implemented using integrated optic circuits, which virtually eliminates the alignment requirements and makes batch fabrication of the sensors possible.

Our goal in this section is to illustrate some of the underlying principles of integrated and fiber optic chemical sensors, to give few working examples, and to discuss positive and negative attributes of these sensors compared to the other more conventional chemical sensors.

Chemical sensors have always been in the realm of the most difficult devices to design. Noting that sensing smell and taste also fall into the category of chemical sensors, the rewards of developing a "perfect" chemical sensor are obvious. One can only hope to have a "virtual reality" environment where all the human sensory factors are activated. At present, only sight, sound, touch, and mechanical motion are present.

Mechanisms through which light is affected by the presence of a chemical substance are absorption, index of refraction, physical path length, and polarization. In the case of chemical sensors, the sensor shell usually plays a very active role. Chemical molecules and atoms interact with the sensor shell either physically or chemically. The distinction between these two methods of interaction in most cases is the energy involved in "un-doing" the interaction. In the case of a physical process, this energy is usually quite low, while in the case of chemical interaction, the energy is high compared to the thermal energy.

When the chemical substance interacts with the sensor shell, it may change its dimensions, it may modify its temperature, it may change its composition in an appreciable manner, or it may modify its luminescence. To design chemical sensors for continuos operation, the absorption/de-absorption of the chemical substance onto the sensor shell should be easily achieved. If the interaction energy between the chemical substance and the sensor shell is too high, the reaction may not be conveniently reversible, rendering the sensor inoperable after the first sensing cycle.

In chemical sensors, the sensor shell itself may be a renewable part of the optical sensor. In this case we call the device extrinsic. In other configurations, where the sensor shell is an integral part of the optical waveguide, the sensor is called intrinsic.

Because of the intimate contact and interaction between the chemical substance and the sensor shell, optical chemical sensors are notorious for their lack of long-term stability. In some cases even short term drifts can make the chemical sensor unreliable. Clearly, there is a non-trivial relationship between the interaction energy of the chemical substance, the sensor shell, and the stability of the sensor.

In the following sections we first discuss electrical gas sensors and then discuss intrinsic and extrinsic optical chemical sensors in detail.

8.4.1 Electrical Gas and Chemical Sensors

Electrical gas sensors are quite simple and are used extensively in safety and process monitoring systems in industry and to monitor CO and other bio-hazardous gases in residential applications. These sensors are usually composed of a sensing method and a gas sensitive film. In the following sections we discuss some typical electrical gas sensors.

Resistivity-Based Gas Sensors. These are probably the simplest forms of electrical gas and chemical sensors. A gas sensitive film is contacted by metallic "read" electrodes. As the sensitive film is exposed to different gases or chemicals, its resistance changes. The using the contacting electrodes the resistance of the film is measured by passing a small amount of current and monitoring the voltage drop across the film.

Air quality sensors usually use a sensing film that can be readily oxidized (by NO_2 or O_2) or reduced (by CO or NH_3). Upon oxidation or reduction, the electrical resistivity (or some other physical parameter) of the film changes and it is detected to infer the gas concentration.

Figure 8.15 shows an example of a Langmuir-Blodgett film [$(C_6H_{12})_3SiOSiPcOGePcOH$] sensor [19]. According to figures 8.15b and 8.15c, it is apparent that both the response time and the final value of the sensor output depends on the gas, its concentration and temperature. Thus, the sensor's selectivity to a particular gas is maximized by appropriately setting its temperature (figure 8.15). Alternatively, a neural network can be used to monitor the response (conductance as a function of time) of the sensor and adjust its temperature to selectively detect different gases. This observation is the base of our proposed intelligent VLSI gas and chemical sensors.

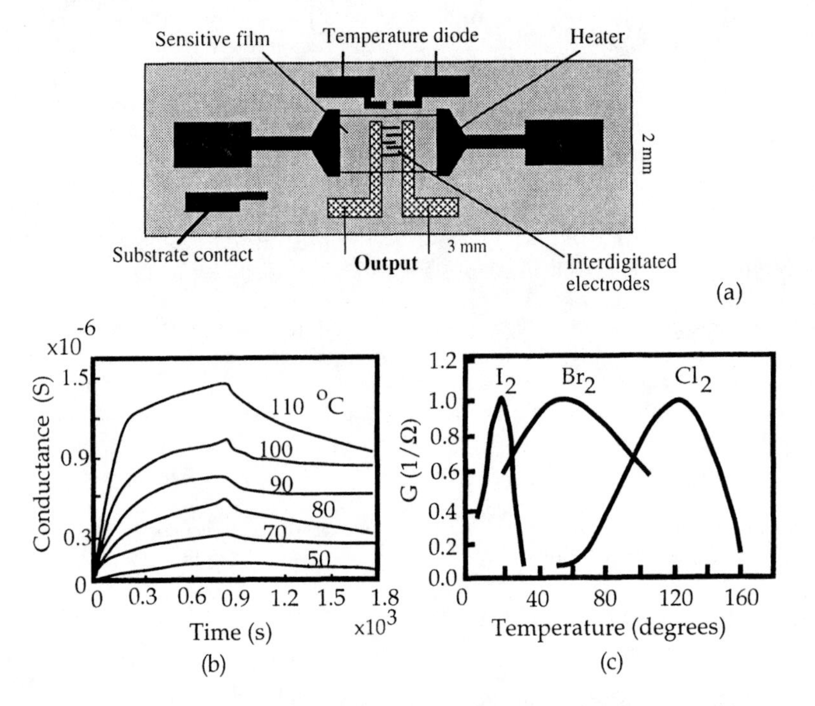

Figure 8.15 (a) A simple sensor structure that uses a gas-sensitive film deposited over interdigitated transducers. The heater is used to increase the temperature of the film. The diode is used to measure its temperature. (b) The response of the gas sensor shown in figure 2 to 111 ppm Cl_2 in nitrogen at different temperatures. (c) Normalized conductance change as a function of temperature for Cl_2, Br_2, and I_2 gases [19].

Barrier-Height-Based Gas Sensors. Palladium is extensively used to detect hydrogen in trace amounts. There are efforts around the world [9,20] to incorporate Pd films with SiC (a wide band gap semiconductor with devices that can operate at elevated temperatures up to 800 °C) and catalitically convert hydrocarbons to hydrogen at high temperatures and measure the fuel quality by measuring the H_2 concentration. Figure 8.16 shows an example of a palladium sensor [21]. In this sensor, the palladium was incorporated in a Schottky diode. Upon exposure to H_2, the physical changes in the Pd result in the reduction of the Schottky barrier height increasing the reverse leakage current of the diode (figure 8.16b). A variety of impurities can be incorporated in Pd to increase its sensitivity (Ag) or to increase its structural stability (Ni, and Cr). The fabricated sensor operated at elevated temperatures and it could detect ppm levels of H_2.

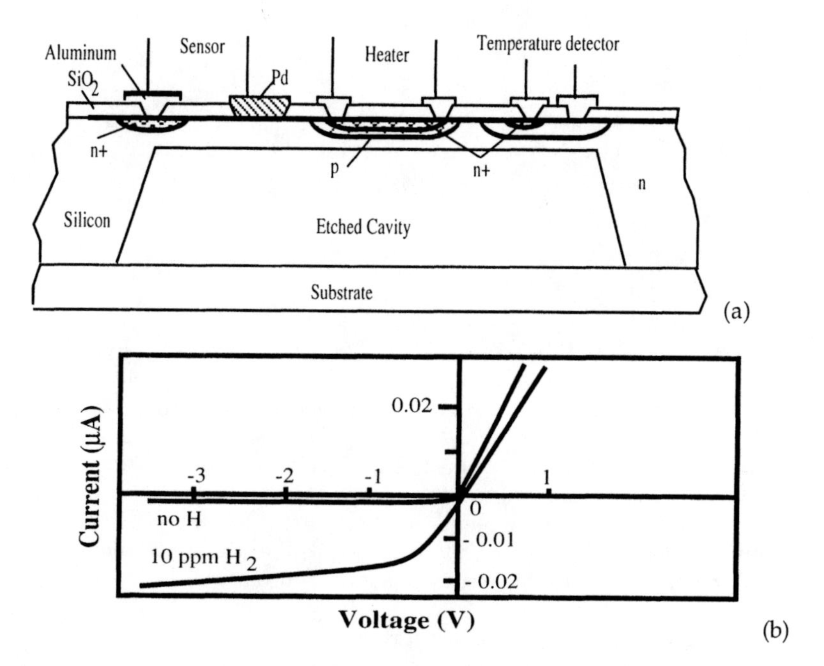

Figure 8.16 a) A Schottky diode palladium H_2 sensor [21]. b) When exposed to H_2, the reverse leakage current of the Schottky palladium diode increases. The anisotropically-etched cavity is to reduce the thermal load and increase the thermal isolation of the sensor.

Ion-Sensitive FET Sensors. Figure 8.17 shows a cross-section of a gas-sensitive field-effect transistor with a Pd gate which is used as H_2 sensor. Hydrogen molecules dissociate at the Pd surface (on two neighboring adsorption sites) and the resulting hydrogen atoms diffuse into the Pd and set-up a potential difference between the Pd and the silicon substrate. The presence of the hydrogen atoms at the interface (H_i) result in the depletion of the p-type silicon channel. Thus the silicon flat band voltage shifts by an amount $-\Delta V_H$ due to the interfacial hydrogen [9,20].

In air 1 ppm H_2 gives a threshold-voltage shift of about 25-50 mV at 150 °C in a depletion mode n-channel FET. In an inert atmosphere less than 1 ppb of hydrogen can be detected [20].

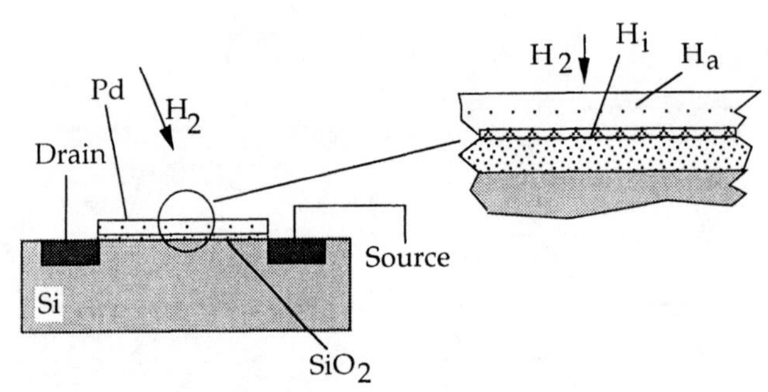

Figure 8.17 A field-effect transistor with Pd gate electrode. As hydrogen atoms diffuse into the Pd, they change the threshold voltage of the FET by inducing charge at the Si/SiO_2 interface [20].

The above device structure is the basis of many ion-sensitive and different types of gas-sensitive sensors. Various gas-sensitive films can be incorporated with silicon to render the sensor selective to a particular gas or chemical ions. References [9,20] reviews all the different methods and structures which are used in gas and chemical sensors.

8.4.2 Guided-Optics Intrinsic Chemical Sensors

These sensor types are based on the fact that chemical species can affect the waveguide properties. Hence, it is not the absorption or emission properties of an analyte that are measured, but rather the effect of the analyte upon the optical properties of the optical waveguide. More specifically, these sensors are based on one or more of the following effects of the analyte:

(a) An increase in the strain/stress of the coating,
(b) Modification of the waveguide temperature,
(c) Attenuation of the guided light amplitude,
(d) Change of the effective refractive index of the mode,
(e) Modification of the polarization of the light.

Except in (a) and (b), in all the above mechanisms, the interaction with the chemical substance takes place through the evanescent waves. In (a) the physical shape of the waveguide may change. In (b), the waveguide parameter changes due to a change in the physical dimension as well as to a change in the refractive index of the waveguide as a function of temperature. In (c) the intensity of the optical field decreases exponentially as it travels in the waveguide. The Kramers-Kronig relationship relates the imaginary part of the refractive index to its real part. Therefore, in attenuation mode chemical sensors, the phase of the guided mode also changes. In (d), the phase of the guided light changes. In (e), the TM and TE mode experience different phase shifts or attenuation. In the following sections we discuss examples of each of these mechanisms are reported in the literature.

All of the above mechanisms can be incorporated in an interferometer, and phase-modulated or interferometric optical sensors offer the highest sensitivity. Interferometric sensor systems typically employ the Mach-Zehnder interferometer configuration shown in figure 8.14. Other configurations such as those of Michelson and Fabry-Perot, have also received some attention.

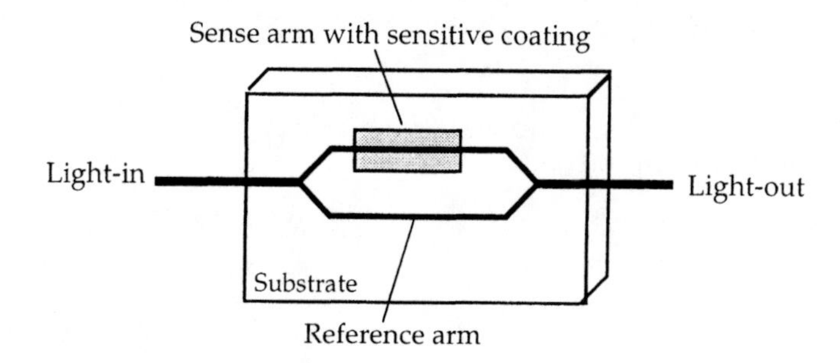

Figure 8.18 A Mach-Zehnder chemical sensor.

The Mach-Zehnder inteferometer is extremely sensitive to changes in optical path length; for time-varying modulations of the input wavelength, the theoretical limit to the sensitivity of an interferometer is on the order of 10^{-13} m (1 part in 10^{14} for a 10 m sensing fiber). The dependence of the interference signal is given by the equation:

$$P = P_o \cos^2(\Delta\phi/2) \tag{8.33}$$

where P is the output light power, P_o is the input light power, and $\Delta\phi = k_o L_{sf} \Delta n_{eff}$ is the relative phase shift between the two arms, k_o is the free space wavenumber, L_{sf} is the length of the sensing arm exposed to the chemical, and Δn_{eff} is the change in the mode refractive index of the sense arm due to the chemicals. The attainable resolution of the measurement of chemical concentration (dC) may be written as:

$$dC = \frac{\partial C}{\partial n_{sf}} \frac{\partial n_{sf}}{\partial n_{eff}} \frac{\partial n_{eff}}{\partial \phi} d\phi \tag{8.34}$$

where n_{sf} is the refractive index of the chemically sensitive film (which is deposited over the sense arm). Since the optical power is mostly confined inside the waveguide and only the evanescent part of it is affected by the change in the refractive index of the sensitive film, dn_{eff}/dn_{sf} can be quite small and the waveguide should be designed carefully. Near cut-off operation will probably result in the best dn_{eff}/dn_{sf}. We note that:

$$\frac{dn_{eff}}{d\phi} = (k_o L_c)^{-1} \tag{8.35}$$

Taking L_c to be 1 cm, the light wavelength to be 830 nm ($k_0 = 2\pi/\lambda_o = 7.57\times10^4$ cm^{-1}), and a minimum detectable phase change of $\delta\phi \approx 10°$ (this corresponds to a 1% change in P), we have $dn_{eff} \approx 10^{-4}$ for a Mach-Zehnder Interferometer with a relatively simple and straight-forward signal detection scheme. Using synchronous detection, dϕ can be easily reduced to 1°. In the literature there has been reports of dn_{eff} as low as 10^{-7}, which is quite impressive. To increase the sensitivity of this sensor, one has to render the reference arm inactive as much as possible. Various inorganic coatings, including SiO_2, have been used to passivate the reference arm and MgF_2 has been found to be the best passivator for most gases.

In the following sections we will discuss different examples of optical sensors reported in the literature.

H_2 Sensors. Incorporation of hydrogen in palladium (Pd) results in a change in the dimensions of the Pd film. Based on this principle, a hydrogen sensor was constructed and implemented using a Mach-Zehnder interferometer [9,10]. The Pd coating of 1.5 μm thick was sputtered onto the sense arm and a 0.5-mW He-Ne laser was used. The movement of the fringe pattern can be observed visually or by use of a simple photodetector and chart recorder.

An example of thermal chemical sensing is provided in the references [22] where solution concentrations of hydrogen peroxide are quantified using the enzyme catalase immobilized to the sensing arm of a fiber optic Mach-Zehnder interferometer. The heat released upon interaction of the reactants causes the index of refraction to change on the waveguide, which in turn generates a measurable phase shift on the guided wave. In [22.b] a fiber optic interferometric hydrogen sensor was developed. In this sensor the heat of chemisorption on a palladium-coated sensing waveguide is monitored and related to the amount of hydrogen present in the surroundings of the palladium.

Humidity Sensors. A humidity sensor, based on the measurement of variations of the transmitted light at 680 nm in a U-shaped unclad silica optical fiber, coated with cobalt chloride, has been developed [23.a]. Recently, a humidity sensor based on the phenol-red dye-doped polymethylacrylate plastic fiber has been reported [23.b]. The moisture in soil can also be measured by using this sensor. The absorption is usually wavelength dependent. Therefore, in designing attenuation-based chemical sensors, spectroscopy is performed to find the optimal wavelength. Intrinsic fiber optic sensors based on the attenuation of light generally rely on evanescent field interaction. This technique is referred to as evanescent field spectroscopy here. One advantage of the evanescent field spectroscopy is that it may be used for opaque samples, while conventional spectroscopy may need adaptation of the sample (further dilution or modification of the sample chamber).

Hydrocarbon Sensor. Using preferential absorption of hydrocarbons by an organo-philic compound (octadecyltrichlorosilane) forming the clad of a silica fiber, a transmission loss measurement-based hydrocarbon sensor is developed [24]. This scheme has been used in several modified forms for constructing hydrocarbon sensors. Schemes employing remote fluorescence analysis of underground waters have also been proposed and constructed [24].

To detect hydrocarbons, a sensitive polymer film that was fabricated by a method similar to Langmuir-Blodgett techniques [16] is used. Polysiloxane was spread on the surface of distilled water, giving a film approximately 1 μm thick. Then the chip was inserted, such that one arm was covered with the oligomer, which was then polymerized by a mercury arc. The set-up for monochromatic and spectral measurements is as shown in figure 8.19 [16].

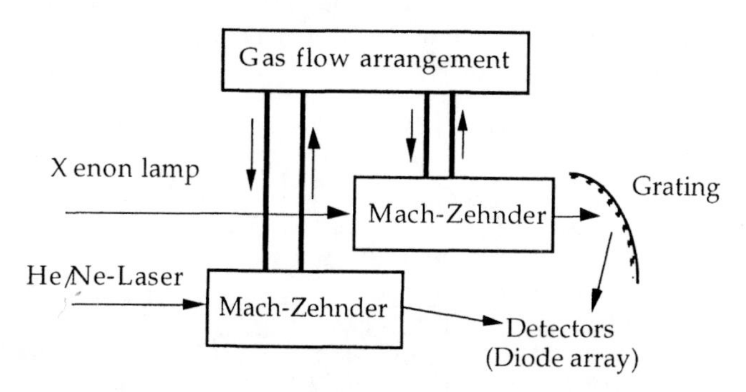

Figure 8.19 The set-up for monochromatic and spectral measurements [16].

Polarization of the light can also be used in sensing applications. An example of a chemical sensor using the polarization of a guided light in an integrated optic interferometer is given in [25]. The sensor consists of an integrated difference (or polarimetric) interferometer that uses only one waveguide. The waveguide is designed to have only two modes of propagation: the fundamental TE and TM modes. A small portion of the waveguide's core where it contacts the measurand (a gas or a liquid) is exposed. The propagation constant of the TE and TM modes is sensitive to the refractive index of the sample. The dependence of each propagation constant on the refractive index of the sample is different and is dictated according to the equations for a three layer waveguide. At the output of the interferometer the light exiting the waveguide is passed through a polarizer at 45° and into a photodetector. After the polarizer, the waves arising from the original TE and TM are polarized in the same direction and may interfere. The measured output power is (assuming no absorption):

$$P=P_0[1+\cos(\theta_0+\Delta\theta_{TE_0}-\Delta\theta_{TM_0})]^2 \tag{8.36}$$

where $\Delta\theta_{TE}$-$\Delta\theta_{TM}$ is the induced phase difference for the TE and TM modes and is related to the change in the corresponding wavenumbers ($\Delta\beta_{TE}$-$\Delta\beta_{TM}$):

$$\Delta\Theta=\Delta\theta_{TE_0}-\Delta\theta_{TM_0} = L(\Delta\beta_{TE_0}-\Delta\beta_{TM_0}) \tag{8.37}$$

where L is the interaction length. In reference [25], $\Delta\Theta$ could be determined with an experimental resolution of $2\pi/1000$. Thus, for an interaction length of L=12 mm, an effective refractive index change of $\sim10^{-7}$ could be resolved. For a 140-200 nm TiO_2 -SiO_2 waveguide, such a resolution in the n_{eff} resulted in a resolution of 10^{-6} in detecting changes in the refractive index of water (n=1.33).

8.4.3 Extrinsic Chemical Sensors

In these sensors, the fiber (or the fiber bundle) acts as a light conduit to remotely probe the spectral properties of the measurand [26.a]. In contrast to intrinsic fiber sensors, the measurand should not affect the waveguide properties of the fiber. Fiber and measurand do not necessarily have to be in intimate contact. A few typical examples of absorption cells are shown in figure 8.20.

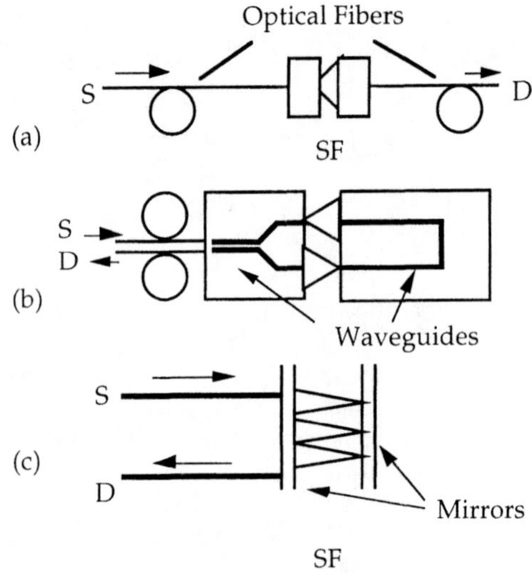

Figure 8.20 A schematic presentation of different extrinsic chemical sensor configurations [26.a].

The simplest case (a) is a flow-through cell whose diameter defines the optical path length, with two fibers attached to the cell at both sides. The fiber optic peers through the cell, and both liquids and gases can be investigated. Another single fiber approach, but with two sample volumes and more versatile geometry with respect to sensor insertion into a sample, is shown in (b). While 0.1 to 10 cm path lengths are sufficient for ultraviolet and visible wavelength absorption measurements, the small molar absorptions in near-infrared have necessitated the path length 1 m in order to improve sensitivity. To increase the path length without increasing the physical size of the device, a multiple reflection chamber, as shown in (c), is used.

The Lambert-Beer law relates analyte concentration and path length, with absorbance. Because of the constant background absorption by fibers at certain wavelengths, or the sample solvent, usually an additional term dealing with losses (exp(-b)) is incorporated into the Lambert-Beer law:

$$I=I_0 e^{-\varepsilon c \ell} e^{-b} \tag{8.38}$$

where I_o is the input light intensity, ε is the absorption coefficient of the chemical per unit normal concentration, c is its concentration, ℓ is the path length, and b is the attenuation of light due to background absorptions determined from blank runs.

Figure 8.21 shows the experimental system arranged for the remote absorption measurement of low-level CH_4, gas in air employing long-distance low-loss silica optical fiber links [26.b].

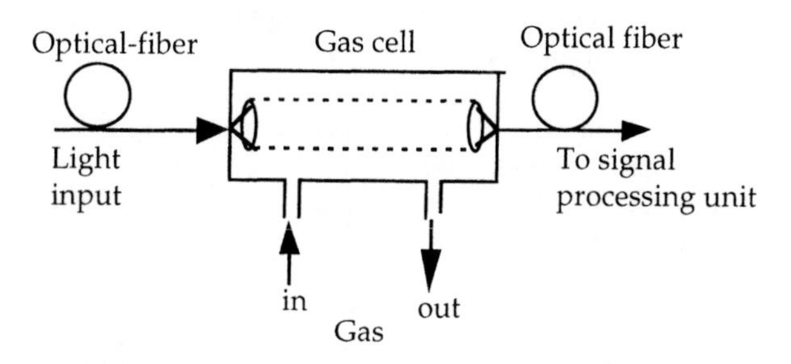

Figure 8.21 A schematic of a remote optical absorption measurement of low-level CH_4.

The light source used for the experiment was either a 1.34-μm InGaAsP laser diode or a 1.64-μm InGaAs laser diode used as a light-emitting diode (LED) with spectral width of about 80 nm [26.b]. Both LEDs were modulated with 90-Hz current square pulses for synchronous detection with a lock-in amplifier. The absorption cell working as the gas sensor head was 50 cm long, in which the partial pressure of CH_4 gas was appropriately changed while the total pressure of the CH_4 air mixture was kept at 1 atm. Absorption spectroscopy was performed using a grating monochromator [26.b].

8.4.4 Polymer Waveguide Chemical Sensors

This is an example of a chemical sensor where both intrinsic and extrinsic sensing mechanisms are present (figure 8.22). In this case, the optical waveguide itself is made of a polymer that interacts with the chemical substance. An example of this type of sensor is given in [27], where polyimide light-guides are used for the detection of n-heptane vapors within a mixture with iso-octane vapors. In this case, the change in propagation constant is due to a real change in the refractive index of the light-guides, arising from the absorption of n-heptane molecules at the exposed waveguides surface, which is later diffused into the bulk of the optical waveguide.

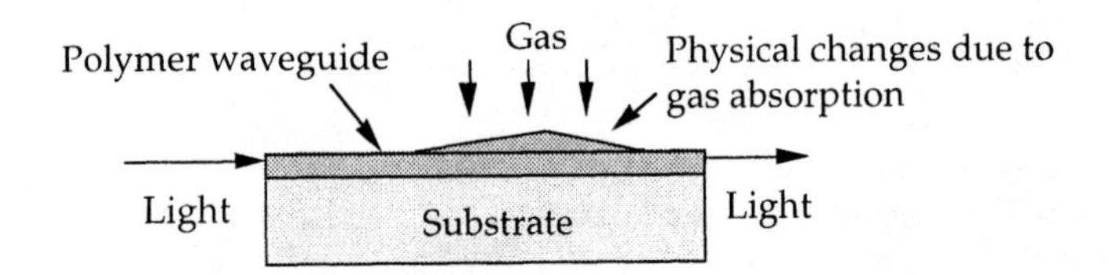

Figure 8.22 In polymer waveguides, gases can interact with the evanescent fields as well as with the waveguide material itself.

Apparently, both alkanes are able to absorb onto the polymer surface. However, only the n-heptane can diffuse into the bulk. By diffusing into the waveguide the n-heptane is separated from the mixture. Thus, in this case, the waveguide is involved in both the separation process and the quantitative sensing of the target chemical.

This type of structure may find widespread application in the future since the optical changes in the waveguides can be quite large, leading to large signals. They can also be potentially inexpensive and fabricated on glass substrates with minimal patterning.

8.4.5 Surface Plasmon Chemical Sensors

An important drawback of the evanescent field sensing scheme is that dn_{eff}/dn_{sf} is quite small, resulting in a low probing efficiency, as mentioned before. To increase the probing efficiency, the optical energy inside the sensitive film should be increased. This can be achieved by operating the waveguide near its cut-off, or, more conveniently, by exciting surface plasmons [13,28]. Figure 8.23 shows the basic principle of operation of chemical sensing using surface plasmons. Light propagating inside the waveguide interacts with the charge density waves of a thin metallic (usually silver) layer. Provided that the guided wave's wavenumber matches that of the surface plasmon, the optical energy is absorbed by the plasmons and is consequently dissipated. The plasmon wavenumber, in turn, depends on the conductivity and on the surface condition of the metallic layer. A thin sensitive polymer film is usually deposited over the metallic film which, upon absorption of chemicals and gases, changes the plasmon wavenumber of the metallic film [13,29].

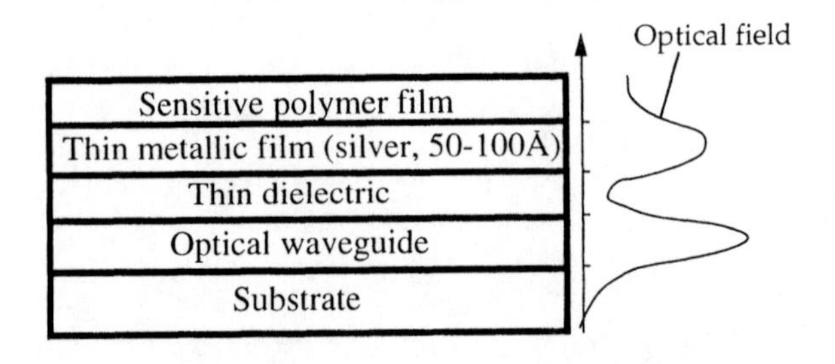

Figure 8.23 A schematic of a surface plasmon integrated optic chemical sensor.

Due to the large attenuation of the optical field that matches the plasmon wavenumber, this scheme cannot be used in interferometry. Using this method, different gases including CH_2I_2 are detected [13]. The surface plasmons are so sensitive to the conditions of the polymer film that very large modulation-index light modulators with electro-optically active polymer films have been proposed and constructed [30].

8.4.6 Indicator-Mediated Extrinsic Sensing

Only a limited number of chemicals have an intrinsic absorption or a related spectroscopic property that can be utilized for direct sensing without compromising selectivity. For several important measurands, including pH, metal ions, and oxygen in water, no direct and sensitive methods are known. In these cases the well-established indicator chemistry can be used [25,31]. By immobilizing a proper indicator on the waveguide, a device is obtained whose spectral properties is sensitive to the measurand. Practically all known indicator-mediated sensors rely on absorption or luminescence measurements. Unlike the case of intrinsic and most cases of extrinsic optical sensors, there is no need for direct physical contact between the sensor shell and the waveguide. This can be important in applications where frequent changes in the reagent phase may be required.

A major advantage of indicator-mediated sensors over other optical sensors is that they are not affected by the refractive index of the medium, reporting only the concentration of a chemical or a physical parameter. However, the reversibility of some of these indicator-mediated sensors is very poor or even nonexistent. Some of these devices are suitable for "single shot" assay in a fashion much like a dipstick test.

There are two principally-different ways to fabricate indicator-mediated optical sensors. In the first approach, the chemistry is produced directly on the waveguide. In the second approach, the sensor chemistry is first built up separately, and the material, in a final step, is placed on the fiber. The latter method appears to give a better reproducibility and has found many more widespread applications. Typical indicator-mediated sensor configurations are shown in figure 8.24 (a)-(e).

In the past 10 years, numerous kinds of oxygen sensors based on fluorescence quenching have been developed. This technique utilizes the ability of oxygen to quench the photo-excited states of other molecules. This is an optical method alternative to standard electrochemical and paramagnetic methods for measuring concentrations of molecular oxygen. The method consists of monitoring the light emission intensity, for example, of a fluorescent p-bonded organic molecule in the absence and presence of molecular oxygen. The quenching of phosphorescent light emissions can also be utilized [32]. Phosphorescence methods have enhanced the sensitivity of oxygen sensing significantly because of the much longer lifetimes of the photo-excited states. This enhances the probability of an encounter between oxygen and an excited molecule.

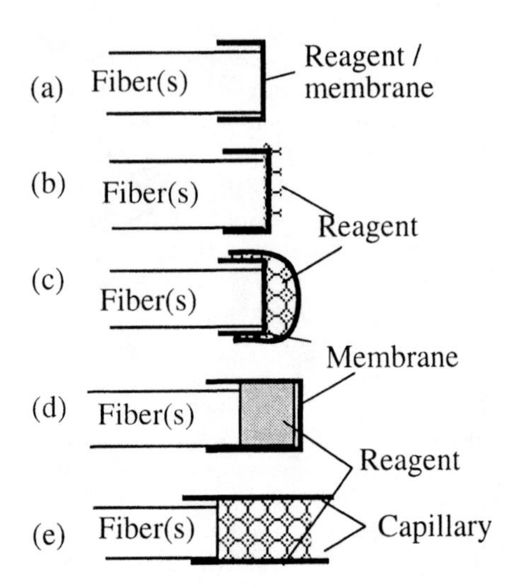

Figure 8.24 A schematic of typical indicator-mediated sensor configurations.

pH Sensors. Optical pH sensors are based on monitoring changes in the spectroscopic properties of an acid-base indicator in the immobilized phase. Both measurement of absorbance or reflectance changes and measurement of fluorescence intensity have been employed. Major efforts, so far, have been to develop pH sensors for physiological applications. A sensor for monitoring sea water pH has been reported by [33]. Flow injection analysis technique was used to monitor rain water pH [34]. A pH sensor which employed Congo red immobilized on a porous cellulose polymer film showed good stability over the range 0-13 and a very short response time (<2 s) [35]. A pH sensor was also developed to measure pH in the 6.0-8.0 range [36]. The development of pH sensors for environmental applications has been rather slow, mainly because of the complex nature of the samples, limited dynamic range offered by the acid-base indicators and low indicator stability. By immobilizing a pH-sensitive dye on a hydrophilic polymer at the end of an optical fiber or on a planar surface, a pH sensor is obtained. Major problems associated with pH optodes result from the ionic-strength sensitivity of all pH indicators, which makes them sensitive to salt effects. On the other hand, pH optodes have distinct advantages over electrodes in case of high pH or when measuring within only a small pH range. The problems associated with optical pH measurement are discussed in [37].

CO_2 Sensors. Carbon dioxide and related acidic or basic gases can be determined indirectly by entrapping a buffer solution plus a pH sensitive dye in a silicone polymer as shown in figure 8.25.

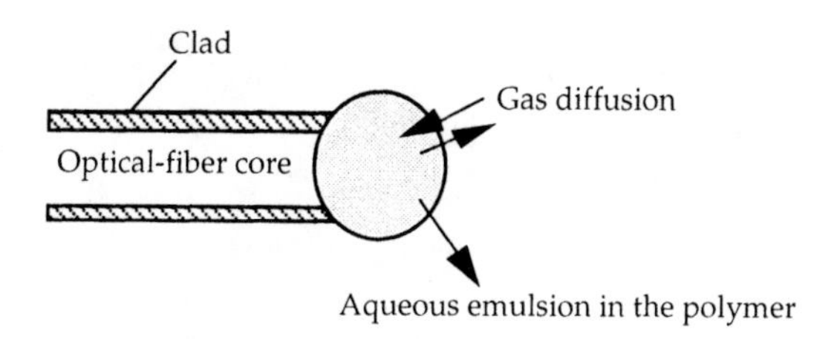

Figure 8.25 A schematic of an optical gas sensor.

CO_2 from the sample passes the polymer membrane and changes the pH of the internal buffer. This is detected by the pH indicator.

All fluoro-sensors based on excited-state interaction of indicator and measurand (such as quenching) are fully reversible. In sensors based on ground state interactions, the reversibility depends on the binding constant (or the Stern-Volmer static-quenching constant), the analytical range, and the relative concentrations (or amounts) of analyte (A) and reagent (R). Considering, for instance, the equilibrium constant for a reaction between A and R, we see that the equilibrium

$$A + R \leftrightarrows AR \tag{8.39}$$

can be driven completely to the right side if [A] (the analyte concentration) is kept low and the equilibrium constant (K_e) favors binding over dissociation. AR is the optically-detected product. In such a "stoichiometric" reaction, the analyte is reacted with an excess of reagent and is completely converted into AR, whose concentration is measured. The signal increases with time until all reagent is consumed. At this point, reaction has gone to completion and the probe has to be disposed or regenerated outside the sample by chemical means [25].

Because it is the total amount of analyte that is determined this way, the volume of the sample needs to be known in order to determine the analyte concentration. Under conditions of unlimited supply of analyte, the relative signal change is proportional to the analyte concentration in case of sensors, while in probes it is the product of analyte concentration and time of exposure.

Ion Sensors. Measurement of ion concentrations in drinking water and in environmental samples is necessary for health reasons. Several optical sensing schemes for both the anions and cations have been reported. An optical sensor for halides, based on dynamic fluorescence quenching has been reported in [38]. The quenching was found to increase in the order Cl < Br < I. Phosphate, perchlorate or nitrate up to 1M concentrations were found not to interfere, but sulfite, isothiocynate and cyanate show interference. Numerous papers have been written on the quenching of various fluorophores by heavy metal ions (well known as heavy metal effect). Sulfide in aqueous solutions has been determined, using the changes in reflectance of immobilized methylene blue, due to the in situ generation of methylene blue during a reaction of N,N dimethyl-p-phenylenediamine hydrochloride in the presence of acidic Fe^{3+}. A fluoride probe based on changes in the reflectance of immobilized alizarin blue-Ce(III) or alizarin blue-La(III) was developed and reported [39].

Measurement of cations is important in several biomedical, environmental and industrial process applications. Several approaches to monitoring these cations have been proposed [40-43]. This includes use of ionophores that bind the cations [40]. A reversible indicator system based on fluorescence and ion-pair extraction has been reported in [41]. Feasibility of constructing optical ion sensors using potential sensitive fluoro-phores has been demonstrated by developing a potassium ion sensor [42].

Organic Contaminants. Due to the difficulty in finding a specific and reversible reaction for selective detection, there has been rather little work on optical sensor development for the measurement of small amounts of organic contaminants in natural samples. The measurement of organic substances in water is of paramount importance, as many of these are known for their high toxicity. A reversible, polar solvent vapor sensor, based on the measurement of reversible de-colorization of blue thermal printer-paper used in graphic plotters, has been developed and reported in [44]. This sensor is specific to polar solvent vapors, such as those of alcohols, ethers, and easters. No interference from hydrocarbons or chlorinated hydrocarbons was found.

Methanol concentrations during the fermentation process have also been measured using fiber optics [45]. Non-invasive, short-wavelength near-infrared spectroscopy (700-1100nm) was used for this purpose. A thin film micromirror sensor for measuring volatile hydrocarbons has been developed and reported in [46].

8.4.7 Optical Biosensors

Most of the above examples dealt with sensors of relatively simple chemicals. Their principles of operations, however, can be extended to construct sensors of biological entities. This field has grown considerably over the past 10 years and it is beyond the scope of this article to give a comprehensive review of this fast-growing field [47-50].

Optical biological sensors are almost exclusively of the extrinsic type. Their transduction mechanisms are identical to simple chemical sensors and they use much more complex "binders" that serve as intermediaries between the optical channel and the biochemicals [47-50]. Due to the complexity of the interaction between the "binders" and the biochemicals, these sensors tend to be even less reliable than other chemical sensors. Here we discuss an example of a biochemical sensor that takes advantage of microstructures to facilitate optical sensing.

Proteins in blood or pollutants in water can be monitored using immunoassay technology, where the highly specific interaction of an analyte (antigen) with its corresponding antibody is quantified. Figure 8.26 shows a schematic of a structure that is used in an optical biosensor to detect anti-immunoglobulin.

Using this sensor, anti-immunoglobulin G(a-IgG) at the level of 1μg/ml could be detected in less than 10 minutes [41].

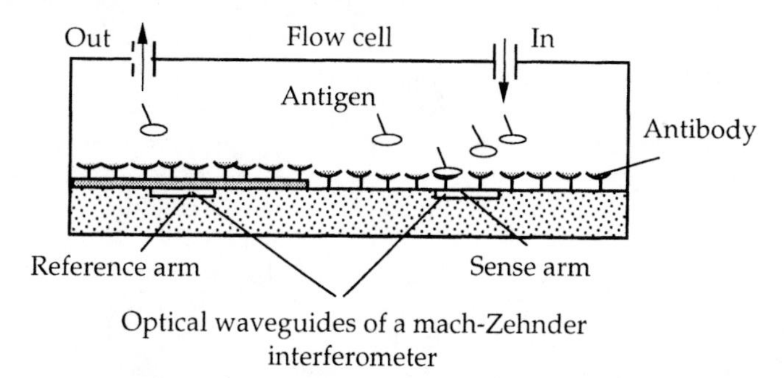

Figure 8.26 A schematic of a biosensor that uses a Mach-Zehnder interferometer to measure changes in the refractive index of an anti-body (IgG) when it is exposed to a flow of antigen (anti-immunoglobulin) [50].

The structure of the above sensor clearly shows that this type of sensors can benefit greatly from the use of microstructure technology. The flow cell can be constructed using silicon bulk micromachining. There may also be a need for cavities situated above the sensor waveguide, where the size of the cavities may be used to adjust the sensor's selectivity.

8.4.8 Ultrasonic Gas and Chemical Sensors

Ultrasonic methods have been extensively used to detect gases and chemicals. In these methods a gas or chemical sensitive layer is deposited over an ultrasonic vibrator. When gases or chemicals are absorbed by the sensitive layer, they change the layer's mechanical properties. These mechanical changes influence the vibration amplitude and phase and can be picked up by monitoring the vibrational characteristics of the oscillator. Two different ultrasonic methods based on bulk and surface waves have been developed in the past.

The bulk wave detection scheme is also called microbalance method since it can be used to detect very small mass loading that occurs when minute quantities of materials are deposited over the surface of the oscillator. This method is the basis of thickness-monitors used in evaporation systems and can easily detect nanogram of materials.

The surface acoustic wave method is even more sensitive than the microbalance method since it directly detect the surface loading. Its only drawback is its cost. It can be used to detect simple mass loading effects and any mechanical changes that may occur in its gas or chemical sensitive layer. This method can detect changes as small as one part in billion.

8.4.9 Intelligent Sensors

As mentioned in the previous sections of this chapter, the output of the chemical sensors is usually collected by a data processing unit, such as a computer, to process the data and calculate concentration and other attributes of the chemicals that are used as measurands. More often, the output of the sensor is not single-valued, as in interferometers, and the data processing unit keeps track of the output cycles. Another problematic issue is that chemical sensors usually respond to more than one type of chemical and, to remedy this "cross sensitivity" problem, multiple sensors having different selectivities are usually employed. In these schemes, the computational data processing performs correlation calculations to estimate the "un-correlated" part of the signal that may be related to the concentration of a particular substance. Correlation calculations can be reduced to problems in pattern recognition.

In optical chemical sensors, multiple wavelengths can be used and processed separately. Thus, in these sensors the sensor shell may be used by different wavelength-sensing means. The response of these different channels to different chemicals can be recorded and processed to eliminate cross sensitivities.

Due to the complexity of the chemical sensor output, the data processing unit is, therefore, of a great importance [51]. Neural network [52] and fuzzy logic [53,54] have emerged as powerful intelligent computational tools. The success of these tools in other fields necessitates a careful evaluation of their application to chemical sensors. The objective of this evaluation is to understand how these "intelligent" data processing schemes can improve the quality of the chemical sensors, their processing speed, and their cost effectiveness. Although we will discuss these issues in relation to chemical sensors, their application to other types of sensors follow almost an identical path. In principle, the data processing itself can be carried out in the optical domain, and there has been some research activity in this area.

8.5 Connections/Links and Wiring

Many different methods can be used to connect sensors and actuators to a processor unit. Communication links should preferably address the following issues:

- Ease of interface with sensors/actuators and processors
- Materials compatibility
- Immunity to electromagnetic interference
- Multiplexing capabilities
- Low losses
- Cost effective

The electrical wiring using twisted pairs or coaxial wires is a familiar example of a communication link. This method may seem to be inexpensive at the first sight but as the number of sensors and actuators increase, the wiring complexity and cost becomes quite high. The well known "spaghetti wire" effect is the outcome of wiring many different sensors and actuators to a central processing unit. In these situation special precautions should be taken to reduce the cross talk which also adds to the system cost and complexity. Non the less, electrical wiring is used in most applications.

Radio links are used to set up communication links between different sensor/actuator sites and processors. In applications where simple electrical wiring is not possible or acceptable, rf links have gained considerable popularity. One such application is in biomedical telemetry where an external control/data processing or power source is linked to a unit inside the human body. Rf links have been used to both deliver power and communicate with implanted units quite successfully. In these applications, electrical wiring is not practical.

Sound and ultrasound can also be used as a communication link. Their main shortcoming is their limited range due to scattering by interfaces and defects that may exist inside the system.

Chemical methods are used in biological organisms as a communication link between brain and other parts. Nervous system, and neuron firing are all chemically induced electrical phenomena.

Free-space or guided-wave optic are extensively used as communication links. Optical wiring using fiber-optics is used in long-haul voice and data communication links around the globe. Free-space optics has a limited use in this context but, non the less, it is extensively used to link the remote control unit to TV and other appliances. For these applications, free-space optics (mostly infrared light) is ideal since its main shortcoming, i.e., short travel distance due to scatterings, etc., is an asset that is used to limit the control range of the device and to prevent cross-talk between different neighbors' remote control units.

Fiber-optic links are quite attractive in the context of smart structures. They are immune to electromagnetic interference, they can use a variety of time-domain or frequency domain multiplexing methods to address different sensor/actuator sites, they are loss-free, they are compatible with a wide range of materials used in smart structures ranging from composites to ceramics, and they can easily interface with devices and processors.

Potential widespread use of smart structure technology in industry requires significant reduction in system cost and complexity. There are many different definitions of smart materials and structures in use today [2]. Here, by smart materials and structures we mean a mechanical structure with imbedded sensors in its different parts so that its mechanical state can be sensed and assessed. Smart materials may also have imbedded actuators that can be used to alter their mechanical properties or to repair them [1-3].

Here, we focus only on addressing and networking the imbedded sensors in these materials. These imbedded sensors can provide essentially two types of different but inter-related information. On one hand, they can report on variations of the external stimuli in different parts of the mechanical structure. On the other hand, they can be used to monitor internal "relaxation" in the stress field or deterioration of the mechanical structure itself. To be able to accomplish these different types of monitoring, the imbedded sensors should affect the mechanical structure minimally. In other words, the imbedded sensors should not alter the strain field distribution inside the mechanical structure nor should they contribute to the failure of the structure. Additionally, these sensors should be addressed and read efficiently and the scheme that is used to interrogate them should be immune to a variety of interfering signals that may be present inside the mechanical structure.

8.5.1 Optical Links

Optical schemes using fiber-optics are very attractive and offer interesting possibilities in interrogating sensors in smart materials and structures. First, these schemes are immune to electromagnetic interferences. Therefore, in industrial environments, where 60 and 120 Hz noise may be bothersome, optical schemes in general, and fiber optic schemes in particular, are ideal. Second, due to the non-interfering nature of the "photons" of different wavelengths that can travel in fiber-optics, one can use parallel schemes to address different sensors using different wavelengths [55,56]. These parallel schemes can be used in addition to time and frequency (base-band) multiplexing schemes. Additionally, fiber optics being composed of SiO_2 does not add to the electromagnetic cross-section of the structure. This may be of value where the smart structure or material is required to offer a certain amount of stealth.

In addition, because of their small diameters (80 to 125 μm), fiber optics can be surface-mounted or embedded with minimal effect in a large number of materials, including metal-matrix composites [57], polymers [58], and concrete [59]. Thus, optical fiber sensing technique is a strong candidate for structural monitoring.

A fiber optic network for monitoring embedded sensors of a smart structure must meet the following requirements: i) Capability of multiplexing a large number of sensors, ii) low cross-talk between sensors, iii) high signal to noise ratio, iv) sufficient sensitivity, v) high dynamic range, vi) light weight and compactness, and vii) low power consumption.

The different methods that can be used to interrogate sensors using fiber optics can be divided to three categories: i) In-line schemes where the fiber optics is spliced and light is made to pass through or reflected from the sensor, ii) evanescent field schemes where the guided light interacts with the sensor through the evanescent fields, iii) mechanical schemes where the sensor interacts with the guided light through the mechanical deformation that is caused in the fiber optics, and iv) these mechanical deformations, such as microbends, affect the guided light through the elasto-optic effects.

In-line methods suffer from many drawbacks, including excessive insertion losses, miss-alignment problems, difficulty of implementation and excessive time required to assemble the sensor/fiber system. However, they offer the simplest possible method of interrogating the sensor.

Evanescent field schemes are based on interaction with the guided light through its evanescent fields which have extensions in the vicinity of the fiber optic or waveguide [55]. Therefore, they require the cladding part of the fiber optic to be modified so that the interaction with the evanescent field can take place. This scheme is very well-suited to interrogating passive sensors without external power sources and actuators.

To make this scheme spatial frequency selective and to improve its sensitivity, the exposed region can be modified by periodic dielectric loading that results in a fiber-optic Bragg diffraction grating (FBG) [55,60]. FBG provides us with many interesting possibilities for sensing. Any "stretching" of the FBG changes its spatial frequency and hence affects its reflection/ transmission spectra.

Deformation of an FBG can also result in a change in its spatial frequency and can be sensed optically. Bragg gratings are extensively used as distributed reflectors in solid-state lasers [61], in light couplers and in waveguide filters [55,61]. FBG's will be used in this work to interrogate sensors. However, another great advantage of FBGs is that they can be used to couple light out of or into the fiber as well. Thus, in more general smart structures or materials, where actuators may also be imbedded along with sensors, FBGs can be used to activate these actuators or deliver other "commands" such as changing the nature of the response of the smart material.

Mechanical schemes, in principle, do not require the modification of the cladding of the fiber optic [55]. In these schemes, however, long interaction lengths are needed and mechanical forces that may be required to induce the necessary changes in the fiber may be too large. To interrogate passive sensors, mechanical schemes are not acceptable. FBGs can be easily fabricated using standard photo-lithography methods. It has been shown that a number of FBGs can be implemented and addressed on a single fiber [62]. For a large sensor network, however, passive FBG fiber-optic networks have the following drawbacks:

1. Number of sensors is limited. To date only up to 20 FBGs multiplex on a single passive fiber has been reported [62] since the available bandwidth from a broadband source is limited to about 36 nm [64].

2. Cross talk among sensors. Experiments indicate this is a serious problem when a large number of sensors are used; the sensed signals would be substantially degraded [65].

3. Low energy efficiency. An unacceptable low signal-to-noise ratio results when the number of sensors is large due to the narrow bandwidth and the narrow pulse-width of light signal. This is because passive multiplexing schemes only use a very small portion (10^{-4} to 10^{-5}) of the energy from a constant broadband light signal.

Figure 8.27 illustrates the concept of the actively addressed sensor network. FBGs, fabricated using standard photolithography. Large number of FBGs, all with the same grating period on a single fiber can be used to address embedded sensors in smart structures. A super fluorescence regenerator can be used as the optical energy source of network.

When the modulation period T_m is set equal to T_i, the light pulses that are reflected by the ith FBG are amplified. With the gain of the amplifier greater than its losses for this i^{th} path, stable output pulses at the resonant wavelength of the addressed sensor are generated. On the other hand, when the round trip-time of an optical pulse that is reflected by the ith FBG matches the optical pulse modulation period, the optical gain of the round-trip path becomes larger than its losses and the ith FBG is addressed. By changing the modulation period, different FBG's on the fiber are addressed.

A sensor with identical characteristics to the imbedded sensor is added to the reference arm to tune the resonant wavelength of the reference FBG (λ_r). When λ_r matches the resonant wavelength of the ith FBG (λ_i), the reference FBG reflects maximum light power to the detector. The signal that is applied to the reference FBG is the required signal proportional to the measurand [66] sensed by the ith sensor.

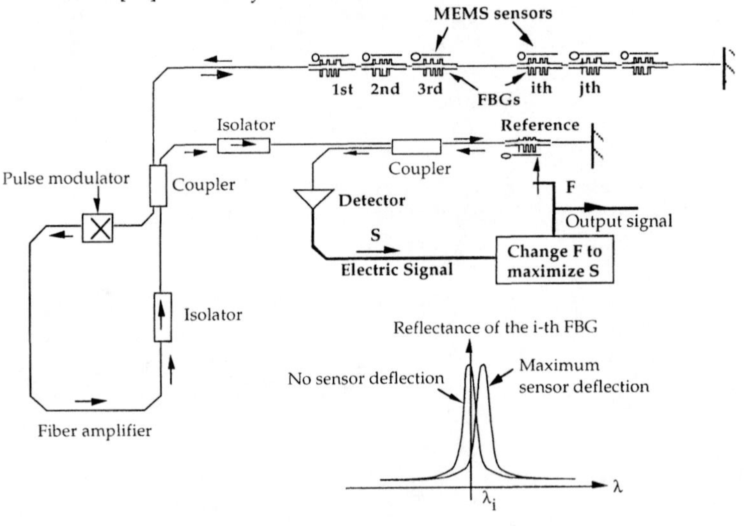

Figure 8.27 A large sensor network using active multiplexing techniques for aircraft structure monitoring.

8.5.2 Requirement on the Processing Unit/Intelligence

As mentioned in the previous sections of this chapter, the output of the chemical sensors are usually collected by a data processing unit, such as a computer, to process the data and calculate concentration and other attributes of the chemicals that are used as measurands.

More often, the output of the sensor is not single-valued, as in interferometers, and the data processing unit keeps track of the output cycles. Another problematic issue is that chemical sensors usually respond to more than one type of chemical and, to remedy this "cross sensitivity" problem, multiple sensors having different selectivities are usually employed. In these schemes, the computational data processing performs correlation calculations to estimate the "un-correlated" part of the signal which may be related to the concentration of a particular substance. Correlation calculations can be reduced to problems in pattern recognition.

In optical chemical sensors, multiple wavelengths can be used and processed separately. Thus, in these sensors the sensor shell may be used by different wavelength-sensing means. The response of these different channels to different chemicals can be recorded and processed to eliminate cross-sensitivities.

Due to the complexity of the chemical sensor output, the data processing unit is, therefore, of great importance [46]. Neural network [51] and fuzzy logic [67] have emerged as powerful intelligent computational tools. The success of these tools in other fields necessitates a careful evaluation of their application to chemical sensors. The objective of this evaluation is to understand how these "intelligent" data processing schemes can improve the quality of the chemical sensors, their processing speed, and their cost effectiveness. Although we will discuss these issues in relation to chemical sensors, their application to other types of sensors follow an almost identical path. In principle, the data processing itself can be carried out in the optical domain and there has been some research activity in this area. In the following sections, we give only a brief review of different computational intelligent schemes without any reference to their implementations.

8.6 Actuators

The previous chapters discussed actuators in some depth. Most of these actuators can be used in smart structures. From the point of view of smart structures with microactuators, it is necessary to consider the following characteristics in addition to other actuator attributes which are discussed in Chapter 1:

* Interfacing with the actuator (both power and data output)
* Efficiency

An efficient interfacing method in structures with many actuators is of outmost importance. The interfacing method should also be reliable and capable of withstanding stress and mechanical deformations. The efficiency of the actuators themselves becomes a very critical issue in structures with large number of these actuators. We discussed some aspects of actuator efficiencies in Chapter 1. The section 8.5 discusses some of the interface and interconnection issues.

8.7 Signal Processing/Computing

It is not possible to cover vast topics of signal processing and computing in a section of a chapter. These are topics at the heart of the high tech revolution of our era, involving information, communications and industrial automation. Thus, we focus our attention on the signal processing needed in simple smart structures and draw a few specific conclusions.

Computation and signal processing were divided into the two categories of implicit and explicit processors as discussed in the following sections.

8.7.1 Implicit Computation

Mechanical Methods. There are a number of functions that can be easily implemented using mechanical devices. Spatial integration, multiplication and division, mapping, filtering, clipping, rectification, sub-harmonic and harmonic generations, among other functions, can be implemented.

Spatial integration is simply achieved by adjusting the active area of the sensor to cover the area over which the integration is sought. Thus, the sensor takes in data simultaneously over the active area covering the desired integration range. Multiplication and division can be achieved by gear ratios or by levers. Mapping can be achieved by applying the input at different points in a deformable mechanical structure and obtaining the output as a magnitude of deformation at a different point. This is schematically shown in figure 8.28.

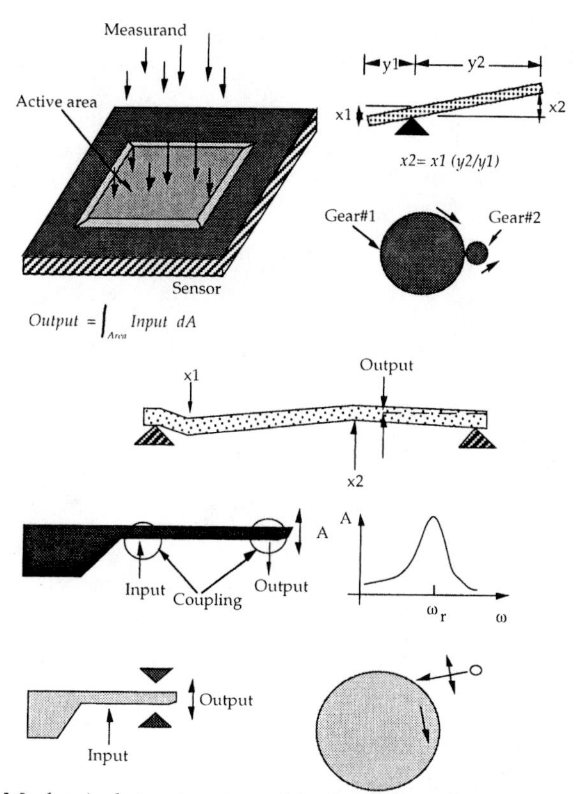

Figure 8.28 Mechanical structures used in implicit analog computers.

Band-pass, low-pass and high-pass filterings can be all implemented using micromechanical structures. Figure 8.28 schematically shows a few different configurations that can be used to achieve these functions. A resonator in the form of a cantilever beam, or any other second-order structure, is used to achieve band-pass filtering. Low-pass filtering is achieved using an inertial mass located strategically on a diaphragm or by increasing the inertial mass of a cantilever beam to make its response sluggish.

High-pass filtering is not as straight forward to achieve. One method is to use a mechanical resonator. Resonators operated in the frequency range below their resonance can be considered a high pass filter. Another idea is to "couple" two deformable mechanical members through surfaces with velocity-dependent friction. At the molecular level, the friction tends to be directly proportional to the velocity, being smaller at lower velocities and increasing as a function of velocity. Thus, as the frequency of oscillation of the input beam increases, its velocity at the contact point increases, in turn increasing the "coupling" to the output device.

Clipping can be easily achieved by restricting the motion of a deformable structure to a range that limits its free motion, as schematically shown in figure 8.28.

Rectification of oscillatory motions using mechanical structures is an interesting area of research still in its infancy. An example of such a device was discussed in section 2.3.3 of Chapter 2 where the oscillatory motion of a diaphragm was rectified by a rotor with fins attached underneath it. Devices that rectify the oscillatory motion exhibit a rich variety of harmonic and sub-harmonic generation phenomena.

Analog Electronic Methods. With simple on-board analog electronics, a variety of functions can be implemented. These include: i) temporal integration, ii) differentiation, iii) summation (weighted and linear), iv) subtraction, v) compression, vi) expansion, vii) level detection, viii) clipping, ix) ac-dc and dc-ac conversion and x) automatic gain control, among other functions. These are schematically shown in figure 8.29. For a comprehensive review of these different functions, interested readers are referred to [68].

A Combination of electronics with mechanical devices can be used to implement local processors in charge of "reflexes". A global processor can be used to set the overall goals, bias these local processors and impart adaptability to them.

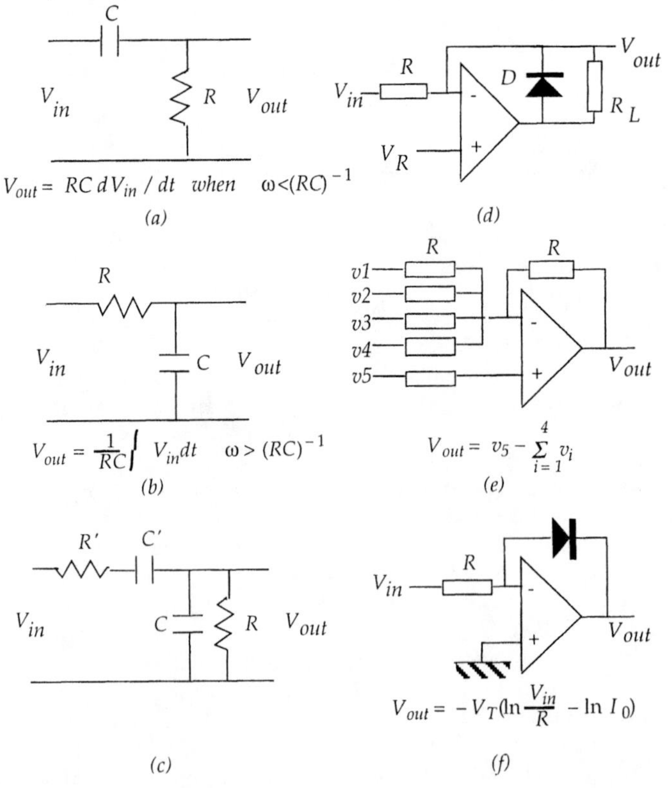

Figure 8.29 Electronic circuits used for implicit computation task.

8.7.2 Explicit Computation

Almost universally, the explicit computations and signal processing is performed by digital microprocessors and specialized signal processing devices. Thus, the signal from the sensor should be amplified and applied to an analog-to-digital (A/D) converter. The output of the A/D is given to a microprocessor which in turn, performs the desired operations on it and transforms it into a digital output. The digital output can be given to an output device in the form of a display or audio signal, or it can be processed further and applied to an actuator to modify the environment.

Inspired by biological systems, one of the interesting trends in robotics and smart structures is to have local processors impart necessary "reflexes" to the structure and have a central processing unit perform tasks that require a higher level and more global knowledge and data bases. This approach establishes a hierarchy in decision-making that allocates simple and localized tasks to local and relatively simple processing units and other more general and universal tasks to a central processing unit. This global processing unit oversees the operation of the local units and sets the agenda for the overall functionality of the structure. There are ample examples in biology that take advantage of such an hierarchy to distribute different tasks to processing units with differing levels of sophistication and power. These hierarchical systems have many interesting characteristics, such as:

* Responsivity
* Adaptability
* Efficiency
* Graceful degradation
* Resilience

It is beyond the scope of this book to discuss all of the above attributes in any depth. But briefly stated, hierarchical systems [3] are quite responsive because most of the decision-making tasks are performed locally, without burdening the global processor with a myriad of data from sensors all over the structure. Hence, smart structures with hierarchical control and processing systems are more efficient because there is no need to transmit all the data to a central location. They degrade gracefully because, as different sensors die, the global processing unit can over-compensate for them by performing extrapolations and other data processing functions. A combination of these and computational methods based on artificial intelligence ideas would make these structures adaptive and resilient.

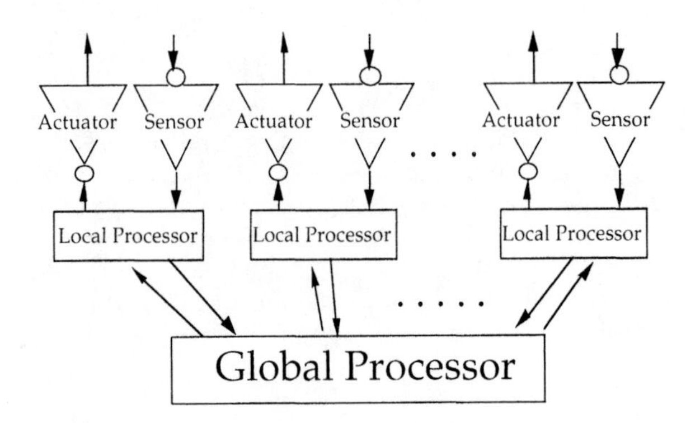

Figure 8.30 Hierarchical processing system for biologically inspired smart structures.

Another important issue in the context of smart structures is control strategies that can be divided to discrete time domain and analog control problems. In the case of having implicit signal processors, probably analog control approach makes more sense while in the cases with explicit processors both analog and discrete time domain approaches may be necessary. However, by performing most of the processing associated with the controller by the processor, the discrete time domain approach may be utilized readily.

For example consider the control issues associated with "smart skins" that can react and respond to pressure build-up at the hull of submarines. These skins can be of great value in reducing acoustic signature and noise of submarines. They can also be used in residential buildings, in cars and in air-planes to reduce the outside noise.

Consider the structure shown in figure 8.31 where an acoustic wave is incident on a structure. The wave is partially reflected (P_r) and partially absorbed by the target which may cause secondary wave generation by the structure (P_s) [69].

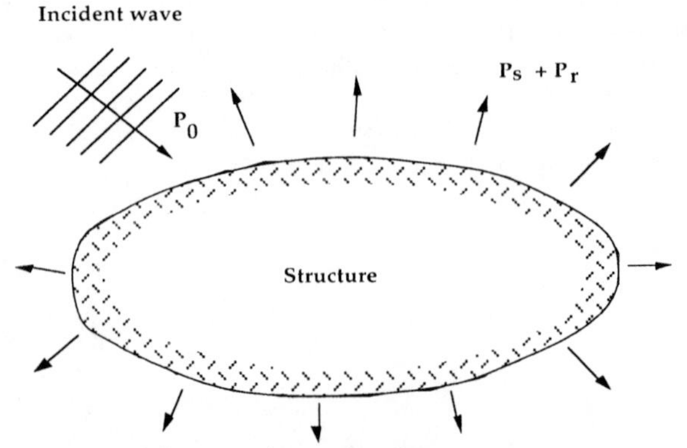

Figure 8.31 Acoustic reflections from a structure.

Passive echo cancellation methods use a coating that absorbs the acoustic energy . These methods work well at high frequencies where the acoustic wavelength is smaller than the thickness of the layer.

Because of the limitations of passive methods, active echo cancellation methods are developed that use an active layer, such as a piezoelectric layer, to respond to the pressure variations caused by the incident acoustic wave. Figure 8.32 schematically shows a structure of a coating with embedded sensors, actuators and localized control/signal processing units that can be used in active echo cancellation. The important requirement of these systems is that even if the control system fails, the structure should remain silent.

A pseudo-active echo cancellation can be achieved by connecting the piezoelectric coating to an appropriate energy dissipating load. Polymers with embedded piezoelectric particles have been designed and fabricated as pseudo-active echo cancellation coating.

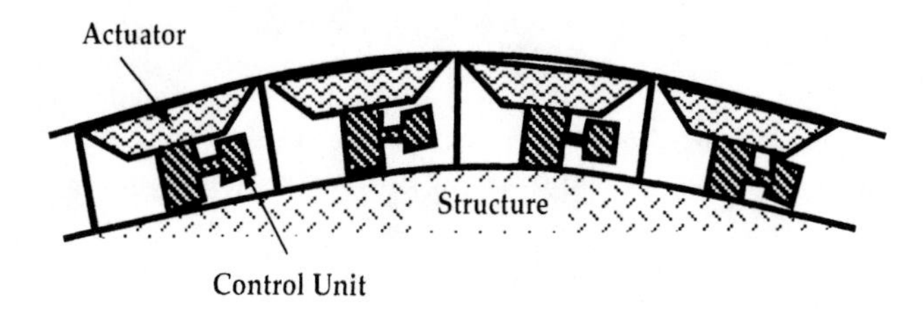

Figure 8.32 A schematic of an active echo-cancellation system [69].

In active echo-cancellation systems, both the incident acoustic wave and the target echo (P_r+P_s) are monitored and using a control system the target echo is reduced. Figure 8.33 shows an active system with hydrophones to sense the incident and target echo acoustic waves. Many of these basic units may be needed in large structures shown in figure 8.32.

Note that the hydrophones should not have a large acoustic signature themselves. The first hydrophone detects the incident acoustic wave without being perturbed by the reflected/scattered wave. Since it takes a finite amount of time for the wave sampled by the first hydrophone to travel to the actuator, the signal from the first hydrophone should be appropriately phase shifted. Then the scattered wave is sensed by the second hydrophone which produces the error signal. The goal is to have this error signal as small as possible. The various time delays in this system are frequency dependent making it only feasible at a single or a narrow band of acoustic frequencies.

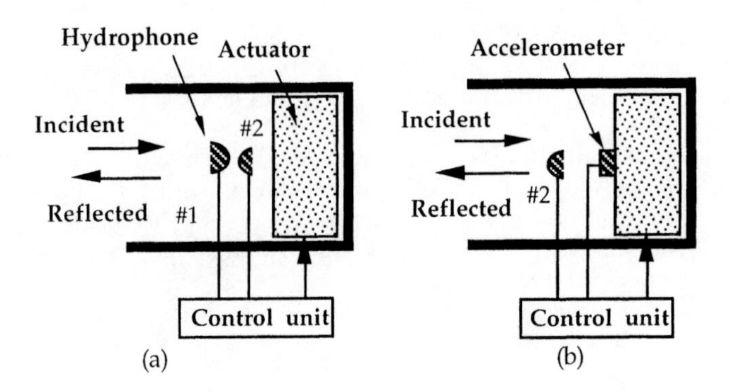

Figure 8.33 Local active cancellation of echo with (a) two hydrophones, and (b) with a hydrophone and an accelerometer [69].

In the second approach, the first hydrophone is replaced by an accelerometer. This approach has a wider operation frequency than the first approach. The accelerometer senses the particle velocity on the actuator (effectively it treats the actuator surface as the sensing surface) and using a digital signal processor minimizes the error signal detected by the hydrophone. Figure 8.34 shows oscilloscope traces of target echo with and without the active cancellation. Echo attenuations as high as 15 dB has been reported [69].

With today's microprocessors, most control issues are addressed by computational by a microprocessor. There are a variety of methods to perform these computations including "rigid" algorithms as well as "adaptive" algorithms. We discuss two examples of "adaptive" methods, namely neural network and fuzzy logic, in the following sections.

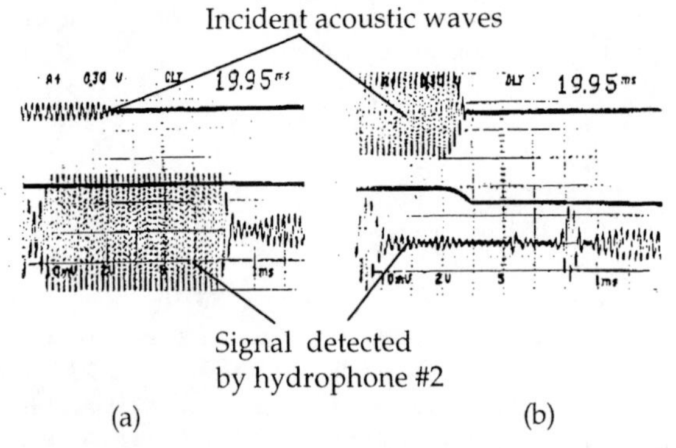

Figure 8.34 (a) Target echo without active cancellation. (b) Target echo with active echo cancellation. The top trace was 0.3 V/div. and the bottom trace was 2 V/div. The time scale was 1ms/div. In (b) the incident wave amplitude was increased [69].

<u>Neural Networks.</u> An important application of neural networks in smart gas sensors, for example, is the detection of multiple components in a gas mixture. In these applications, an array of gas sensors, with each sensor

sensitized to a specific chemical substance, is usually used. The task of separation of different chemical elements from the data from the sensor array is also known as multi-component analysis. Traditional processing methods employ feature extraction and pattern classifier algorithms for the separation of different constituents in the gas mixture. Neural networks are relatively new computational paradigms which emulate the functioning of the brain (figure 8.36) [51].

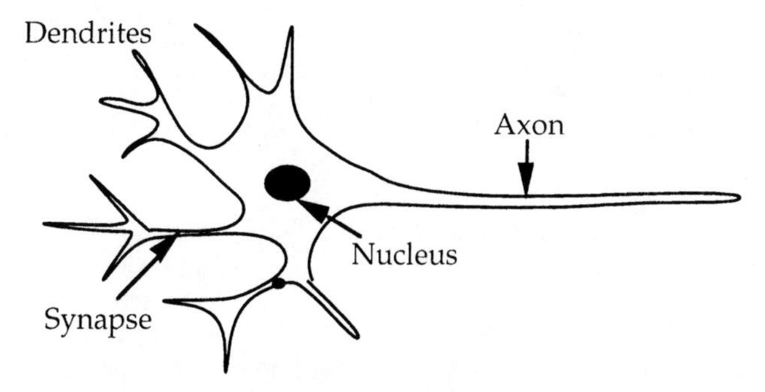

Figure 8.36 Structure of a biological neuron.

A specific category of neural network known in the literature as the supervised learning network has the powerful ability of discovering complex non-linear mappings from recorded data samples. Recently there have been proposals to approach the multi-component analysis task using neural networks [51-54]. Another possible use of neural networks is in sensor drift correction. Sensors, in general, are affected or influenced by external variables, such as temperature, pressure, etc. If the disturbing variables can be measured, we can compute a drift-free measurement signal by learning an inverse mapping

$$x_f = f(x_d, y_1, \ldots y_n) \tag{8}$$

where x_f is the fault-free measurement, x_d is the disturbed measurement signal and $y_1, \ldots y_n$ are the disturbing external variables. The mapping can be learned by exposing the sensor to a gas of known concentration and recording the disturbing variables and the output of the sensor. This data can be used by a neural network to learn the unknown mapping.

Using the above scheme, we have implemented a neural network to process the output of a displacement and force sensor that is based on a speckle pattern detection [70]. The schematic of the sensor is shown in figure 8.37.

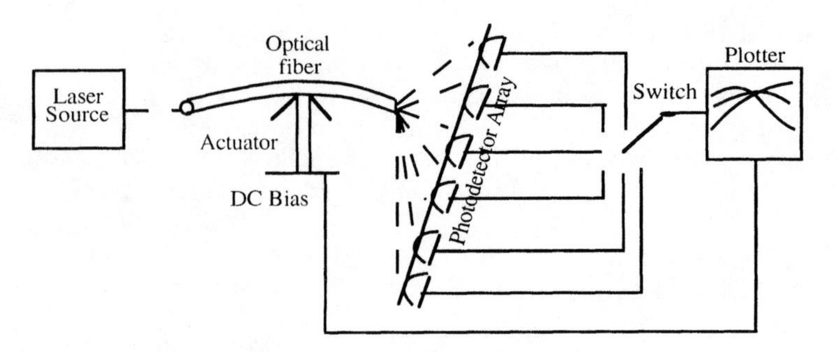

Figure 8.37 The experimental set-up of a fiber optic displacement and force sensor based on speckle pattern detection.

The displacement sensor uses a multi-mode fiber optic as the central part of its sensing scheme. The laser light excites the propagation modes of the fiber. The output of the fiber is projected onto a screen, where, due to interference between the different modes, a field of light dots or speckles can be observed. Since the distribution of the speckle pattern is directly related to the power distribution among the different modes, any perturbation of the fiber results in a change in the speckle pattern [70].

Due to the very large sensitivity of this scheme, "cross-talk" between the different types of perturbations are quite troublesome. However, it was observed that the change in the speckle pattern due to temperature variations in the fiber was different from the speckle pattern change when the fiber was deformed mechanically. We developed a neural net to differentiate among the different patterns of the speckle field [70].

The speckle configuration depends not only on the amplitude of the deformation of the actuator, but also on the mean position of the fiber. It is found that the output of a photodetector, placed so as to measure a small region of the speckle configuration, follows a non-linear function as the mean position of the fiber is varied. The neural network trained for displacement values within a 1.5 μm range is found to generalize with less than $\pm$ 0.015 μm error in the input range. It is found that a functional-link net (figure 8.38) with 11 functional links and 4 neurons can be trained for a convergence criterion of 5×10^{-6} in less than 1000 iterations. The error in the individual targeted output during training was less than $\pm$ 0.01 μm [70].

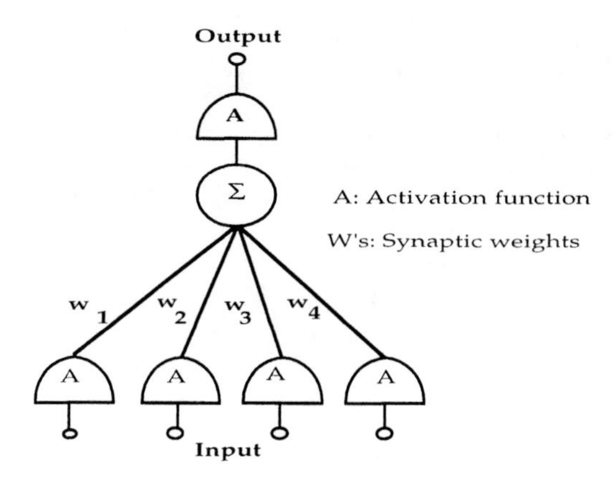

Figure 8.38 The structure of an artificial neural network parallels the structure of natural neurons.

The above example discusses the application of neural networks to improve the performance of a displacement sensor and its application in optical chemical sensors follows almost identical steps.

Fuzzy logic. Another tool which can be used for detection of multiple constituents in a gas mixture is fuzzy logic processing. Fuzzy logic attempts to deal with imprecise information and it defines the concept of degree of membership of an entity in a set. This approach differs from the absolute nature of Boolean logic, which assigns true or false to every entity to which it is applied.

Neural networks have a strong link with the fuzzy logic method since they share the same representation space. In fuzzy logic, an entity may belong to different sets with varying degrees. An application where this is well-suited is in the detection of a gas mixture using an array of sensors [43,46]. In [49], the concentration of each component is given a membership value associated with the linguistic sets described as 'very high', 'medium, and 'low'. Using fuzzy logic, therefore, one can process the output of a gas sensor in a manner that is much closer to that of human decision-making. The concept of membership function in fuzzy logic represents the gray nature of the processes that occur in nature. Applied to chemical sensors, fuzzy logic deals with two types of uncertainty which include the uncertainty in sensing procedure and in that of the process itself.

We conclude this chapter by noting that in the task of constructing smart structures require knowledge drawn from many different fields of knowledge as far apart as physiology, behavioral sciences, cognitive sciences, large-scale system concepts, device physics, microfabrication technologies and good understanding of environment of the smart system. This chapter covered some of the different components of smart structures in a very sketchy manner. I have intentionally left out examples of smart structures in use today. These includes smart building that using an active control system can withstand earthquakes, bridges that change their load-bearing characteristics depending on the traffic patterns and many other systems that are in use today. These examples are covered in ref. [2] thoroughly.

My intentions in covering materials of this chapter was to give the reader a flavor of different components that may go into a smart structure. Issues associated with many variable-control systems are quite difficult and many research effort are on the way to address them. The important development in recent years has been in the area of miniaturization of actuators and their integration with sensors and signal processing units that may pave the way to the solution of some long-standing problems in control of large systems.

8.8 References

1. J. Holland, Adaptation in Natural and Artificial Systems. The MIT Press, Cambridge, MA (1995).
2. Brian Culshaw, Smart Structures and Materials. Artech House, Boston MA, 1996.
3. a) J. H. Holland, Hidden Order. Addison-Wesley Publishing Company, Inc., Reading, MA, (1995).
   b) H. A. Simon, The Sciences of the Artificial. 3rd Ed. The MIT Press, Cambridge, MA (1996).
4. P. de Latil, "Thinking by Machine." Houghton Mifflin Company, Boston, MA, (1957).
5. M. Tabib-Azar, Integrated Optics and Microstructure Sensors. Kluwer Academic Publishings, Boston (1995).
6. A. Garcia, and M. Tabib-Azar, "Sensing Means and Sensor Shells: A New Method of Comparative Study of Piezoelectric, Piezoresistive, Electrostatic, Magnetic, and Optical Sensors." Sensors and Actuators A. Physical Vol. 48 (2), pp. 87-100 (1995).
7. G. Beheim, K. Fritsch, and M. Tabib-Azar, "A Sputtered Thin Film Fiber-Optic Temperature Sensor." Sensors, pp. 3743 (1990).
8. G. Beheim, H. Sotomayor, J. Flatico and M. Tabib-Azar, "Fiber Optic Thermometer Using Fourier Transformation Spectroscopy." Presented in SPIE's 1991 Boston Symposium Proc.
9. W. Gopel, "Ultimate Limits in the Miniaturization of Chemical Sensors." Sensors and Actuators A 56, pp. 83-101 (1996).
10. J. N. Zemel, "Future Directions for Thermal Information Sensors." Sensors and Actuators A 56, pp. 57-62 (1996).
11. A. Sharma, "Optical Sensors in Environmental Applications." edited by Tuan Vo-Dinh. Proc. of SPIE Environment and Process Monitoring Technologies, Pub.# 1637, p.p. 270-279 (1992).
12. J. Janata, M. Josowicz, and M. DeVaney, Chemical Sensors. Analytical Chemistry Vol. 66, p.p. 207R-228R (1992).
13. P. V. Lambeck, "Integrated Opto-Chemical Sensors.", Sensors and Actuators B, Vol. 8, p.p. 103-116 (1993).
14. A. D. Kersey and A. Dandridge, "Applications of Fiber-Optic Sensors." Vol. 13(1), p.p. 137-143 (1992).
15. K. Chan, H. Ito, and H. Inaba, "All-Optical-Fiber-Based Remote Sensing System for Near Infrared Absorption of Low-Level CH_4 Gas." Journal of Lightwave Technology, LT-5(12), p.p. 1706-1710 (1987).

16. N. Fabricius, G. Gaughitz, and J. Inganhoff, "A Gas Sensor Based on an Integrated Optical Mach-Zehnder Interferometer", Sensors and Actuators, Vol. 7, p.p. 672-676 (1993).
17. A. Mendelis and C. Christofides, Physic, Chemistry and Technology of Solid State Gas Sensor Devices. John Wiley & Sons, N Y (1993).
18. O. S. Wolfbeis, Novel Techniques and Materials for Fiber Optic Chemical Sensing. Optical Fiber Sensors, Springer Proceedings in Physics Vol. 44, p.p. 417-424 (1989).
19. a) H-Y. Wang, and J. B. Lando, "Gas-Sensing Mechanism of Phthalocayanine Langmuir-Blodgett Films." Langmuir Vol. 10, pp. 790-796 (1994).
    b) H. Y. Wang, C. W. Chiang and J. B. Lando, "Structural Investigation of Gas Sensing Langmuir-Blodgett Films of Phthalocyanine ($(C_6H_{13})_3SiOSiPcOGePcOH$))." Thin Solid Films, Vol. 273, p. 90 (1996).
20. I. Lundstrom, "Why Bother About Gas-Sensitive Field-Effect Devices." Sensors and Actuators A 56, pp. 75-82 (1996).
21. Q. Wu, "Characterization of Pd Schottky Barrier Hydrogen Detector." M. S. Thesis, CWRU (1997).
22. a) S. J. Choquette, and L. Locascio-Brown, "Thermal Detection of Enzyme-Labeled Antigen-Antibody Complexes Using Fiber-Optic Interferometry." Sensors and Actuators B Vol. 22, p.p. 89-96 (1984).
    b) M. A. Butler, "Optical Fiber Hydrogen Sensor." Appl. Phys. Lett. Vol. 45 (10). p.p. 1007-1009 (1984).
23. a) H. E. Posch, and O. S. Wolfbeis, "Fiber Optic Humidity Sensor Based on Fluorescence Quenching." Sensor and Actuator Vol. 15, p.p. 77-83 (1988).
    b) K. Ogawa, S. Tsuchiya, H. Kawakami, T. Tsutsui, "Humidity Sensing Effects of Optical Fibres with Microporous SiO_2 Cladding." Electronics Lett. Vol. 24, p.p. 42-43 (1988).
24. B. S. Matson, J. W. Griffin, "Infrared Fiber Optic Sensors for the Remote Detection of Hydrocarbons Operating in the 3.3 to 3.6 μm Region." SPIE Proc. of Chemical, Biochemical and Environmental Sensors, editors: R. A. Lieberman, and M. T. Wlodarczyk, Vol. 1172, p.p. 13-26 (1989).
25. Ch. Stamm and W. Luckosz, "Integrated Optical Difference Interferometer as a Refractometer and Chemical Sensor." Sensors and Actuators B Vol. 11, p.p. 177-181 (1993).
26. a) Otto S. Wolfbeis (1991), Fiber Optic Chemical Sensors and Biosensors, Vol. I and II, CRC Press, New York (1991).
    b) K. Chan, H. Ito and H. Inabo, "All-Optical-Fiber-Based Remote Sensing System for Near Infrared Absorptions of Low-Level CH_4 Gas." Journal of Lightwave Technology LT-5(12), p.p. 1706-1711 (1987).
27. R. P. Podgorsek, H. Franke, and C. Feger, "Selective optical detection of n-heptane/iso-octane vapors by polyimide lightguides." Opt. Lett. Vol. 20(5), p.p. 501-503 (1995).
28. E. Kretchmann, "The Determination of the Optical Constants of Metals by Excitation of Surface Plasmon." Z. Phys., Vol. 241, p.p. 313-324 (1971).
29. O. Solgaard, F. Ho, J. I. Thackara, and D. M. Bloom, "High Frequency Attenuated Total Internal Reflection Light Modulator." Appl. Phys. Lett. Vol. 61, p.p. 2500-2502 (1992).

30. C. Jung and S. Yee, "Feasibility of an Integrated Optics Surface Plasmon Modulator." Proc. of SPIE International Conference on Integrated Optics and Microstructures II, editors: M. Tabib-Azar, D. L. Polla, and Ka-Kha Wong, Vol. 2291, pp. 361-370 (1994).
31. Ph. M. Nellen, and W. Lukosz, "Integrated Optical Input Grating Couplers as Chemo- and Immuno- Sensors." Sensors and Actuators Vol. B1, p.p. 592-596 (1990).
32. J. M. Charlesworth, "Optical Sensing of Oxygen Using Phosphorescence Quenching." Sensors and Actuators B Vol. 22, p.p. 1-5 (1994).
33. M. Monici et al, "Fiber Optic pH Sensor for Sea Water Monitoring, Proc. of SPIE Vol. 798, p.p. 294-300 (1987).
34. B. Woods, A. Analyst, p.p. 113-301 (1988).
35. T. P. Jones, and M. D. Porter, "Optical pH sensor based on the chemical modification of a porous polymer film." Anal. Chem. Vol. 60, p.p. 404-411(1988).
36. D. H. Jorden et al, Anal. Chem. Acta. Vol. 59, p. 437 (1984).
37. J. Janata, "Do optical sensors really measure pH." Anal. Chem. 59: 1351-1356 (1987).
38. E. Urbano, et al, "Optical sensor for continuous determination of Halides." Anal. Chem. 56: 427-429 (1984).
39. R. Narayanaswamy, et al, "Optical fiber sensing of fluoride ions in flow stream." Talanta. Vol. 35, p.p. 83-88 (1985).
40. R. M. Izatt, et al, "Thermodynamic and Kinetic Data for Cation-Macrocycle Interaction." Chem. Rev. Vol. 85, p. 271 (1985).
41. Z. Zhujun, J. L. Mullin, and W. R. Seitz, "Optical sensor for sodium based on ion-pair extroetian and flurosence." Anal. Chem. Acta. Vol. 184, p.p. 251-258 (1986).
42. Schaffer, et al. Analyst. Vol. 113, p. 693 (1988).
43. J. Van Gent et al., "Chromoionophores in Optical Ion Sensors." Sensors and Actuators, Vol. 17, p.p. 297-305 (1989).
44. H. E. Posch, et al, "Optical and Fiber Optic Sensors for Vapors of Polar Solvents." Talanta Vol. 35, p. 89 (1989).
45. A. G. Cavinato et al. Anal. Chem. Vol. 62, p. 1977 (1990).
46. M. A. Butler, A. J. Ricco, and R. J. Buss, R. J., J. Electrochemical Society Vol. 137(4), p. 1325 (1990).
47. V. Kasche et al., "Principles of Signal Generation and of Coupling to Optical Fibers: Dynamic Fluorescence Biosensors." In: Biosensors Applications in Medicine, Environmental Protection and Process Control. Edited by: R. D. Schmid and F. Scheller, GBF Monographs, Volume 13, VCH Publishings, Germany, p.p. 233-242 (1989).
48. J. S. Schultz, "Biosensors." Scientific American, p.p. 64-69, August (1991).
49. U. Noack et al., "Algae Toximeter - A Biosensor for Water Monitoring." in: Biosensors Applications in Medicine, Environmental Protection and Process Control. Edited by: R. D. Schmid and F. Scheller, GBF Monographs, Volume 13, VCH Publishings, Germany, p. 243 (1989).
50. A. A. Boiarski et al., "Integrated-Optic Sensor with Macro-Flow Cell." Proc. of SPIE International Conference in Integrated Optic and Microstructures I, edited by M. Tabib-Azar, and D. Polla, Vol. 1793, p.p. 199-211 (1992).

51. Yoh-Han Pao, Adaptive pattern recognition and neural networks. Addison-Wesley Publishing Company (1989).
52. H. Sundgren, F. Winquist, I. Lukkari, and I. Lundstrom, "Artificial neural networks and gas sensor arrays: quantification of individual components in gas mixture." Meas. Sci. Technol. Vol. (2), p.p. 464-469 (1991).
53. J. A. de Agapito, et al. ,"Fuzzy logic applied to gas sensors." Sensors and Actuators B Vol. 15-16, p.p. 105-109 (1993).
54. K. Takahashi and S. Nozaki, "From intelligent sensors to fuzzy sensors." Sensors and Actuators A Vol. 40, p.p. 89-91 (1994).
55. See articles in: Active Materials and Adaptive Structures. Edited by G. J. Knowles, Institute of Physics Publishing, Bristol, UK (1991).
56. A. D. Kersey, T. A. Berkoff and W. W. Morey, "Multiplexed fiber Bragg grating sensor system with a fiber Fabry Perot wavelength filter," Opt. Letter, 18, pp. 1370-1372 (1993).
57. J. Qiu, J. Tani, T. Takagi, "Smart Composite Material Without Thermal Bending Deformation." in Smart Materials, Edited by V. K. Varadan, Proceedings of SPIE, Vol. 1916, pp. 22-27 (1993).
58. D. Bullock, J. Dunphy and G. Hufstetler, " Embedded Bragg grating fiber optic sensor for composite flexbeams" SPIE 1798, pp. 253-261 (1992).
59. R. Maaskant, T. Alavie, R. M. Measures, M. Ohn, s. Karr, d. Glennie, c. Wade, G. Tadros and s. Rizkalla, "Fiber optic Bragg Grating sensor network installed in a concrete road bridge," SPIE 2191, 2191, pp. 457-465 (1994).
60. G. Meltz, W. W. Morey and W. H. Glenn, " Formation of Bragg gratings in optical fibers by a transverse holographic method" Opt. Lett. 14 823-825 (1989).
61. A. Yariv, Quantum Electronics, John Wiley and Sons, Third Edition, pp. 611-620 (1987).
62. K. Takahashi, "Sensor Materials for the Future: Intelligent Materials." Sensors and Actuators, Vol. 13, pp. 3-10 (1988).
63. B. Malo, D.C. Johnson, F. Bilodau, J. Albert, and K. O. Hill, "Single-Excimer Pulse Writing of Fiber Gratings by Use of a Zero-Order Milled Phase Mask: Grating Spectral Response and Visualization of Index Perturbations." Opt. Lett. Vol. 18 (15), pp. 1277 (1993).
64. J- M. P. Delavaux, C. Flores, Y. K. Park, S. Y. Huang, and C. Jack, "A Packaged Er-Doped Fiber Amplifier Module at 980 nm." SPIE Proceedings, Vol. 1789, Fiber Laser Sources and Amplifiers IV, pp. 164 (1992).
65. S. A. Al-Chalabi, B. Culshaw, D. E. N. Davies, "Partially Coherent Source in Interferometric Sensors." Proc. 1st Intern. Conf. on Optical Fiber Sensor, London, pp. 132-135 (1983).
66. S. Wu, S. Rajan, S. Yin, and F. T. S. Yu, "Multiple Channel Sensing with Fiber Specklegram," Applied Optics, 31, 28, 5975 (1992).
67. a) J. A. de Agapito, et al. ,"Fuzzy logic applied to gas sensors." Sensors and Actuators B Vol. 15-16, p.p. 105-109 (1993).
    b) K. Takahashi and S. Nozaki, "From intelligent sensors to fuzzy sensors." Sensors and Actuators A Vol. 40, p.p. 89-91 (1994).
68. P. Horowitz and W. Hill, The Art of Electronics. 2nd ed., Cambridge University Press, New York (1989).

69. C. Audoly, "Review of Active Methods for Acoustic Echo Cancellation." in Smart Materials, Edited by V. K. Varadan, Proceedings of SPIE, Vol. 1916, pp. 156-167 (1993).
70. a) A. Garcia-Valenzuela, and M. Tabib-Azar, "Fiber Optic Force and Displacement Sensor based on Speckle Detection with 0.1 Nano-Newton and 0.1 Angstrom Resolution." Sensors and Actuators A. Physical, Vol. 36 (3), p.p. 199-208 (1993).
b) U. K. Rao, A. Garcia-Valenzuela, and M. Tabib-Azar, "Smart Integrated-Optics Displacement Sensor Based on Speckle Pattern Detection Using Neural-Net with 0.1 Å Resolution." Sensors and Actuators A: Physical, Vol. 39, p.p. 37-44 (1993).

Index